9th edition

AN INTRODUCTION TO ASTRONOMY

9th edition

AN INTRODUCTION TO ASTRONOMY

Laurence W. Fredrick
Director, Leander McCormick Observatory
University of Virginia

Robert H. Baker

D. VAN NOSTRAND COMPANY

New York • Cincinnati • Toronto • London • Melbourne

Cover photo courtesy of National Aeronautics and Space Administration

D. Van Nostrand Company Regional Offices:
New York   Cincinnati

D. Van Nostrand Company International Offices:
London   Toronto   Melbourne

Library of Congress Catalog Card Number: 78-069754
ISBN: 0-442-22422-2

Published by D. Van Nostrand Company
135 West 50th Street, New York, NY 10020

10 9 8 7 6 5 4 3 2 1

Preface

The importance of astronomy as a broadening subject for college students cannot be overestimated. It is the one subject that forces us to look well beyond our immediate environment, sometimes to the point of being overwhelmed. In order to introduce the student logically and gradually to the vastness of the Universe, *An Introduction to Astronomy* begins with the familiar—the earth—and progresses outward, step by step, to the stars and beyond.

The Ninth Edition represents an extensive rewriting and rearranging of the preceding edition. The treatment of the planets, as well as of the sun, is greatly expanded, based on the results of recent studies of the planets from space. The chapters covering star formation and stellar evolution have been rewritten to incorporate, in a natural way, our latest discoveries concerning white dwarfs, pulsars, and the like. The chapters on galaxies and the origin and evolution of the Universe, currently the most active area of astronomy, are substantially revised. The now tantalizing subject of extraterrestrial intelligence has also been expanded. The many pedagogical features that made this text popular have been retained and improved. All important terms that are defined are set in bold type. There is a Glossary of further terms used in the book and an expanded and updated Appendix. Reading and source materials appear at the end of each chapter.

For those wishing to shorten the course material, we suggest dropping the first four chapters. A few subjects that bear upon the remainder of the book, namely, the equatorial coordinate system, the earth as a planet, and the basics of telescopes, can be covered in an introductory lecture or two with the appropriate sections being pointed out to the student.

The aim of this book is to provide a stimulating adventure into the study of the cosmos for students of every academic background. While most of the mathematics prerequisites have been eliminated, the approach is still logical and infused with the scientific spirit.

An Instructor's Manual is available from the publisher. It contains answers to the text questions, additional questions, up-to-date comments, and a few sample tests.

I wish to thank my many colleagues from all over the world for discussions, detailed critiques, and original photographs and diagrams. I especially thank Professors John Ray, Clemson University, J. M. Malville, University of Colorado, Frank Edmonson, Indiana University, Stephen Hill, Michigan State University, and David Gray, Golden West College, for their helpful reviews. As usual, I express my appreciation to Frances, who completely typed all three drafts of the manuscript.

Laurence W. Fredrick

Contents

chapter 1

THE EARTH AND THE SKY   1
The Globular Earth   2
The Conventional Globe of the Sky   8
Effects of the Atmosphere   11

chapter 2

THE EARTH'S DAILY ROTATION AND TIME   21
Effects of Rotation on the Earth   22
Apparent Rotation of the Heavens   27
Time   34

chapter 3

THE EARTH'S ANNUAL REVOLUTION AND THE CALENDAR   44
The Earth Revolves   45
The Seasons   52
Calendars   56

chapter 4

TELESCOPES AND INSTRUMENTATION   61
Optical Telescopes   62
Radio Astronomy   73
Other Types of Telescopes   76
The Spectrograph   82
Photoelectric Instruments   87

chapter 5

THE MOON IN ITS PHASES   91
Motions of the Moon   92
The Ocean Tides   98
The Features of the Moon   101
The Physics and Origin of the Moon   112
Eclipses of the Moon   117
Eclipses of the Sun   120

chapter 6

THE PATHS OF THE PLANETS   129
Motions of the Planets   130
The Law of Gravitation   133
The Planetary System   139

chapter 7

THE TERRESTRIAL PLANETS AND THE ASTEROIDS   147
Mercury   148
Venus   154
Mars, the Red Planet   165
The Asteroids   184

chapter 8

THE GASEOUS PLANETS   188
Jupiter, the Giant Planet   189
Saturn, the Ringed Planet   199
Uranus and Neptune   204
Pluto, the most Remote Planet   205

chapter 9 ✓

OTHER FEATURES OF THE SOLAR SYSTEM   208
Comets   209
Meteors and Meteor Streams   215
Meteorites and Meteorite Craters   220
The Origin of the Solar System   225

chapter 10 ✓

THE SUN WITH ITS SPOTS   231
Observing the Sun   232
The Photosphere   235
Sunspots   239
The Solar Atmosphere   244

chapter 11

THE STARS IN THEIR SEASONS   256
The Constellations   257
Maps of the Sky   261
Planetarium   275

chapter 12

THE STARS AROUND US   277
Distances of the Stars   278
The Stars in Motion   281
Stellar Spectra   285
Magnitudes of the Stars   295

chapter 13

DOUBLE AND VARIABLE STARS   306
Binary Stars   307
Variable Stars   313
Eclipsing Variables   314
Pulsating Variables   316
Eruptive Variables   323
Pulsars   328
Flare Stars and Planetary Nebulae   333

chapter 14 ✓

COSMIC GAS AND DUST   338
Diffuse Nebulae   339
The Interstellar Medium   346

chapter 15

THE LIVES OF THE STARS   356
Youthful Stars   357
The Stars in Middle Age   361
The Declining Stars   371

chapter 16

STAR CLUSTERS   384
Galactic Clusters   385
Distance of the Cluster   390
Globular Clusters   393
Stellar Populations   397

chapter 17

THE GALAXY   403
The Milky Way   404
Structure of the Galaxy   407

chapter 18

GALAXIES   418
Structural Features of Galaxies   419
Distribution of Galaxies   432
Spectra of Galaxies   440

chapter 19

COSMOGONY   450
Theories of the Universe   451
Origin and Evolution   460
Astronomy from Space   469
Astronomy from the Earth   473

GLOSSARY   481

APPENDIX   491

INDEX   495

chapter 1

The Earth and the Sky

A relatively small planet attending the sun, which itself is an average star among the multitude of stars making up the Milky Way galaxy, the earth owes its importance to the fact that we live here. It is from the earth and its immediate environs that we view the celestial scene around us. Therefore, in order to interpret the scene correctly, we must first consider the earth that looms large in the foreground. Our study of astronomy begins with the globe of the earth at the center of the apparent globe of the heavens and encompassed by the atmosphere through which we look out at the celestial bodies.

The Globular Earth

One of the greatest achievements of man has been the observation of the earth from space where we see the earth as the celestial body that it is. From this perspective our planet appears as a globe with visible surface features. Seas and large lakes contrast markedly with the land while bright snowfields and drifting clouds add variety to the scene. Observed from the moon with the unaided eye, the earth appears in the lunar sky as a distinctly marked disk four times larger than the moon appears to us. The markings clearly reveal the earth's rotation while the earth goes through the whole cycle of phases from new to full and back to new again. On the daylit side, the only evidence of man is marked by the line of the Great Wall of China. By night, cities and street lighting speckle the globe with a patterned array.

With increasing distance the earth becomes less and less remarkable as a celestial body. From Mars it would be a fine evening and morning star accompanied by the moon as a fainter star, both showing cycles of similar phases with the aid of a telescope. From the outermost planets the earth would usually be lost in the glare of the sun while from the nearest star the earth and all the other planets would be invisible even with the largest telescopes. At this distance the sun would be an ordinary star indistinguishable from the countless other stars dotting the sky. The earth, however, is unique in many ways, most notable of which is the fact that it is inhabited by human beings.

Why are we situated here, as opposed to all the other planets in the solar system? First of all, the earth is just the right distance from the sun to enable water to exist as a liquid. Had our planet been closer to the sun, all the water would have boiled away, making it a torrid pressure cooker like Venus. On the other hand, if the earth were a little further away from the sun, the water would be bound up in frost and ice, leaving only permafrost and polar ice caps as on Mars. The earth is also unique among the planets of the sun's family in having an abundance of free oxygen in its atmosphere. This fortuitous combination of steady moderate temperatures and hospitable atmospheric composition has enabled our planet to spawn and nurture life on its surface. Although probably singular in the solar system in this regard, it is anticipated that many other stars in our galaxy possess planets such as the earth which might very well support life in a form resembling our own. We will return to this topic in Chapter 19.

1.1 The planet earth

The earth **revolves** around the sun once each year and, at the same time, **rotates** on its axis once each day. Due to its rotation, the solid

Figure 1.1 *The planet earth as viewed from space. The earth and moon are seen at crescent phase. (National Aeronautics and Space Administration photograph.)*

earth bulges slightly at the equator and is flattened at its poles, assuming the shape of an oblate spheroid. It is 12,700 kilometers in diameter,* 43 kilometers greater at the equator than from pole to pole. But these mild departures from a perfect sphere are almost imperceptibly small. The earth is actually rounder than a good ball bearing or an excellent bowling ball.

The earth does not generate its own light, and were it not for the sun, it would be immersed in darkness. The half of the planet facing the sun remains constantly illuminated while the opposite side is only dimly brightened by reflected light from the moon and the faint backdrop of starlight. The earth therefore has **phases** just as the moon does. However, its surface displays a continually changing appearance due to the earth's rotation and the complex interplay of clouds which cover its surface (Fig. 1.1).

The majority of the earth's surface is covered by water and it is enveloped in an **atmosphere** several hundred kilometers thick. Beneath the oceans, vast mountain chains ring the planet. These rectilinear features are known as mid-ocean ridges. On the margins of the ocean basins deep semicircular trenches are found, while extensive mountain ranges form

* 1 kilometer (km) = 1000 meters = 0.62 miles
  1 meter (m) = 100 centimeters = 3.28 feet
  1 centimeter (cm) = 10 millimeters (mm) = 0.4 inches

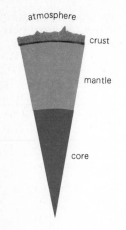

atmosphere

crust

mantle

core

Figure 1.2 *A schematic slice through the earth showing the relative thicknesses of the crust, mantle, and core.*

the backbone of the continents. In spite of its rough topography, however, the earth is exceedingly smooth in comparison to its size.

The interior of the earth consists of essentially three parts: the **crust,** the **mantle,** and the **core** (Fig. 1.2). The crust ranges from a few kilometers in thickness under the ocean floor to over 70 kilometers in depth beneath the largest continental mountain ranges. The upper levels of the crust are composed of **igneous rocks** (those formed from the solidification of molten rock), such as granite and basalt, generally overlain by a veneer of **sedimentary rocks** (those formed by the settling of particles held in suspension by water), such as sandstone, shale, and limestone. The continents are composed predominantly of rock with a granitic composition while the oceans are underlain by basalt. Because of the greater thickness and lower density of the granitic continents, they float above the level of the basaltic ocean basins. The whole crust floats on the mantle. Early stress caused the crust to crack and split into large plates which move with respect to each other and form the present continents and subcontinents. The study of the motions of these plates is called **plate tectonics.**

The mantle is composed of heavier silicates of magnesium and iron. Small pieces of the upper mantle are found in **kimberlite pipes** (where diamonds occur) which were blown to the surface by violent eruptions early in the earth's history. In addition, large slices of the upper mantle have been thrust onto the continental margins. These unusual areas allow the boundary of the crust and upper mantle to be observed at the surface of the earth. However, for an understanding of the deep interior of the planet, no direct observations are possible. Instead, the nature of the earth's deep interior must be inferred from indirect observations.

One of the most valuable means of determining the structure of the interior of the earth is to analyze the way **seismic waves,** generated by earthquakes, are propagated through the body of the earth. By measuring the amount of time required for the seismic waves to be transmitted from their source to distant seismographs stationed all over the world, models of the various layers within the earth have been constructed. At a distance of 3400 kilometers from the center of the earth (2900 kilometers beneath the surface), the seismic waves are observed to undergo a sudden change in velocity and direction. This change defines the boundary between the mantle and the core.

The core of the earth consists of two parts, a liquid outer core and a solid inner core. This conclusion is based on the way in which different types of seismic waves are transmitted through the earth. **Longitudinal waves** (Fig. 1.3), that is, waves which propagate by vibrating in the direction in which they travel, are able to pass through the core and back to the earth's surface. On the other hand, **transverse waves,** which vibrate at right angles to the direction in which they travel, are stopped when they reach the outer core. Since transverse waves cannot travel through liquid,

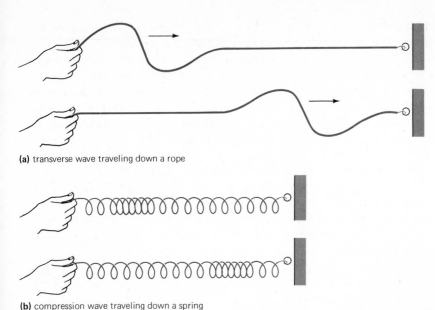

Figure 1.3 *Schematic presentation of a transverse wave (a) and a compression wave (b).*

(a) transverse wave traveling down a rope

(b) compression wave traveling down a spring

whereas longitudinal waves can, the outer core of the earth is inferred to be in a molten state. This is consistent with the high temperatures expected to occur in the earth's interior. However, the behavior of longitudinal waves in the core indicates that at a radius of 1250 kilometers from the center, the earth becomes solid again.

The earth's core is composed predominantly of an iron-nickel alloy and smaller amounts of a lower-density component, probably sulfur or silicon. This composition is deduced from the distribution of density within the earth and the elemental abundances found in the sun and meteorites (Section 9.8). Because iron is an excellent conductor of electricity, currents are maintained in the earth's core which give rise to the earth's magnetic field.

1.2 The earth's magnetic field

The earth's **magnetic field** resembles the field of a bar magnet with a north and south pole. It is therefore referred to as a **dipole field.** The axis of the field passes through the earth's center but is inclined at an angle of 11° to the rotational axis (Fig. 1.4). The north **geomagnetic pole,** toward which the north-seeking pole of a magnetic compass needle is directed, is located in northern Canada, several hundred kilometers from the earth's pole of rotation.

The magnetic field of the earth extends for great distances into space (Fig. 1.5). In the direction of the sun it is detected for a distance of about ten to fifteen earth radii, while in the opposite direction it extends

Figure 1.4 *A schematic drawing showing the earth's magnetic field.*

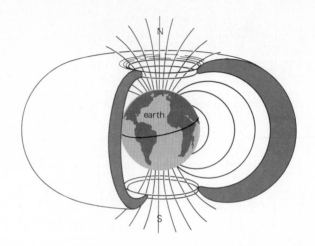

to a much greater distance where it is referred to as the earth's **magnetic tail.** Charged particles emitted by the sun (the solar wind, see Section 10.9) are captured by the magnetic field and concentrated at heights of approximately 3200 and 16,000 kilometers. These two regions of high-energy charged particles are known as the inner and outer **Van Allen belts** (Fig. 1.6) after the scientist who discovered them in 1958.

1.3 Positions on the earth

One way of denoting positions on the earth's surface is with reference to natural or conventional areas. It is often satisfactory to the inquirer

Figure 1.5 *A schematic drawing of the earth's magnetic field in the direction of the sun and away from the sun. The units are in earth radii.*

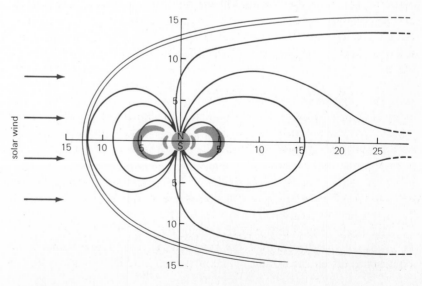

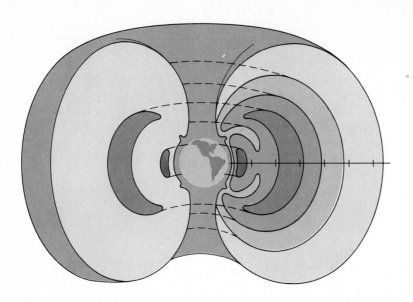

Figure 1.6 *A drawing showing the relative locations of the Van Allen belts.*

if we say, for example, that Cleveland is in Ohio. A second way, especially where positions are required more accurately, is with reference to circles imagined on the conventional terrestrial sphere (Fig. 1.7) that best represents the earth's surface. These familiar circles of our globes and maps are mentioned here so that the resemblance to systems of circles imagined in the sky may be noted later.

The earth's **equator** is the great circle* halfway between its rotational north and south poles. **Parallels of latitude** or **latitude circles** are small circles† parallel to the equator. **Meridians** pass from pole to pole and are accordingly at right angles to the equator; they are slightly elliptical but are considered as half-circles for our purposes. The meridian of Greenwich, or **prime meridian,** passes through the original site of the Royal Observatory at Greenwich, England. It crosses the equator in the Gulf of Guinea at the point where the longitude and latitude are zero.

The **longitude** of a place is the angular distance of its meridian measured along the equator east or west from the Greenwich meridian, from 0° to 180° either way. If the longitude is 60°W, the place is somewhere on the meridian 60° west of Greenwich meridian. The latitude of a place is its angular distance in degrees north or south from the equator, from 0° to 90° either way. If the latitude is 50°N, the place is somewhere on the parallel of latitude 50° north of the equator (Fig. 1.7). When the

* This is a circle going around the globe, on the largest circumference of the globe, that is formed by a plane going through the center of the globe. Because a great circle passes through the center of the earth, it divides it into two equal hemispheres.

† A small circle is any other circle on the surface of the globe.

Figure 1.7 *The earth's coordinate system of longitude and latitude.*

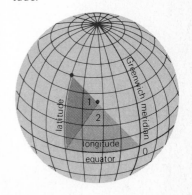

longitude and latitude are given, the position of a place is uniquely defined. As an example, the longitude of Yerkes Observatory at Williams Bay, Wisconsin, is 88°33′W, and the latitude is 42°34′N.

Here we are using the customary symbols for angular measure: degrees (°), minutes (′), and seconds (″). It should be clear when the symbol (°) is used to indicate degrees of temperature because we will always use it in conjunction with a capital letter. Thus °F, °C, and °K (or K) are degrees of temperature on the Fahrenheit, Celcius (or centigrade), and Kelvin or absolute scales, respectively. The symbols for feet (′) and inches (″) will never be used in this text.

The longitude of a place is often given in hours, minutes, and seconds east or west of the Greenwich meridian. The equator is divided into 24 hours, 12 east and 12 west. Since the full circle of the equator is 360°, one hour of longitude equals 15°. In our earlier example, the longitude of Yerkes Observatory would be 5ʰ 46ᵐ west.

The Conventional Globe of the Sky

A very early picture of the earth and sky represents their appearance so well that it is often employed in diagrams, as shown in Fig. 1.8. Here the earth is imagined as a circular plane on which the sky rests like an inverted bowl. A somewhat later picture, proposed by Greek scholars in the fifth century B.C., is useful in other diagrams (Fig. 2.9). It shows the sky as a spherical shell surrounding a spherical earth at its center. The earth is relatively so small that it may be considered to be only a point at the center of the celestial sphere.

1.4 The celestial sphere

Although the stars are scattered through space at various distances from the earth, the difference in their distances is not perceptible to ordinary observation. All the stars seem equally remote. As we view the evening

Figure 1.8 *The horizon coordinate system of altitude and azimuth.*

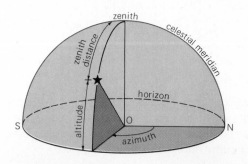

sky, we may imagine that the celestial bodies are set like jewels on the inner surface of a vast spherical shell. This **celestial sphere,** long regarded as a tangible surface, survives only as a convenient means of representing the heavens for many purposes. By this convention the stars can be shown on the surface of a globe or in projection on a plane map. Their positions are then denoted in the same ways that places are located on the globe of the earth.

The center of the celestial sphere may be the center of the earth, the observer's place on the earth's surface, the sun, or anywhere else we choose. The size of the sphere is as great as we care to imagine it. Because the sphere is infinitely large, parallel lines, regardless of their actual distance apart when nearby, will appear to converge at some point a great distance away. This is a familiar axiom from our high-school geometry studies. It is clearly demonstrated by the parallel railroad tracks, which seem to converge in the distance, or the parallel contrails of a high-flying aircraft. Similarly, parallel lines in space, regardless of their distance apart, are directed toward the same point on the remote celestial sphere. This concept will be referred to several times in the text.

The **apparent place** of a star is its position on the celestial sphere. It denotes the star's direction, and nothing else about its location in space. Where two stars have nearly the same direction, although one may be more remote than the other, they have nearly the same apparent place. In addition, the apparent places of the sun, moon, and planets refer to their positions among the stars projected on the celestial sphere, as though all were at the same distance from us.

The **apparent distance** between two celestial bodies is accordingly their difference in direction; it is often called the **distance** (i.e., the angular distance) between them when there is no chance for ambiguity. Such distances are expressed in angular measure. The distance between the pointers of the Big Dipper is somewhat more than 5°; it is a convenient measuring stick for estimating angular distances in the sky (Fig. 1.9).

How may the place of a star be described so that other people will know where to look for it? One way is to specify the constellation in which the star appears. If the star is in the constellation Perseus, anyone who can recognize the different constellations knows approximately where this star is situated. A second way of denoting the apparent place of a star is with reference to circles of the celestial sphere, such as the horizon.

1.5 The horizon

A cord by which a weight is suspended provides a vertical line when the weight comes to rest. This line leads upward to the **zenith,** the point directly overhead in the sky, and it leads downward through the earth to the **nadir,** the point directly underfoot (Fig. 1.10).

Figure 1.9 *The Big Dipper's pointer stars are very nearly 5° apart and make an excellent measuring reference as shown.*

Figure 1.10 *A plumb bob defines the direction of the zenith and the nadir.*

zenith

nadir

The **celestial horizon,** or simply the horizon, is the great circle of the celestial sphere that is halfway between the zenith and nadir, and therefore 90° from each. The direction of the horizon is observed by sighting along a level surface perpendicular to the zenith line. Obviously the positions of the zenith, nadir, and horizon among the stars are different at the same time in different parts of the world.

The **visible horizon,** the line where the earth and sky seem to meet, is rarely the same as the horizon of astronomy. On land the visible horizon is usually irregular and above the celestial horizon. At sea in calm weather it is a circle that lies below the celestial horizon; this **dip** of the visible horizon at sea increases with increasing height of the observer's eye above the level of the sea, caused by the curvature of the earth's surface.

If the observer is at sea and his eye is at sea level he sees the true horizon. If he is above the level of the sea he sees a depressed horizon caused by the dip of the horizon. Further, the observer's visible horizon recedes as he increases his height above the sea. We will note later that local sunrise, sunset, and so on, are affected by the dip of the horizon, that is, the observer's height.

1.6 The celestial meridian

Vertical circles are great circles of the celestial sphere that pass through the zenith and nadir, and accordingly across the horizon vertically. The most useful of these circles is the observer's **celestial meridian,** the vertical circle passing through the north and south poles of the heavens (Section 2.6). The meridian determines the positions of the four cardinal points of the horizon: north, east, south, and west.

North and south are the opposite points where the celestial meridian crosses the horizon. The east and west points are midway between them; as we face north, east is to the right and west is to the left. When the cardinal points are already located, it is proper to define the celestial meridian as the vertical circle passing through the north and south points of the horizon and the zenith.

1.7 Azimuth and altitude

The **azimuth** of a star can be measured in degrees along the horizon from the north point eastward to the foot of the vertical circle through the star. Thus the azimuth of a star is 0° if the star is directly in the north, 90° in the east, 180° in the south, and 270° in the west (Fig. 1.8).

The **altitude** of a star is its distance in degrees from the horizon, measured along the vertical circle of the star. The altitude is 0° if the star is rising or setting, 45° if it is halfway from the horizon to the zenith, and 90° if it is in the zenith. The zenith distance, or the star's distance in degrees from the zenith, is the complement of the altitude, i.e., 90° minus the altitude.

This is one way of denoting positions on the celestial sphere. If, for example, a star is in azimuth 90° and altitude 45°, the star is directly east and halfway from the horizon to the zenith; if it is in azimuth 180° and altitude 30°, the star is directly south and a third of the way from the horizon to the zenith. Certain instruments operate in this system based on the horizon. The engineer's transit is an azimuth–altitude instrument.

Positions of stars denoted in this way are correct only for a particular instant and for a particular place on the earth. The azimuths and altitudes of the celestial bodies are always changing with the daily motion of the heavens, and they are different at the same time in different parts of the world. More permanent positions relative to the celestial equator rather than the horizon are defined in Chapter 2. Meanwhile, we turn our attention to the earth's atmosphere and note particularly the effect it has on our view of the sky.

Effects of the Atmosphere

1.8 The region of the clouds

The earth's atmosphere is a mixture of gases surrounding the earth. From its pressure at sea level (1033.23 g-wt/cm^2) the mass of the entire atmosphere is calculated to be somewhat less than a millionth of the mass of the earth. The air becomes rarefied so rapidly with increasing elevation that half of it by weight is within 5.6 kilometers from sea level. Three-quarters of the atmosphere lies below the level of an airline passenger jet crossing the country at a typical cruising altitude of 11 kilometers.

The composition of the lower atmosphere is remarkably uniform. Its major constituents are nitrogen and oxygen molecules, found in proportions of 4 parts nitrogen to 1 part oxygen by volume. It also contains water vapor, carbon dioxide, and other gases in relatively small amounts, as well as dust in variable quantity.

When sunlight strikes the earth, a portion of the radiation is absorbed by the ground. The surface is thereby warmed and it reradiates this heat as longer wavelength infrared radiation. However, since carbon dioxide and water vapor, which permit the sunlight to reach the ground, absorb the longer wavelengths that are re-emitted, the temperature of the air rises near the surface. A similar phenomenon is observed in a greenhouse where the glass roof is transparent (like carbon dioxide and water vapor) to visible light but opaque to the infrared radiation which the ground emits. Heat is therefore retained instead of being radiated into space (Fig.

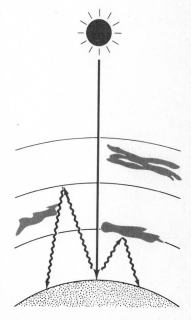

Figure 1.11 *A schematic effort to represent the greenhouse effect.*

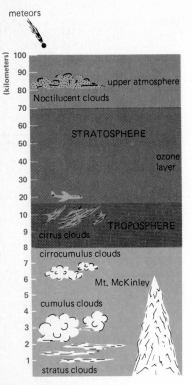

meteors

Figure 1.12 *A schematic representation of the earth's atmosphere.*

1.11). This process is known as the **greenhouse effect.** It serves to maintain the temperature of the lower atmosphere.

The lower atmosphere is divided into two layers (Fig. 1.12): the **troposphere** and the **stratosphere.** The troposphere extends from the surface to heights ranging from 16 kilometers at the equator to 8 kilometers at the poles. It is characterized by an abundance of water vapor and decreasing temperature with increasing altitude due to the greenhouse effect. Since warmer air lies beneath cooler denser air, the atmosphere near the surface is unstable and convection currents form, resulting in a verticle mixing of the air. The troposphere is therefore a region of turbulence where nearly all weather phenomena occur. (Tropo comes from a Greek word meaning change.)

The stratosphere extends from the troposphere to a height of 70 kilometers. It is characterized by dry air rich in ozone. Ozone, composed of molecules containing three atoms of oxygen instead of two, is formed mainly by the action of the sun's ultraviolet radiation on ordinary oxygen molecules. The absorption of the sun's ultraviolet rays by ozone warms the air in the stratosphere so that temperatures gradually increase with increasing height in this layer. As a result, the denser cooler air lies below the warmer air and the stratosphere is stably stratified. (Strato comes from a Greek word meaning layer.)

Ozone is most abundant at elevations from 15 to 30 kilometers. It helps to protect us from the ultraviolet rays of the sun, which would be injurious to life of all kinds if they could penetrate to the earth's surface. Until rockets could transport instruments above the ozone layer, the ultraviolet part of the spectrum was inaccessible to observation. Now that these wavelengths can be photographed, a whole new field of astronomy has opened up.

Clouds form in the lower atmosphere from the condensation of water vapor. Foglike stratus clouds begin to form at an average elevation of 1 kilometer. Cumulus clouds rise from flat bases about 2 kilometers above the ground. The cirrocumulus clouds of the "mackerel sky" (Fig. 1.13) have an average height of 6 kilometers, whereas the filmy cirrus clouds of ice crystals may be as high as 11 kilometers or more. Even higher clouds, called noctilucent clouds because they are observed after sunset and before sunrise, occasionally occur at heights of 90 kilometers. These clouds are composed of dust formed from the disintegration of meteors.

1.9 The upper atmosphere

The upper atmosphere reaches from the stratosphere to an altitude of more than 800 kilometers. Here the rarefied gases are most exposed to impacts of high-frequency radiations and high-speed particles from outside. The molecules are largely reduced to separate atoms, and the atoms

Figure 1.13 *A mackerel sky showing a well-defined pattern.*

themselves are shattered into electrically charged components. The **iono-sphere,** the region ranging from an altitude of 70 to 320 kilometers, contains at least four fluctuating layers (D, E, F_1, F_2) (Fig. 1.14) where the ionized gas is concentrated. By successive reflections between these layers and the ground, radio waves can travel long distances before they are dissipated. By the same token, the ionosphere limits the wavelengths of cosmic radio radiation that can reach the ground. When the layers are disrupted during a geomagnetic storm, communication by radio in the higher frequencies is disturbed.

The impact of particles from space on the gases of the upper atmosphere makes these gases luminous in the varied colors of the aurora and in the airglow. In the lower ionosphere the resistance of the denser air to the swift flights of meteors (Section 9.4) heats these intruders to incandescence, so that they produce bright trails across the sky.

1.10 The daytime sky

The stars are invisible to the unaided eye in the daytime because the atmosphere scatters the intense sunlight so that it reaches us from all parts of the sky. The sunlit air therefore conceals most celestial bodies by outshining them. Other than the sun, the moon is the only heavenly body ordinarily conspicuous in the daytime. The planet Venus near the times of its greatest brilliancy can be seen without a telescope as a star in the blue sky. The bright planets and stars, however, are easily visible

Figure 1.14 *A schematic draw-ing showing the relative levels of the ionosphere.*

250 km F_2

170 km F_1

120 km E

70 km D

Figure 1.15 *Bright planets and stars are visible in daytime with a telescope and can even be photographed near the edge of the sun.*

in the daytime sky with a telescope, and with special devices they can be photographed even near the edge of the sun (Fig. 1.15).

Why is the clear sky of the daytime blue, whereas the sunlight itself is yellow? Sunlight is composed of light of many colors, as we observe when the light passes through a prism, or through raindrops or the spray of a waterfall; it contains all the colors of the rainbow. As sunlight comes through the atmosphere, the violet and blue light is most easily reflected, or scattered, by the air molecules, and the red light is least affected. Hence on a clear day the sky takes on the blue color of the light that is scattered down to us preferentially.

When the sun is near the horizon, most of the blue of its direct light is scattered away before it can reach us through the greater thickness of air that then intervenes. Thus the sun appears reddened at its rising and setting.

1.11 Apparent flattening of the sun

Another aspect of the sun near the horizon is as familiar as its reddened color. There the sun sometimes appears so noticeably flattened at the top and bottom that its disk appears elliptical. This appearance is caused by refraction of the sunlight in the atmosphere. This is also true for the moon (Fig. 1.16). Refraction of light is the change in the direction of a

Figure 1.16   *Atmosphere refraction as one views the moon through the earth's atmosphere. (National Aeronautics and Space Administration photograph.)*

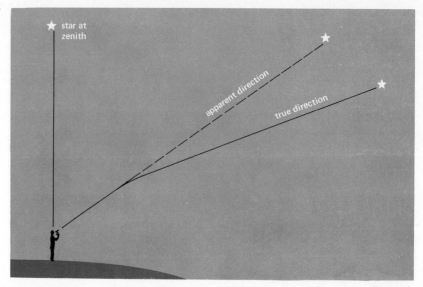

ray of light when it passes from one medium into another, as from rarer into denser air. The effect is described in Chapter 4 in connection with the operation of the refracting telescope.

As the light of a celestial body comes through the air, the rays are bent downward toward the earth by refraction (Fig. 1.17), so that the object appears higher in the sky than it actually is. The amount of the apparent elevation increases with increasing zenith distance of the object, but is so gradual at first that for more than half the distance to the horizon the difference is not enough to be detected without a telescope. Near the horizon the increase is rapid, until the effect becomes conspicuous to the unaided eye. An object on the horizon is raised above its true place more than half a degree, or more than the apparent diameter of the sun or full moon. Because of the rapid increase in the amount of refraction as the horizon is approached, the circular disk of the sun appears elliptical; the lower edge is raised considerably more than the upper edge.

In addition to its flattened appearance the sun's disk seems to be larger near the horizon than when it is higher in the sky. This is an illusion having nothing to do with refraction. By the same illusion, the moon and the star figures, such as the Big Dipper, seem magnified near the horizon.

1.12 Twinkling of the stars

A familiar feature of the clear night sky is the **twinkling,** or **scintillation,** of the stars; their light is unsteady, and the fluctuations in apparent brightness are especially noticeable near the horizon. The lower air is often turbulent; warmer (low density) air currents are rising, cooler (high density) air currents are descending, and horizontal movements of layers of different densities add to the confusion. Viewed through this turmoil the stars twinkle because of variable refraction and interference of their light, just as the landscape seems to be affected by the "heat waves" over the highway on a summer day.

The bright planets usually do not twinkle. They are disks instead of point sources of light, as a telescope shows. Each point of the disk may twinkle like a star, but the different parts do not do so in unison because their rays take slightly different paths through the disturbed air. Similarly, the moon shines with a steady light.

Viewed with a telescope, the stars appear blurred when the air is especially turbulent. The rays in each beam of starlight that enter the telescope have been diverted from perfect parallelism, so that they are not brought to the same focus by the lens. Features of the moon and planets also appear blurred in these conditions. When this occurs the astronomical **seeing** is said to be bad, and there is not much to be done about it except to wait for better seeing. With small apertures the image is less blurred but may shift about somewhat.

1.13 Lunar and solar halos

Bright rings, or **halos,** that sometimes appear around the moon and sun have no special astronomical significance; they are noticed and commented on by watchers of the skies and are fine examples of refraction effects in the atmosphere. The rings are produced by ice needles and snowflakes in the cirrus (thin, featherlike clouds that are high in the troposphere) and cirrostratus clouds. These six-sided crystals refract the moonlight and sunlight, concentrating it in certain directions. Although many effects are possible, the most common one is a single ring (Fig. 1.18) having a radius of 22° around the moon or sun. The ring often shows rainbow colors, with the red on the inner, sharper edge the most important.

"Moon dogs" and "sun dogs" are two enlargements of the ring on opposite sides of the moon or sun. These appear when many snowflakes in the clouds float with their bases in a horizontal position. A second ring having a radius of 46° and parts of other rings are seen less frequently (Fig. 1.18). The impression that the appearance of a ring around the moon or sun gives warning of an approaching storm has some factual basis because the filmy clouds that form the halos are likely to be blown ahead of storm clouds.

1.14 The aurora and the airglow

Two natural illuminations of the night sky in addition to the light of the celestial bodies are the aurora and the airglow. In the northern

Figure 1.18 *Solar halos: Common ring halo (bottom). The lower portion of an unusual complex halo (top). (Photographs by E. Everhart.)*

hemisphere the **aurora** is characterized by a luminous arch across the sky, with its apex in the direction of the geomagnetic pole. In the southern hemisphere they are visible in the southern sky in the direction of the south geomagnetic pole. Rays like searchlight beams rise above the arch, drifting, dissolving, and reforming, and draperies may appear in other parts of the sky. The rays are usually green, but may be red and yellow.

Figure 1.19 *A looped curtain aurora over Alaska. (Geophysical Institute, University of Alaska photograph.)*

Especially in latitudes farther north and south ribbons spread across the sky (Fig. 1.19). The aurora is referred to as the **Aurora Borealis** in the northern hemisphere and as the **Aurora Australis** in the southern hemisphere. Geomagnetic storms result from the influx of charged particles from unusual solar activity being trapped in the earth's magnetic field (Section 1.2). Where the magnetic field dips down into the ionosphere at the magnetic poles, these particles interact with the atmosphere, causing the aurora. The displays are the most intense in two zones centered around 23° from the magnetic poles. South of latitude 35° in the northern hemisphere, or south of a line through San Francisco, Memphis, and Atlanta, they appear only during geomagnetic storms of considerable intensity.

The **airglow** is an illumination suffused over the sky. Its light is also caused by energy coming primarily from the sun. It is faintest overhead and brightest not far from the horizon, which shows that the glow is from the atmosphere. It is undetected by the eye but is effectively studied with the photoelectric cell and color filters. Because the airglow is always present, it is the greater menace to celestial photography. The airglow fogs astronomers' photographs, placing a severe limit to the faintest objects that can be reached by increasing the exposure times. Only by carefully selecting a photographic plate and filter combination that eliminates the light from airglow or by using very high orbiting telescopes can this problem be circumvented.

The airglow should not be confused with the zodiacal light and gegenschein discussed in Section 9.6.

Questions

1 If the earth were represented by a globe 1 meter in diameter at the equator, how much shorter would the polar diameter be? What would the height of the highest mountains be on this scale?

2 If you wished to drill through the earth's crust to the mantle, where would you suggest drilling in order to do the least amount of drilling?

3 What type of seismic wave passes through the earth's core?

4 Draw up a list of useful studies from earth satellites. Discuss those that effect your everyday life directly.

5 What is the celestial sphere?

6 Define the celestial meridian.

7 Why is the daylight sky blue? Why is the sun generally red at sunrise and sunset?

8 Why does the sun (and moon) appear to be flattened at rising and setting?

9 The planets normally do not appear to twinkle. Why?

10 What is meant by the term airglow? Why is it a nuisance to astronomers?

Other Readings

Carrigan, C. R. and Gubbins, D., "Source of the Earth's Magnetic Field," *Scientific American,* **240(2)**, 118–130 (1979).

Jacchia, L. G., "The Earth's Upper Atmosphere," *Sky & Telescpe,* **49,** 155–159, 229–232, 294–299 (1975).

Lynch, D. K., "Atmospheric Halos," *Scientific American,* **238(4)**, 144–152 (1978).

Siever, R., "The Earth," *Scientific American,* **233(3)**, 82–91 (1975).

chapter 2

The Earth's Daily Rotation and Time

The distinction between rotation and revolution is more precise in astronomy than other sciences. Rotation is turning on an axis, whereas revolution is motion in an orbit. Thus, the earth rotates daily and revolves yearly around the sun. In this chapter we consider the earth's rotation and some of its effects as well as its relation to time.

Effects of Rotation on the Earth

Even the most casual observer of the heavens cannot escape from the observation that the celestial objects traverse the sky daily; the sun crosses the sky from east to west every day and the stars make the same trek every night. Because the earth is a seemingly stationary object, it was natural for people to assume that the stars were revolving around the earth and that the earth was fixed in space at the center of the universe. This concept of the cosmos, called the **geocentric theory,** was enshrined by the works of the Greek philosopher Aristotle around 350 B.C. and for almost two thousand years there seemed to be no reason to suppose that things might be otherwise. It was not until the Renaissance that evidence was presented which proved that it was the earth that was rotating and not the objects in the sky. Today we recognize many effects of the earth's rotation; the trajectory of rockets, the circulation of the atmosphere and oceans, and even the direction in which rivers flow.

2.1 The Foucault pendulum

A convincing proof of the earth's rotation, independent of any external reference point, was first demonstrated to the public by Jean Foucault in 1851. Under the dome of the Panthéon in Paris, Foucault freely suspended a heavy iron ball by a wire 61 meters long and started it swinging. The audience saw the plane of the oscillation slowly turn in a clockwise direction (Fig. 2.1). What they observed was the changing direction of the meridian due to the earth's rotation. This occurs because a **Foucault pendulum,** once set in motion, always swings in the same direction. On the rotating earth, however, the direction of the lines of longitude (the meridians) are changing continuously because the earth is spinning.

The rate of change in the direction of the swing of a pendulum depends on the latitude. At the equator the direction of the meridian in space is not altered by the earth's rotation because it is parallel to the earth's rotational axis and the direction of swing of a pendulum remains the same. At higher latitudes the meridians become more inclined to the earth's rotational axis. At the latitude of Baton Rouge the meridian is inclined 30° to the earth's rotational axis and it takes two full days for the pendulum to rotate through 360°. At the poles, where the meridians are perpendicular to the rotational axis, the pendulum rotates through 360° in a day.

2.2 The Coriolis effect

When standing on the earth we are carried along by its rotation. At the equator our speed is nearly 1670 kilometers per hour. At pro-

Figure 2.2 *The circumference of the circle traveled in one rotation period is less at high latitudes than at the equator, thus the speed is less.*

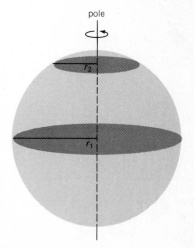

gressively higher latitudes our velocity due to the rotation of the earth decreases because our distance from the earth's axis of rotation decreases (Fig. 2.2). At the latitude of New York our rotational velocity is reduced to 1280 kilometers per hour and in southern Alaska it is only 800 kilometers per hour. At the poles, where we stand right on the earth's rotational axis, we would not move at all but simply spin around at a rate of one time each day.

The course of objects moving north or south over the earth's surface is influenced by the changing speed of the rotating earth beneath them. As an example, consider a rocket launched from the equator toward the north pole. When the rocket takes off it is traveling eastward with a velocity of 1670 kilometers per hour along with the ground underneath it. If the effects of air resistance are neglected, the rocket would still have this eastward velocity by the time it got to the latitude of New York. This means that the rocket, which was initially moving north with an eastward velocity equal to the ground beneath it, is now moving eastward with a velocity 390 kilometers per hour faster than the ground below. A rocket launched north from the equator would therefore appear to veer eastward in its flight over the earth; it would be deflected to the right relative to

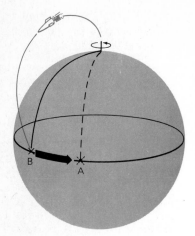

Figure 2.3 *A rocket launched from the north pole directly at its intended landing point will land behind that point because of the Coriolis effect.*

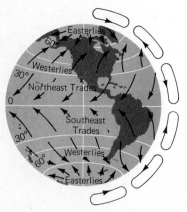

Figure 2.4 *Prevailing surface winds. The moving air is deflected to the right in the northern hemisphere and to the left in the southern hemisphere.*

the ground below. If a rocket were launched from the latitude of New York toward the equator the same thing would happen; it would be deflected to the right. In this case the rocket is moving from a place of slower rotational speed to an area of greater rotational speed. The result would still be a deflection to the right, i.e., westward (Fig. 2.3). In the southern hemisphere the result is just the opposite; whether moving north or south, the rocket would be deflected to the left. This deflection, known as the **Coriolis acceleration** after the scientist who demonstrated it, arises because the earth is not an inertial system of reference. That is, it is not a fixed reference frame but a rotating one.

Although most people are unaware of it, the Coriolis acceleration is constantly affecting us because it has a major effect on the world's weather patterns. The global circulation of the atmosphere is generated by the energy in sunlight, which heats the air most intensely in the tropics. The warmed air rises there and flows toward the poles. Cooled at higher elevations the air descends, notably at latitudes 30° where it flows north and south over the surface. From here the north-flowing winds move to 60°N where they rise and flow north and south again. The high-altitude air cools at the pole and descends. The same is true in the southern hemisphere. Because of the Coriolis acceleration, the surface winds are deflected to the right in the northern hemisphere and to the left in the southern hemisphere (Fig. 2.4). Thus we have the easterly trade winds of the tropics and the prevailing westerly winds of the temperate zones. The easterly winds of the frigid zones are an associated effect.

Ocean currents follow the prevailing winds in a general way but are complicated by the land barriers. We see clearly the effect of our rule in the Gulf Stream, which flows initially from the southwest.

Eddies in the air circulation, such as cyclones and hurricanes, show the Coriolis effect in the directions of their whirling. Consider a cyclone in the northern hemisphere, where the air is locally rising. The surface currents flowing into this area of low pressure are deflected to the right and reach its center indirectly. Thus the whole area, perhaps 2400 kilometers in diameter, is set whirling in a counterclockwise direction. In the anticyclones, or "highs" of our hemisphere, the air is descending and flowing out over the surface; these accordingly whirl in a clockwise direction. The directions of cyclones in the southern hemisphere are clockwise, and those of anticyclones are counterclockwise. The Coriolis effect can also be seen in the global circulation on the planets Mars and Jupiter.

2.3 Centrifugal effect of the earth's rotation

The equator is more than 21 kilometers farther from the earth's center than are the poles; in this sense it is downhill toward the poles. Why then doesn't all the water of the oceans assemble in these lowest regions around the poles? Why does the Mississippi River flow "uphill" toward the equator? The reason is found in the earth's rotation.

All parts of the rotating earth have a tendency to move away from the axis. It is the same **centrifugal effect** that urges a stone to fly away when it is whirled around at the end of a cord. Part of the effect of the earth's rotation on an object at its surface is to slide the object toward the equator (Fig. 2.5). The earth has adjusted its form accordingly so that the upslope toward the equator offsets the tendency to slide toward it.

The second component of the centrifugal effect lifts the object, so that its weight on a spring balance is diminished. An object weighing 86 kilograms at the pole, where there is no centrifugal effect, becomes almost 0.5 kilogram lighter at the equator because the distance from the center of the earth is greater. If the period of the rotation of the earth should decrease to 1 hour and 24 minutes by our present clocks, an object at the equator would weigh nothing at all when placed on an ordinary scale. At this excessive rate of rotation the earth would be unstable and liable to disruption.

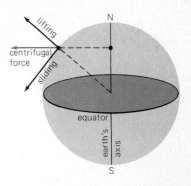

Figure 2.5   *The earth's rotation tends to diminish the weight of an object (except at the poles) and slide it toward the equator.*

2.4 Wanderings of the poles

When the positions of places on the earth are determined repeatedly with high accuracy, it is found that their latitudes do not remain the same. This **variation of latitude** has been observed for many years at a number of international latitude stations distributed over the world, and these results tell us how the equator and therefore the poles have been wandering. This causes slight changes in the apparent positions of celestial objects.

The wanderings of the poles called the **Chandler wobble** are caused by shifts of the earth relative to its axis, so that locations of the poles change slightly on the surface (Fig. 2.6). One variation is seasonal, having a period of a year; the other is an oscillation in a period of 14 months. Various contributions to the wandering of the poles are the seasonal variations of ice, snow, and air masses and elastic effects in the nonsolid earth due to rotation. The resulting motion of the poles is irregular and limited. Neither pole is withdrawing much more than 12 meters from its average place; all of the wanderings are now confined to an area smaller than that of a baseball diamond. Some scientists suggest the possibility of wider migrations in the past, which might have caused the marked changes in climate in geological times. Similar changes may have caused dramatic changes on Mars.

Figure 2.6   *The trace of the earth's north pole from 1972 to 1975. The year mark is at the beginning of that year.*

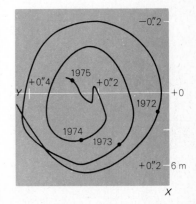

2.5 Changing period of the earth's rotation

The earth's rotation has long set the standard for our timekeeping. The consequent daily rotation of the heavens has provided the master clock by which all other clocks have been corrected. The earth-clock was formerly considered entirely reliable; the period of the earth's rotation was supposed to be uniform, until it was finally proved otherwise when more accurate timekeeping methods were developed.

Suppose that someone begins with the idea that his watch is always right. As the days go by he is surprised to find that everything is getting ahead of time by his watch. He misses trains that seem to depart too early; the sun rises before it should; the town clock runs faster and faster. Presently he decides that his watch must be running slow. It is so with the earth's rotation. Periodic occurrences, such as the revolution of the moon around the earth, are forging ahead of regular schedules timed by the earth-clock. We conclude that the period of the earth's rotation is increasing, and also that the increase is not perfectly regular (Fig. 2.7). This is discussed again in Section 5.7.

By comparing the recorded times and places of early eclipses with the calculated places (the times at which they would have occurred if the earth had been rotating uniformly), astronomers have concluded that the length of the day is increasing at the rate of 0.0016 second in a century. This would mean that the earth-clock has run slow by more than 3.25 hours during the past 20 centuries. The high- and low-water tides in the oceans are regarded as the brakes responsible for reducing the speed of the rotation. The tides are caused chiefly by the moon's attraction and the bulges of water, which produce the tides, follow the moon around the earth once in a month, while the earth itself rotates under the tidal bulges once a day. We discuss the tides in Chapter 5. Much uncertainty, however, remains in the size of the increase of the period. This uncertainty should be reduced with the present, more precise, means of making observations.

There are also sudden and not as yet accurately predictable variations in the length of the day. The earth's rotation has run off schedule, both

Figure 2.7 *A graph of the earth's rotation period referred to a constant time base. The earth's period of rotation is lengthening.*

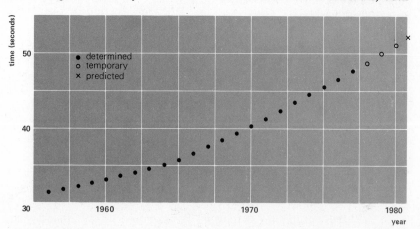

fast and slow, as much as half a minute after allowance for the tidal retardation is made. Small periodic variations, mainly annual and semiannual, are detected by the best clocks; these changes are ascribed to winds and tides, including tides raised in the solid earth, and they are reliably repeated.

Apparent Rotation of the Heavens

The stars rise and set daily, circling westward and keeping precisely in step as they go around. The patterns of stars, such as the Big Dipper, look the same night after night and year after year. It is as though the stars were set on the inner surface of a rotating hollow globe. This celestial sphere seems to turn daily from east to west around an axis which is the axis of the earth's rotation projected to the sky.

2.6 The equatorial coordinate system

Astronomers need to be able to describe the position of a star exactly so that other astronomers may find it. But the rotation of the earth causes the position of stars to change continuously. At different times of the year, different stars occupy the sky, and depending on the location of the observer on the earth, different stars occupy his field of vision. As a result, if the position of a star had to be described according to the observer's local horizon, the exact time of the night, the day of the year, and the observer's position would all have to be taken into account. It would obviously be impractical to use such a system. To circumvent this problem, astronomers have devised various coordinate systems which are fixed on the celestial sphere and rotate with the stars. These coordinate systems are used for special purposes, but the equatorial coordinate system is in general use because of its utility to the observational astronomer. We will introduce the galactic coordinate system later on.

In the **equatorial coordinate system** the celestial sphere is mapped out on a grid like the coordinates of latitude and longitude on earth. The **celestial poles** are the two opposite points on the celestial sphere toward which the earth's axis is directed and around which the stars circle. The north celestial pole is directly in the north, from a third to halfway up in the sky for observers in different parts of the United States. The south celestial pole is similarly depressed below the south horizons of these locations.

The pointers of the Big Dipper (Fig. 2.8) direct the eye to Polaris, the pole star, at the end of the Little Dipper's handle. This moderately bright star is within 1°, or less then two moon diameters, of the pole itself. It is also the north star, showing the approximate direction of north.

Figure 2.8 *The Big Dipper's pointers show the way to the north celestial pole.*

The south celestial pole is not similarly marked by any bright star in its vicinity.

Just as the earth's equator is halfway between the terrestrial poles, so the celestial equator is halfway between the north and south celestial poles (Fig. 2.9). This circle crosses the horizon at its east and west points at an angle that is the complement of the latitude. Thus in latitude 40°N, or the latitude of Philadelphia, the celestial equator is inclined 50° to the horizon and has an altitude of 50° at its highest point in the south.

Hour circles in the sky are like longitude circles on the earth. They are half circles that connect the celestial poles and are therefore perpendicular to the equator. Unlike the circles of the horizon system, which are stationary relative to the observer, these circles are to be considered as sharing in the rotation of the celestial sphere. If 24 hour circles are imagined equally spaced, they will coincide successively with the observer's celestial meridian at intervals of 1 hour. With reference to the celestial equator and its associated circles, the position of a celestial body is given by its right ascension and declination, which resemble terrestrial longitude and latitude.

The **right ascension** of a star is its angular distance measured eastward along the celestial equator from the vernal equinox to the hour circle through the star. (The **vernal equinox** is the point where the sun's center

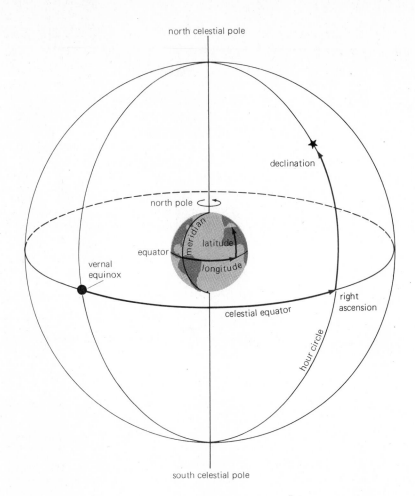

north celestial pole

declination

north pole

meridian

latitude

equator

longitude

vernal
equinox

celestial equator

right
ascension

hour circle

south celestial pole

Figure 2.9 *The equatorial coordinate system. Right ascension is measured eastward from the vernal equinox along the celestial equator (i.e., counterclockwise as viewed from the north celestial pole). Declination is measured north and south from the celestial equator.*

crosses the celestial equator at the beginning of spring; that is, the intersection of the ecliptic and the celestial equator where the sun passes from the southern hemisphere to the northern hemisphere.) Right ascension is expressed in time more often than in angular units. Because a complete rotation of the heavens, through 360°, is made in 24 hours, 15° is equivalent to 1 hour. Thus a star's right ascension may be given as 61° or 4 hours and 4 minutes.

The **declination** of a star is the star's distance in degrees north or south from the celestial equator, measured along an hour circle through the star. The declination is marked either N or with a plus sign if the star is north of the equator, and S or with a minus sign if it is south.

As an example, the right ascension of the star Arcturus is 14^{h}15^m and its declination is 19°18'N. The star is accordingly 213°75' east of the vernal equinox and 19°18′ north of the celestial equator. Notice that right ascension is measured only eastward, whereas terrestrial longitude is measured both east and west. The approximate right ascensions and declinations of the brighter stars can be read from the star maps in Chapter 11.

The local **hour angle** of a star is reckoned westward along the equator from the observer's celestial meridian through 360° or 24 hours. Unlike right ascension, which remains nearly unchanged during the day, the hour angle of a star increases at the rate of 15° per hour and, at the same instant, has different values for observers in different longitudes.

The equatorial coordinate system is most widely used for locating the position of a star with a telescope and tracking it across the sky. This is because a telescope mounted with its axis parallel to the earth's axis of rotation need only rotate around this axis in order to remain pointing at the same star. The driving mechanism for a telescope mounted in the equatorial coordinate system is therefore very simple.

2.7 Latitude equals altitude of celestial pole

In Fig. 2.10 we see that the observer's latitude is the same as the declination (degrees north or south of celestial equator) of his zenith. This angle equals the altitude of the north celestial pole, because both are complements of the same angle between the directions of the zenith and pole. Thus, *the latitude of a place equals the altitude of the celestial pole at that place.*

Figure 2.10 *The latitude of a place on the earth equals the altitude of the celestial pole at that place.*

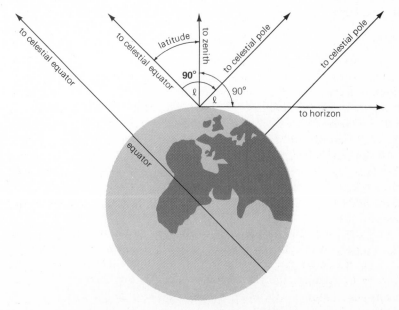

The rule above determines the **astronomical latitude.** This observed latitude depends on the vertical line and is affected by anything that alters the vertical. A mountain near the place of observation would cause the plumb line to incline a little in its direction. Such "station errors" may make a difference of nearly a minute of arc in the latitude, although they are usually much smaller. **Geographical latitude** is the observed latitude corrected for station error; it is the latitude that would be observed if the earth were perfectly smooth and uniform.

The latitude of a place on the earth is determined by the rule mentioned above. If there were a bright star precisely at the celestial pole, the latitude could be found simply by measuring the altitude of that star. The pole star itself may be employed if correction is made for its distance from the pole. More often the observations are made to determine the declination of the zenith, which we have seen is the same as the latitude. The latitude is then simply the co-altitude of the celestial body plus its declination (Fig. 2.11).

Suppose that a navigator sights the sun at its crossing of the meridian south of the zenith and determines its true altitude as 51°10′, so that its zenith distance is 38°50′. The sun's declination obtained from an almanac for the time of the sight is 22°0′N. The latitude equals the sun's zenith distance plus its declination or 60°50′N (Fig. 2.11). If the sun had crossed the meridian the same distance north of the zenith, the zenith distance would receive the minus sign, and the latitude would have been 16°50′S.

Let us now turn the latitude rule around: when the latitude of a place is given, we know the altitude of the north celestial pole at that place and how the daily courses of the stars are related to the horizon.

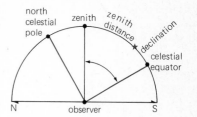

Figure 2.11   *The latitude of a place equals the zenith distance of a celestial body at its upper transit plus its declination at that time.*

2.8 Stars observed from earth

At the north pole, latitude 90°, the north celestial pole is in the zenith, and the stars go around it daily in circles parallel to the horizon. These circles are called **diurnal circles.** Stars north of the celestial equator never set, while those south of the equator never come into view (Fig. 2.12). In this statement we avoid confusion by referring to the true positions of the stars and not as they are elevated by atmospheric refraction (Section 1.11). The sun, moon, and planets rise and set as seen from the north pole whenever they cross the celestial equator. The sun rises about 21 March and sets about 23 September. The moon rises and sets about once a month.

At the south pole everything is reversed. There the south celestial pole stands in the zenith, and the stars of the south celestial hemisphere never set. The long period of sunshine begins with the rising of the sun about 23 September and ends about 21 March.

At the equator, where the latitude is zero, the altitude of the celestial

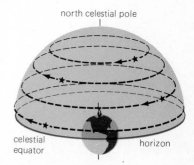

Figure 2.12   *At the pole, stars circle parallel to the horizon.*

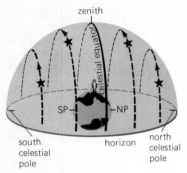

Figure 2.13 *At the equator the stars rise and set perpendicular to the horizon.*

pole is zero by our rule. The north celestial pole is situated at the north point of the horizon, and even the pole star now rises and sets as it circles near this pole (Fig. 2.13). The south celestial pole is at the south point. The celestial equator passes directly overhead from east to west.

At the equator, all parts of the heavens are brought into view by the apparent daily rotation. All stars rise and set. Their courses cross the horizon at right angles and are bisected by it, so that every star is above the horizon 12 hours daily if we neglect refraction. This applies to the sun as well; the duration of sunshine is 12 hours at the equator throughout the year.

The heavens turn obliquely for those who live between the pole and equator. Suppose that we are observing from latitude 40°N. Here the north celestial pole is 40° above the north point of the horizon. The celestial equator arches across from the east point to the west, inclined so that its highest point is 40° south of the zenith. The daily courses of the stars are likewise inclined toward the south. Half the celestial equator is above the horizon (Fig. 2.14). North of the equator the daily courses of the stars come up more and more above the horizon until they are entirely above it. Southward from the celestial equator they are depressed more and more until they disappear completely.

In this oblique arrangement the celestial sphere is divided into three parts: (1) a circular area around the elevated celestial pole contains the stars that never set; (2) a similar area around the depressed pole contains the stars that never rise; (3) the remaining band of the heavens symmetrical with the celestial equator contains the stars that rise and set. Hence, in latitude 40°N the area of the heavens that is always above the horizon is within a radius of 40° around the north celestial pole, and the area that never comes into view is within a radius of 40° around the south celestial pole. The remainder of the heavens, a band extending 50° on either side of the celestial equator, rises and sets.

2.9 Circumpolar stars

Circumpolar stars (Fig. 2.15) go around the celestial poles without crossing the horizon. These stars either never set or never rise, because they are closer to one of the celestial poles than the distance of that pole from the horizon, a distance that is equal to the observer's latitude. We accordingly make this rule about them for observers in the northern hemisphere.

A star having a distance from the north celestial pole (90° minus the star's declination) less than the latitude of a place never sets at that place. A star having a distance from the south celestial pole less than the latitude never rises.

Suppose that the observer is in latitude 40°N, and consider as an

Figure 2.14 *For observers between the pole and the equator, the stars rise and set obliquely.*

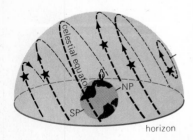

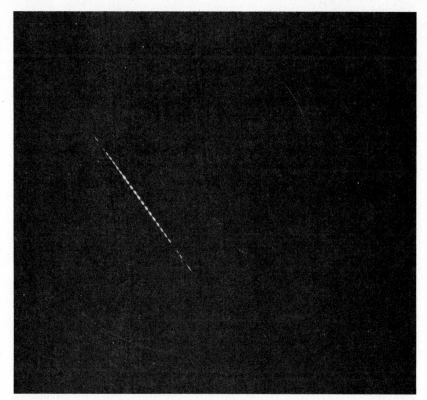

Figure 2.15 *Circumpolar star trails taken with a meteor camera. The north star is the short bright trail near the center. A bright meteor is seen as dashes across the picture.*

example the bowl of the Big Dipper, declination about 58°N. It never sets here because its north polar distance of 32° is less than the latitude. As a second example, consider the star Canopus, declination 53°S. This brilliant star is not visible in latitude 40°N because its south polar distance of 37° is less than the latitude.

If we travel north, the celestial poles move farther from the horizon. The polar areas grow larger until they come together when the pole is reached; there as we have seen no stars rise and set. If we travel south, the celestial poles approach the horizon. The polar areas grow smaller until they disappear when the equator is reached, where all stars rise and set.

2.10   The duration of twilight

The gradual transition between daylight and the darkness of night, which we call **twilight,** occurs during the time that the sun, after its setting and before its rising for us, can shine on the atmosphere above us. What is said here about evening twilight applies in reverse to the morning twilight.

Civil twilight ends when the sun's center has sunk 6° below the horizon. Then it is no longer possible without artificial illumination to continue outdoor operations that require good light. **Nautical twilight** ends when the sun's center is 12° below the horizon. Then the sea horizon is likely to be too dim for the navigator's sextant sights. **Astronomical twilight** ends when the sun's center is 18° below the horizon; by that time the fainter stars have become visible overhead. The times of sunset and sunrise and the duration of twilight can be found in some of the almanacs for any date and latitude. The minimum times for the three critical twilights on clear days can be estimated in a rough way from the fact that the sun, by definition, moves 15° per hour giving 24, 48, and 72 minutes, respectively.

The duration of twilight varies with the time of year and the latitude. Twilight is shortest at the equator where the sun descends vertically and so reaches the limiting distance below the horizon in the shortest time. Astronomical twilight lasts somewhat more than an hour at the equator, and an hour and a half or more in the latitude of New York. On 22 June it does not end at all north of 48°N, about the latitude of Victoria, British Columbia. Civil twilight persists from sunset to sunrise on this date from latitude 60°, or the latitude of Oslo, Norway, to nearly 66°, where the midnight sun is seen. On this same date Tierra del Fuego, Argentina, catches only a brief glimpse of the sun.

Time

Our watches are likely to be more reliable timepieces if they are compared frequently to some common time standard. Previously our time standard was a clock at an observatory which was checked frequently against the master clock in the sky which itself is operated by the rotation of the earth. The present time standard is a set of atomic clocks stationed around the world which are compared frequently. This is because the rotation of the earth is not constant enough to act as a standard for modern timing requirements. Time broadcast from the standard clocks is periodically adjusted so that civil time does not get out of step with the day. While the determination of time is taken for granted because time seems so obvious and natural, we will see that the determination of time is actually rather complex.

2.11 The clock in the sky

Any one of the stars or any other object on the celestial sphere might be chosen as the time reckoner, which can be placed at the end of an hour hand of the clock in the sky. The hour hand would be that part of an hour circle connecting the time reckoner and the celestial pole. This hour hand would go around once each day, telling the time of day to those who can read it. The following definitions are true for any point that is selected as the time reckoner.

We define the **day** to be the interval between two successive **transits** (crossings) of the time reckoner over the same portion of the celestial meridian. Since a star does so twice a day, the distinction is made between **upper transit,** over the half of the circle through the celestial poles that includes the observer's zenith, and **lower transit,** over the other half that includes the nadir. It is noon when the time reckoner is at upper transit. The local time of day is the hour angle (Section 2.6) of the time reckoner if the day begins at noon, or it is 12 hours plus the hour angle if the day begins at midnight.

2.12 Sidereal time

Instead of selecting one of the stars as the time reckoner for **sidereal time,** or "star time," astronomers chose the vernal equinox. The **sidereal day** is accordingly the interval between two successive upper transits of the vernal equinox. This interval is 0.008 second shorter than the period of the earth's rotation, because of the precession of the equinox. Sidereal time, which might have been more correctly called equinoctial time, is the hour angle of the vernal equinox. We should be careful to note that sidereal time takes into account general precession and is not the interval between successive upper transits of a given star but rather those of the vernal equinox.

Sidereal time is kept by special clocks in the observatories, which used to be set directly from the clock in the sky. The principle is as follows: the sidereal time at any instant is the same as the right ascension of a star that is at upper transit at that instant. Evidently what is needed is something to show the place of the celestial meridian precisely, so that the instant of the star's crossing can be correctly observed. The astronomical transit instrument is used for this purpose.

Suppose that the sidereal clock reads $3^h42^m22.7^s$ at the instant the star is at upper transit and that the star's right ascension as given in a catalog is $3^h42^m26.2^s$, which by the rule is the correct sidereal time at that instant. The sidereal clock, therefore, is 3.5 seconds slow. By comparison of clocks and a simple calculation, the error of a standard time clock can be found as well.

Recording devices are employed to time the transits of stars because they are more accurate than direct visual observations. The **photographic zenith tube** has replaced the simple transit instrument at the U.S. Naval Observatory and elsewhere. It is a fixed vertical telescope that photographs stars as they cross the meridian nearly overhead. With this device an

Figure 2.16 *The solar day is longer than the sidereal day because the earth must rotate farther to bring the sun back on the meridian.*

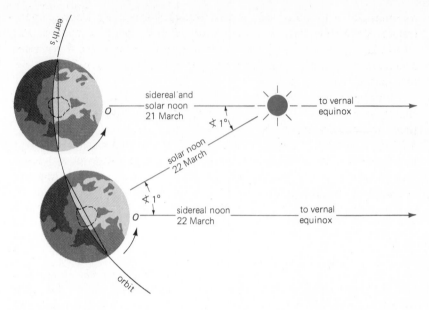

error of only 0.003 second is expected in a time determination from a set of 18 stars.

An even more accurate instrument which we mention in passing is the **modern astrolabe.** It makes observations east and west of the meridian in such a way that the meridian and zenith are well defined. Observations with this instrument have an error of only 0.001 second.

2.13 The solar day is longer than the sidereal day

Suppose that the sun's center has just reached the vernal equinox and they both are at upper transit on the celestial meridian of the observer at *O* (Fig. 2.16). It is sidereal noon and also solar noon, that is, noon by the sun. The sidereal day will end when the earth has made a complete

Figure 2.17 *The earth revolves faster during the northern winter because it is nearer the sun.*

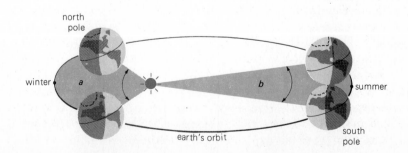

rotation relative to the vernal equinox, bringing the equinox again to upper transit. Meanwhile, the earth has advanced in its orbit around the sun an average of 360°/365.25, or a little less than 1° or about 4 minutes. The earth must turn this much farther, thus the solar day is longer than the sidereal day by about 3^m56^s.

Since the apparent rotation of the celestial sphere is completed in slightly more than a sidereal day, a star rises at nearly the same sidereal time throughout the year. On mean solar time, however, it rises almost 4 minutes earlier from night to night, or 2 hours earlier from month to month, so that at the same time on successive nights the star appears a little farther west than on the previous night. Thus at the same hour by our watches the stars march slowly westward across the sky as the year advances (Section 3.1). Each season, therefore, brings its own display of constellations.

2.14 Apparent and mean solar time

Since our activities are regulated by the sun and not by the stars, we prefer solar time rather than sidereal time for civil purposes. Sidereal noon, for example, comes at night during half of the year. **Apparent solar time** is time by the apparent sun, that is, the sun we see; it is reckoned from local apparent midnight through 24 hours to the next midnight. The apparent sun, however, is not a uniform timekeeper. The apparent solar day, measured by the sundial, varies in length for two principal reasons:

1 **The earth's revolution around the sun is not at a uniform rate.** The earth revolves in its elliptical orbit in accordance with the results of the Newtonian laws of motion (Section 6.3). From these we learn the earth's velocity in its orbit is greatest when it is nearest the sun (Fig. 2.17). Accordingly, the earth revolves farther in a day from November through January and must rotate farther after the ending of the sidereal day to bring the sun again to the observer's meridian. For this reason the apparent solar day is longer in our winter than in summer.

2 **The ecliptic is inclined to the celestial equator.** When the sun is near a solstice, where the ecliptic is parallel to the celestial equator, its daily eastward displacement by the earth's revolution has the full effect in delaying the sun's return to the observer's meridian. When it is near an equinox, however, part of the sun's daily displacement is north or south, and this part does not delay the return. For this reason the apparent solar day is longer in summer and winter than in spring and fall. Both effects conspire to make the day by the apparent sun longer during winter in the northern hemisphere.

Figure 2.18 *A cesium-beam atomic clock. This clock is a major advance in timekeeping and is accurate to better than a millionth of a second over a period of three weeks. (Official U.S. Navy photograph.)*

**Table 2.1
Equation of time**[a]

1 Jan.	− 3^{m}22^s	
1 Feb.	− 13	35
1 Mar.	− 12	29
1 Apr.	− 4	02
1 May	+ 2	54
1 Jun.	+ 2	21
1 Jul.	− 3	39
1 Aug.	− 6	16
1 Sept.	− 0	06
1 Oct.	+ 10	12
1 Nov.	+ 16	22
1 Dec.	+ 11	05

[a] Apparent solar time is faster (+) or slower (−) than mean solar time.

Mean solar time is time determined by the mean sun; it is reckoned from local mean midnight through 24 hours to the next midnight. The **mean sun** may be regarded as a point moving eastward along the celestial equator at a rate equal to the average rate of the apparent sun's motion along the ecliptic. This conventional sun would be a smooth running time-keeper if the earth's rotation were precisely uniform. Its day, which is the average of all apparent solar days through the year, would then have constant length.

The **equation of time** is the difference at any instant between apparent and mean solar time; its value can be found from tables in astronomical almanacs. Table 2.1 shows how much the apparent time is fast or slow with respect of mean solar time on the first of each month. The values are for midnight at the Greenwich meridian averaged over 18 years. The table holds for any other meridian, but in any given year the actual value on the day given may differ by as much as 10 seconds either way except on 1 November, when the equation of time changes very little. The maximum discord in time occurs on 3 or 4 November and may reach 16^{m}25^s. In 1979, for example, the discord was greatest (16^{m}24^s) on 2 November.

The rapid change in the equation of time near the end of one year and the beginning of the next has an effect that is noticed by everyone. At this time of year the earth is nearest the sun and is accordingly revolving fastest. The apparent sun is then moving eastward fastest along the ecliptic, delaying its rising and setting as timed by the mean sun. For this reason the sun does not begin to rise earlier in the morning by our watches, which keep mean time, until about 2 weeks after the date of the winter solstice, although it begins to set later in the evening about 2 weeks before that date.

2.15 Universal time and ephemeris time

The international unit of time is the **atomic second** defined by transitions between the two hyperfine-structure, ground-state levels of the cesium-138 atom in the absence of a magnetic field. What this means is that the cesium atom has two energy levels in its ground, or lowest, state. When the atom goes from one level to the other it will absorb or emit a photon of a very precise wavelength or frequency, and the atomic second is defined in terms of this frequency (Section 2.16).

Universal time is the local mean solar time at the Greenwich meridian. It is the basis of all ordinary timekeeping and is likely to continue to be, because observatories that transmit or monitor mean solar time signals keep their clocks corrected by frequent checks of the master atomic clocks and the meridian transits of the stars. These observations are said to be **universal time zero (UT 0).** For certain purposes this time must be corrected for the variation of the observer's meridian resulting from the motion of the poles (Section 2.4). This time is called **UT 1.** A third universal time is **UT 2,** which is UT 1 corrected for variations in the rate of rotation of the earth (Section 2.5). The time actually broadcast (Section 2.18) and hence used in civil affairs is **UT C (coordinated universal time),** which is a smoothed approximation of UT 2. One can clearly see that the matter of timekeeping is not to be taken lightly. In forecasting the universal times of astronomical events, the irregularities in the earth's rotation make precise predictions difficult. For this purpose we define a more uniform time.

Beginning with the year 1960, the *American Ephemeris and Nautical Almanac* and the *British Astronomical Ephemeris,* which now conform in other respects as well, have tabulated the fundamental ephemerides* of the sun, moon, and planets for intervals of ephemeris time. **Ephemeris time** goes on uniformly without regard for irregularities in the earth's rotation. Its constant arbitrary unit equals the length of the tropical year at the beginning of the year 1900 divided by 31,556,925.9747, which was the number of mean solar seconds in the year at that epoch.

Ephemeris time has run increasingly fast with respect to universal time during the present century. The difference between the two was 50 seconds in 1979.

---

* Ephemeris is derived from the Greek word meaning diary and is a table of predicted positions of a planet, the moon, or the sun.

2.16 Atomic clocks

We have seen (Section 2.5) that the earth's rotation is not exactly uniform and that the day is increasing in length. While the small changes cause no inconvenience to our everyday affairs, they do make it difficult to time intervals with precision. As the need for greater and greater accuracy in time measurements grew, it became necessary to rely on the inherent accuracy of atomic clocks. Clocks based upon the vibration of the ammonia molecule have proven highly satisfactory for the measurement of precise intervals of time. In this clock, microwaves from a well-controlled and correctable oscillator are transmitted into ammonia gas and are absorbed if they vibrate at the exact frequency of the gas. If the frequency of the microwaves drifts, the amount absorbed decreases and this is sensed by a circuit that then corrects the oscillator. Such clocks are accurate to within a few seconds in 50 years.

Even more accurate atomic clocks use a beam of cesium atoms (Fig. 2.18) sorted out by a series of magnets and irradiated by microwaves of the proper frequency. In nature the spin axes of cesium atoms are distributed at random. A first set of magnets serves to align the atoms, one-half spinning one direction and the other half spinning the other direction. The microwave radiation then serves to flip the atoms over. If the microwaves are at the exact frequency all flip over and the numbers with each spin are the same, but if the microwave frequency drifts even the least amount more atoms with one spin flip over than do those of the other. A second series of magnets serves to separate cesium atoms of one spin to one detector and those of the opposite spin to another detector. Thus if the microwave frequency has drifted, one detector or the other gets an increased number of cesium atoms and the microwave generator must be corrected accordingly. Such clocks are accurate to within a few seconds over several centuries.

Atomic time referred to as **A 1** is defined in terms of the vibrations of the cesium atom. The second of atomic time is agreed to be 9,192,631,770 vibrations of the cesium atom. Atomic time agreed exactly with UT 2 at midnight on 1 January 1958 by international agreement. At that time UT 2 differed from ephemeris time by 32.15 seconds, so ephemeris time differs from atomic time by this amount and the difference remains constant.

The **atomic second,** like the second of ephemeris time, is a precise interval of time. If we were to keep counting the same number of seconds per year, civil time as kept by atomic clocks would get out of step with civil time as determined by the rotation of the earth (Fig. 2.7). Since we adjust our living patterns to the latter, an extra second, or **leap second,** is placed in the time that is broadcast for civil purposes. This is done whenever atomic time gets ahead of civil time by ½ second. For example, the last minute of 1977 contained 61 atomic seconds instead of the normal 60.

2.17 Time zones

Unlike sidereal time and apparent solar time, which are always local time, mean solar time is most often employed in the conventional forms of zone time and standard time now to be considered.

The difference between the local times of two places at the same instant is equal to the difference between their longitudes expressed in time. The rule applies to any kind of local time, whether it is sidereal, apparent solar, or mean solar, so long as the same kind of time is denoted at the two places. When the local time at one place is given and the corresponding local time at another place is required, add the difference between their longitudes if the second place is east of the first, subtract if the second place is west.

In the time diagram of Fig. 2.19 we are looking at the earth from above its south pole and are projecting the sun onto the equator in the longitude where the sun is overhead; the east to west direction is counter-clockwise. The observer on the earth is in longitude 60°W, or 4^{h}W. The **local mean time (LMT)** there is 9^h and the **Greenwich mean time (GMT)** or universal time is 13^h, so that the difference of 4^h in the local times of the two places is the difference of their longitudes. In this case the time of day is counted through 24 hours continuously: 9^h is 9 A.M. and 13^h is 1 P.M.

Conversely, the difference between the longitudes of two places is equal to the difference between the local times of the places at the same instant. This is the basic rule for determining longitudes.

The local mean solar times at a particular instant are the same only for places on the same meridian. The time becomes progressively later toward the east and earlier toward the west; the rate of change at latitude 40° is 1 minute of time for a distance of a little more than 21 kilometers. The confusion that would ensue if every place kept its own local time is avoided by use of time zones and is called **zone time.** In its simplest form the plan is as follows:

Standard meridians are marked off around the world at 15° intervals, or 1-hour intervals, east and west from the Greenwich meridian. Accordingly, the local times on these meridians differ successively by whole hours. The time to be kept at any other place between the meridians is the local mean time of the standard meridian nearest that place. Thus the world is divided by boundary meridians into 24 equal time zones, each 15° wide and having one of the standard meridians running centrally through it. The time is the same throughout each zone; it is 1 hour earlier than the time in the adjacent zone to the east and 1 hour later than the time

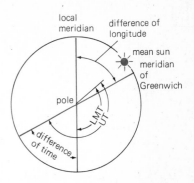

Figure 2.19 *Difference of local time between two places equals their difference in longitude.*

in the zone to the west. The rule to be followed when crossing a boundary between zones on an eastward voyage is to set your watch ahead 1 hour, the minutes and seconds remaining the same as before, and on a westward voyage to set your watch back 1 hour.

Zone time is used in the operation of ships at sea. This uniform plan is occasionally modified near land so that the ship's clock may agree with the standard time kept ashore. Because of the speed of modern aircraft it has become the practice to use Universal Time as the standard operations time. This is so even within the continental United States, although the prevailing local time is used for public schedules.

Standard time divisions on land generally follow the pattern of the time zones at sea. Their boundaries are often irregular, they are affected by local preference, and are subject to change. In some areas the legal time differs from the times in adjacent belts by a fraction of an hour. There is also the arbitrary and non-uniform practice of setting the clocks ahead of the accepted standard time for part of the year, called **daylight saving time** and often referred to as "fast" time. In the United States daylight saving time has been made uniform, beginning at 2 A.M. on the fourth Sunday in April and ending at 2 A.M. on the fourth Sunday in October and has been adopted on a state-by-state basis.

Four standard times are employed in the greater parts of the continental United States and Canada, namely, **Eastern, Central, Mountain,** and **Pacific** standard times. The standard time in Newfoundland and Labrador is 3½ hours instead of 4 hours slow, and in Alaska is generally 10 hours slower than universal time. Apart from Quebec, Ontario, and the Northwest Territories, each province of Canada has adopted a single standard time.

Clearly, if we work our way around the earth the date must change abruptly by one day somewhere. By international agreement a line approximating the meridian 12 hours east (and west) of the Greenwich meridian is referred to as the **international date line** and is the meridian where the date changes by one day.

2.18 Radio time service

Time signals from the *National Bureau of Standards* (NBS) transmitters at Fort Collins, Colorado and the *Dominion Observatory* in Ottawa supply convenient standard time service to all of North America. Throughout North America the time from NBS can be obtained by telephone as well by dialing 303–499–7111. The time given is UT C (Section 2.15). These signals are monitored by the U.S. Naval Observatory and compared with observations of the stars. The time broadcast is allowed to accumulate an error of 0.7 second and then the signals are corrected by a whole second. These changes are announced in advance and detailed corrections for the entire year past are published.

Questions

1 Study Fig. 2.1. Suppose you are suspended by a boom from the top of the Empire State building above 34th Street and drop a steel ball: will it fall east or west of the spot exactly below?

2 If a washbowl were perfectly hemispherical and the drain exactly at the bottom, in what direction should the water swirl as it drains out?

3 What is the sliding force at the north pole? At the equator?

4 What is the lifting force at the north pole? At the equator?

5 What are some of the causes of the changing period of the earth's rotation?

6 Describe how to find Polaris on a clear night.

7 Give the rule for finding your latitude when you are in the northern hemisphere.

8 The declination of the star Canopus is $-52°41'$; at what latitude should you see this star just peak above the horizon? At what latitude will Canopus just be circumpolar?

9 Why does astronomical twilight end so much later in the evening than the other named twilights?

10 What is meant by apparent solar time?

Other Readings

Markowitz, W., "Polar Motion: History and Recent Results," *Sky & Telescope,* **52,** 99–103, 108 (1976).

Newton, R. R., "The Crime of Claudius Ptolemy," Johns Hopkins University Press, Baltimore, 1978.

Reid, F. and K. Honeycutt, "A Digital Clock for Sidereal Time," *Sky & Telescope,* **52,** 59–63 (1976).

Seidelmann, P. K., "A Design for an Analemmatic Standard-Time Sundial," *Sky & Telescope,* **50,** 368–369 (1975).

Shrader, W. W., "A Sundial on an Office Ceiling," *Sky & Telescope,* **49,** 217–218 (1975).

chapter 3

The Earth's Annual Revolution and the Calendar

The earth revolves around the sun once a year. This annual revolution causes the constellations to systematically shift their positions during the course of the year. Because the earth's equator is inclined to the plane of its orbit, the sun also appears to change its path across the sky. In this chapter we will consider the effect of the earth's revolution on the apparent motion of the sun and stars. We will show how the changing path of the sun causes the cycle of the seasons. Then we consider the calendar which is tied to the earth's annual revolution.

The Earth Revolves

While the stars are circling westward around us as though they were set on a rotating celestial sphere, the sun seems to lag behind gradually. The sun shifts slowly toward the east against the turning background of the heavens. If we could readily view the stars in the daytime sky, we would see that the sun moves eastward about twice its breadth in a day, and that it circles completely around the heavens in this direction in the course of the year. Although it is not possible to observe the sun's progress directly, this movement was recognized and charted by early watchers of the skies, for it is clearly revealed by the steady procession of the constellations.

3.1 Annual motion of the constellations

At the same hour each night as the seasons go around, the constellations move slowly across the evening sky and down past the sun's position below the horizon. If we look at a particular group of stars at the same time night after night, we eventually observe their westward movement. We notice, for example, Orion's part in the unending parade (Fig. 3.1); it is the brightest and among the most familiar of the constellations.

Orion, a tilted oblong figure with three stars in line near its center, comes up over the east horizon early in the evening in the late autumn. At the approach of spring this constellation appears upended in the south. As spring advances Orion comes out in the twilight farther and farther west, with its oblong figure now inclined the other way, until it follows the sun too closely to be visible. Then at dawn in midsummer Orion appears in the east again, on the other side of the sun.

This westward procession of Orion and other constellations past the sun's position shows that the sun is moving toward the east among the stars. Its motion is a consequence but not a proof of the earth's annual revolution around the sun. The same effect would occur if the sun revolved around the earth, as most people before the time of Copernicus thought.

Everyone knows today that the earth revolves around the sun. We learn this fact at an early age and accept it as an item of common knowledge. We are told, too, that the sun is approximately 146.9 million kilometers away, so that the earth must be speeding along at the average rate of 29.8 kilometers per second to travel completely around it in a year. There is nothing in our everyday experiences, however, to convince us that the earth is moving in this way. The apparent annual displacements, or parallax, of the stars that prove the earth's revolution are too minute to be detected without a telescope. Another proof is the aberration of starlight, but it also requires a telescope to detect it.

Figure 3.1 *Orion in the evening at different seasons.*

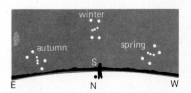

3.2 Aberration of starlight

Raindrops fall vertically when there is no wind, yet they seem to slant if one runs instead. As the observer's speed increases, the place from which the raindrops seem to fall shifts farther in the direction he is going. This is the aberration of raindrops (Fig. 3.2).

There is a similar effect on the rays of light from a star. If a telescope is pointed toward the true position of a star, the light from the star will travel down the tube of the telescope at an angle because the telescope is moving along with the earth as the light travels down the tube. As an analogy, think of running in the rain with an open tin can held vertically. While standing still, the rain would fall straight down and hit the bottom of the can. However, if you ran fast enough, the rain would enter the can at an angle and would strike the sides before reaching the bottom. If you wanted the rain to strike the bottom of the can while running very fast, you would have to point the can toward the direction in which you were running. Similarly, in order to see a star, the telescope must be pointed slightly ahead of the true position of the star toward the direction in which the earth is moving. Since the earth travels in a circular orbit around the sun, the apparent displacement of the star toward the direction in which the earth is moving is the **aberration of starlight.**

This effect only occurs because the speed of light is finite (light does not travel from place to place instantly). As a result, the amount of aberration of starlight is dependent upon the ratio of the earth's velocity around the sun to the speed of light. It is a much smaller displacement than occurs with raindrops; starlight comes down so much faster than the rain that its direction is not appreciably altered by ordinary speeds. Even the earth's swift flight around the sun of 29.8 kilometers per second displaces the stars only 20.″5 at most, an amount too slight to be detected by the unaided eye. This displacement of the stars from their true positions is readily observed with the telescope, and here we have convincing evidence of the earth's revolution.

In 1727, nearly two centuries after the death of Copernicus (who correctly predicted the earth's movement around the sun, but had no proof for his theory), James Bradley first observed the aberration of starlight at Oxford and explained its important meaning. Henceforth, there could be no reasonable doubt that the earth revolves around the sun. In a later chapter (Section 12.1) we consider another proof of the earth's revolution, the parallax effect on starlight, which is even smaller and depends upon the distance to each star. The first stellar parallax was not measured until 1837 when F. G. W. Struve at Dorpat measured the parallax of Vega. This was followed by the parallax of 61 Cygni in 1838 by F. W. Bessel at Königsberg.

Figure 3.2 *Aberration of rain-drops and starlight. Just as rain-drops come down slanting to one who is running, so the stars are apparently displaced ahead of us as we go around the sun. Each star seems to describe a small or-bit around its true position.*

3.3 The earth's orbit

If the earth's orbit around the sun were a circle with the sun at its center, the earth's distance from the sun would remain the same throughout the year, so that the sun would always appear to be the same size. Although the difference is not noticeable to the unaided eye, the sun's apparent diameter does vary during the year. It is greatest early in January and smallest early in July; the difference between the two is $\frac{1}{30}$ of the average diameter. Thus the earth is nearest the sun in early January and farthest from the sun in early July, which might seem surprising to someone who thought that the seasons are caused by the earth's varying distance from the sun. The earth's distance from the sun averages about 1.496×10^8 kilometers; it varies from 1.469×10^8 to 1.52×10^8 kilometers.

The earth's orbit is an ellipse of small eccentricity with the sun at one focus. An **ellipse** is a plane curve in which the sum of the distances from any point on its circumference to two points within, called the foci, is constant and equal to the longest diameter, or major axis, of the ellipse. The definition suggests a way to draw an ellipse (Fig. 3.3), and the drawing of several ellipses in this way makes clear the significance of the eccentricity.

The **eccentricity** *(e)* of the ellipse denotes its degree of flattening. It is determined by the ratio of the distance from the center of the ellipse and the focus to the length of the semimajor axes. Alternately, it is the square root of 1 minus the square of the ratio of the semiminor axis to the semimajor axis (Fig. 3.4). If the eccentricity is zero, the foci are together at the center and the curve is a circle. The ellipse flattens more and more as the eccentricity increases, until at an eccentricity of 1 the curve becomes a parabola. The earth's orbit is nearly a circle; its eccentricity is only 0.017. If Fig. 3.5 were drawn to scale, the sun's center would be 0.5 millimeter from the center of the ellipse, and the ellipse itself could be scarcely distinguished from a circle.

Perihelion and **aphelion** are the points on the earth's orbit which are, respectively, nearest and farthest from the sun; they are at opposite ends of the major axis. The earth arrives at perihelion in early January and at aphelion in early July, as we have noted before. These times vary a little in the calendar and also advance an average of 25 minutes a year, because the major axis of the earth's orbit moves slowly around the sun in the direction of the earth's revolution. The earth's mean distance from the sun is the average of the perihelion and aphelion distances. The earth is at this distance from the sun in early April and again in early October; this distance is referred to as the **astronomical unit (AU)**.

3.4 The ecliptic

The earth revolves eastward around the sun once in a year and at the same time rotates on its axis in the same direction once in a day.

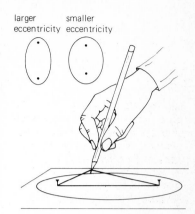

larger   smaller
eccentricity eccentricity

Figure 3.3 *An ellipse can be drawn by looping a string around two nails as shown. The nails serve as the foci of the ellipse.*

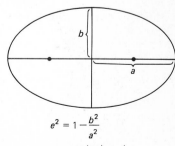

Figure 3.4 *The eccentricity* (e) *of an ellipse is the measure of its shape. As the ratio of* a *to* b *increases, the eccentricity increases and the ellipse becomes flatter.*

$$e^2 = 1 - \frac{b^2}{a^2}$$

a = semimajor axis
b = semiminor axis

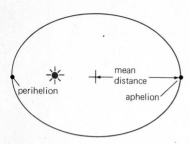

Figure 3.5 *The earth's orbit is an ellipse of small eccentricity (much exaggerated in the diagram) having the sun at one focus.*

The earth's axis is inclined 23°.5 from the perpendicular to the plane of its orbit, or we may say that the earth's equator is inclined 23°.5 to the plane of the orbit. This determines the relation (Fig. 3.6) between the celestial equator and the eastward path the sun seems to describe around the heavens as we revolve around the sun.

The **ecliptic** is the sun's apparent annual path around the celestial sphere; it is a great circle inclined 23°.5 to the celestial equator. The ecliptic is in the plane of the earth's orbit, and the celestial equator is in the plane of the earth's equator. The **north** and **south ecliptic poles** are 90° from the ecliptic, and are, respectively, 23°.5 from the north and south celestial poles. The north ecliptic pole is in the constellation Draco (Map 1, Chapter 11).

The **equinoxes** are two opposite points on the celestial sphere where the ecliptic crosses the celestial equator. They are so named because days and nights are said to be equal in length when the sun arrives at an equinox, although atmospheric refraction makes the duration of sunlight slightly the longer on such occasions. The **solstices** are two opposite points midway between the equinoxes, where the ecliptic is farthest north or south from the celestial equator. Here the sun "stands," so far as its north and south motion is concerned, as it turns back toward the equator.

The equinoxes and solstices are points on the celestial sphere; their positions in the constellations are shown in the star maps in Chapter 11. The **vernal equinox** (R.A. 0^h, Decl. 0°) is the point where the center of the sun crosses the celestial equator on its way north, about 21 March. The **summer solstice** (R.A. 6^h, Decl. 23°.5N) is the northernmost point of the ecliptic; the center of the sun arrives here about 22 June. The **autumnal equinox** (R.A. 12^h, Decl. 0°) is the point where the sun crosses

Figure 3.6 *The celestial equator is inclined to the ecliptic by the same amount that the earth's equator is inclined to the plane of its orbit around the sun. The equinoxes occur where the ecliptic crosses the celestial equator. The solstices are the two points on the ecliptic which are farthest from the celestial equator.*

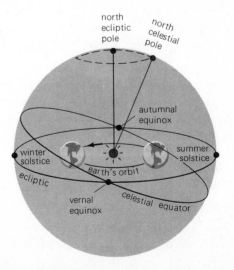

the celestial equator on its way south, about 23 September. The **winter solstice** (R.A. 18ʰ, Decl. 23°5S) is the southernmost point of the ecliptic; the sun arrives here about 22 December. These dates vary a little from year to year owing to the plan of leap years.

The celestial equator keeps the same position in the sky throughout the year. It is inclined to the horizon at an angle that is the complement of the latitude. The ecliptic, however, takes different positions in the evening sky during the year.

Because the ecliptic is inclined 23°5 to the celestial equator, its inclination to the horizon can differ as much as 23°5 either way from that of the equator. At sunset at the beginning of autumn (Fig. 3.7) in middle northern latitudes the ecliptic is least inclined to the horizon; the moon and bright planets that may be visible at the time are seen rather low in the south. At sunset at the beginning of spring the ecliptic is most inclined to the horizon; the moon and planets are then crossing more nearly overhead.

A number of features familiar to watchers of the skies are affected by the varying angle between the ecliptic and horizon. Among these are the harvest moon, the direction of the horns of the crescent moon, the favorable times for seeing the planet Mercury as evening or morning star, and the favorable seasons for viewing the zodiacal light (Section 9.6).

3.5 Celestial longitude and latitude

The moon and bright planets never depart very far from the sun's path around the heavens. For this reason, astronomers of early times who were especially interested in the motions of the sun, moon, and planets referred the places of these and other celestial bodies to the ecliptic, and they named the coordinates celestial longitude and latitude. **Celestial longitude** is measured in degrees eastward from the vernal equinox along the ecliptic. **Celestial latitude** is measured north or south from the ecliptic

Figure 3.7 *Relationship between ecliptic and horizon for an observer at 40° N. (a) The ecliptic is least inclined to the horizon in our latitudes at the beginning of autumn. (b) It is almost inclined at the beginning of spring.*

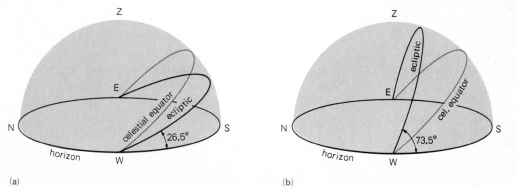

(a)

(b)

along a great circle at right angles to it and passing through the ecliptic poles.

The ecliptic coordinates have only limited use today. For most purposes the position of a celestial body is now referred to circles based on the celestial equator by means of right ascension and declination, which closely resemble terrestrial longitude and latitude. These newer coordinates themselves would doubtless have been called celestial longitude and latitude if the terms had not already been assigned.

3.6 The earth's precession

If a top is spinning with its axis inclined to the vertical, the axis moves around the vertical line in the direction of the spin. The rotation of the top resists the effort of gravity to tip it over, and the conical motion of the axis results, until the spin is so reduced by friction that the top falls over.

The earth is rotating similarly on an axis that is inclined to the ecliptic plane, which is nearly the plane of the moon's motion around us (Fig. 3.8). The attractions of the moon and sun on the earth's bulging equator tend to bring the earth into the plane with themselves, and thus to straighten up the earth's axis relative to its orbit. Their efforts are resisted by the rotation, so that the earth's axis moves slowly around the line joining the ecliptic poles in the direction opposite to that of the rotation, once around in approximately 26,000 years at the present rate. This is the **earth's precession.** It was first noted and measured quite accurately by Hipparchus in 125 B.C.

As the earth's axis goes around in the precessional motion, the celestial poles, toward which the earth's rotational axis is directed, move among the constellations. The poles describe circles 23.°5 in radius around the ecliptic poles (Fig. 3.9), bringing successively to bright stars along their paths the distinction of being the pole star for a time. Thus alpha Draconis was the pole star in the north 5000 years ago, the predecessor of our present pole star.

As shown in Fig. 3.9, the north celestial pole is now approaching Polaris and will pass nearest it about the year 2100 at half its present distance or within about 25 minutes of arc. Thereafter, Polaris will describe larger and larger daily circles around the pole, and its important place will at length be taken by stars of Cepheus in succession. In the year 7000, alpha Cephei will mark the north pole closely. In 14,000, Vega in Lyra will be a brilliant, although not very close, marker.

In our skies the constellations are slowly shifted by precession relative to the circumpolar areas, while the poles remain in the same places at the distance from the horizon equal to the observer's latitude. About 6000 years ago the Southern Cross rose and set everywhere in the United States.

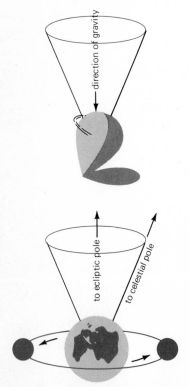

Figure 3.8 *The earth resembles a spinning top. The pull of the moon on the bulging equator of the rotating earth is the chief cause of the precessional motion.*

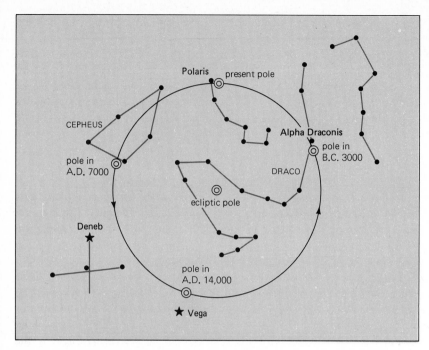

Figure 3.9 *Precessional path of the north celestial pole. The celestial pole describes a circle of 23°.5 radius around the ecliptic pole. It takes 26,000 years for the celestial pole to complete its journey along this circular path.*

Now it is not visible except in the southernmost parts of the country.

The earth's precession, as we have seen, causes the line joining the celestial poles to describe the surface of a right circular cone around the line joining the ecliptic poles. This motion is counterclockwise as we face north. Meanwhile the celestial equator slides westward along the ecliptic, keeping about the same angle between them. Consequently, the equinoxes slide westward along the ecliptic. This is the **precession of the equinoxes.** Their motion, including smaller effects of planetary attractions of the earth, is at the rate of 50 seconds of arc a year. Precession including all effects is referred to as the **general precession.** Precession due to the moon and sun as described earlier is called the **lunisolar precession.**

The precession of the vernal equinox (Fig. 3.10) causes continuous variations in the right ascensions and declinations of the stars. It accounts for the lack of agreement between the constellations and signs of the zodiac, and it makes the year of the seasons shorter than the period of the earth's revolution around the sun.

3.7 The zodiac: its signs and constellations

The *zodiac* (from Greek, meaning "band of animals") is the band of the heavens 18° wide, through which the ecliptic runs centrally. It contains the sun, moon, and bright planets at all times, with the occasional

Figure 3.10 *Precession of the equinoxes. The westward motion of the vernal equinox from V₁ to V₂ causes the signs of the zodiac to slide westward away from the corresponding constellation of antiquity. The coordinates of each star are consequently altered by the precession.*

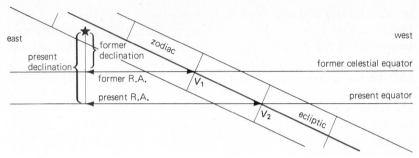

exception of the planet Venus. This is the reason for the special importance of the zodiac in the astronomy of early times and in the ancient pseudoscience of astrology, in which the places of the planetary bodies had great significance.

Twelve constellations of the zodiac lie along this band of the heavens. Because these constellations are of unequal size, the **signs of the zodiac** were introduced in early times in the interest of uniformity; these are 12 equal divisions (corresponding to 12 months) of the zodiac, each 30° long, marked off eastward from the vernal equinox. Each block, or sign, of the zodiac has the name of the constellation it contained 20 centuries ago. The names of the 12 constellations and signs of the zodiac are Aries, Taurus, Gemini, Cancer, Leo, Virgo, Libra, Scorpius, Sagittarius, Capricornus, Aquarius, and Pisces.

During the past 20 centuries, the vernal equinox has moved westward among the stars of the zodiac, and the whole train of signs has followed along, because the signs are counted from the equinox. Each sign has shifted out of the constellation for which the sign was named and into the adjoining figure to the west. Thus, when the sun arrives on 21 March at the vernal equinox, or "first of Aries," the sun is entering the zodiacal sign of Aries. The sun is then in the constellation Pisces, however, and will not enter the constellation Aries itself for another month.

Figure 3.11 *In the northern hemisphere the weather is warmer in summer because the daily duration of sunshine is greater and the sun's rays are more direct.*

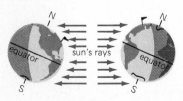

The Seasons

Why is the weather warmer in summer than in winter? It is not because we are nearer the sun in summer, for we have seen that the earth is more than 5 million kilometers farther from the sun in summer in the northern hemisphere than in winter. The cause of the seasons is found in the inclination of the earth's equator to the plane of its orbit (Fig. 3.11).

In the summer the northern end of the earth's axis is inclined toward the sun. Not only is the daily duration of sunlight longer at any place in our northern latitudes than it is in winter, but the sun also climbs higher, so that its rays are more nearly vertical and more concentrated on our part of the world. In winter the northern end of the axis is inclined away from the sun. The daily duration of sunlight is then shorter for us and the sun is lower at noon. The sun's rays are more slanted in the winter, so that they are more spread out over the ground and are also obstructed more by the greater thickness of the intervening air. Thus the weather is warmer in summer than in winter because the sun shines for a longer time each day and reaches a greater height in the sky (Fig. 3.12).

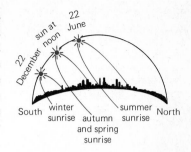

Figure 3.12 *The daily circles of the sun for a northern hemisphere location are farthest north and south in the summer and winter.*

3.8 The year of the seasons

The **sidereal year** is the interval of time in which the sun appears to perform a complete revolution around the heavens with respect to the stars. This is the true period of the earth's revolution. The length of the sidereal year is 365^{d}6^{h}9^{m}10^s of mean solar time.

The **tropical year** is the interval between two successive arrivals of the sun's center at the vernal equinox. Because the equinox is shifting westward to meet the sun, this **year of the seasons,** from the beginning of spring to the beginning of spring again, is shortened. The length of the tropical year is 365^{d}5^{h}48^{m}46^s. This is about the average length of the present calendar year.

There are two other years commonly used in astronomy, the anomalistic year and the eclipse year. The **anomalistic year** is the interval between successive passages of the earth past perihelion. Its length is 365^{d}6^{h}13^{m}53^s. The **eclipse year** is the interval of time between two successive passages of the sun through the same node of the moon's orbit (Sections 5.3 and 5.17). The length of the eclipse year is 346^{d}14^{h}52^{m}51^s.

3.9 The lag of the seasons

By definition the first day of summer in our northern latitude begins about 22 June, when the sun arrives at the summer solstice and turns back toward the south. The duration of sunlight is longest on that day, and the sun climbs highest in the sky. Yet the hottest part of the summer is likely to be delayed several weeks until the sun is well on its way south. Why does the peak temperature of the summer come after the time of the solstice?

As the sun goes south, its rays bring less and less heat to our part of the world from day to day. For some time, however, the diminishing receipts still exceed the amounts of heat we are losing by the earth's radiation into space. Summer does not reach its peak until the rate of heating is reduced to the rate of cooling. Similarly our winter weather is

Figure 3.13 *Insolation and mean temperature curves for Concordia, Kansas and San Luis Obispo, California. The maximum insolation is on 21 June, but the maximum mean temperature lags depending upon geographical circumstances. (Data from U.S. Department of Agriculture.)*

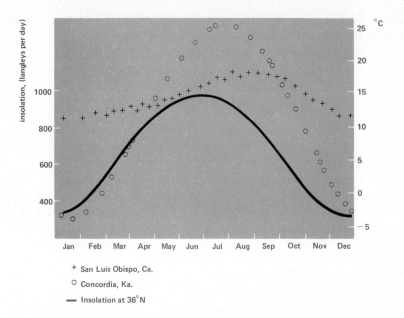

+ San Luis Obispo, Ca.

○ Concordia, Ka.

— Insolation at 36°N

likely to be more severe several weeks after the time of the winter solstice. After the sun turns north, about 22 December, the temperature continues to fall generally until the heat we receive from the sun becomes as great as the daily loss.

Heat loss from the earth lags behind the heat input because the surface of the earth is heated by the sun and then only slowly releases this heat into space. Water is a particularly good retainer of heat. Thus after the summer solstice the oceans continue to warm the atmosphere for many months (Fig. 3.13). By the time of the winter solstice the oceans are still releasing a little warmth into the atmosphere and at the summer solstice the oceans are still warming up after having relinquished their heat in the winter. This lag in the ocean temperature is what is primarily responsible for the lag of the seasons.

Figure 3.14 *The orbit of the earth showing the relationship between the summer and winter solstices and perihelion and aphelion.*

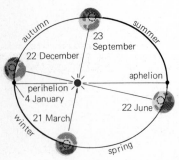

3.10 The seasons in the two hemispheres

Although the seasons are not caused by the earth's varying distance from the sun, the earth is farther from the sun in our summer and nearer the sun in our winter than at similar seasons in the southern hemisphere (Fig. 3.14). Summers in the northern hemisphere might well be a little cooler than southern summers, and northern winters might be milder than southern winters. Thus the northern hemisphere might seem to have the more agreeable climates. Yet the variation in the earth's distance from the sun is only 3% of the distance itself, and there is more water in the southern hemisphere to modify extremes of temperature.

It would not be valid to make a general comparison of the weather in corresponding latitudes north and south of the equator. Differences in elevation and in the effects of air and ocean currents would have to be taken into account.

3.11 The climate zones

The positions and widths of the climatic zones are determined by the inclination of the earth's equator to the plane of its orbit. Because this inclination is 23°5, the torrid zone extends 23°5 north and south from the equator; everywhere in this region the sun is directly overhead on some days of the year. The **Tropic of Cancer** forms the northern boundary of the torrid zone. Here the sun is overhead at noon on 22 June, when it arrives at the summer solstice. (It was called the Tropic of Cancer because about 3000 years ago the sun was in the constellation Cancer at the summer solstice.) The **Tropic of Capricorn** is the southern boundary of the **torrid zone.** Here the sun is overhead at noon on 22 December. (For the ancients, the sun was in the constellation Capricornus at the winter solstice.)

Similarly, the north and south **frigid zones** extend 23°5 from the poles and are bounded by the Arctic and Antarctic Circles, respectively. Although they are modified a little by the effect of refraction near their borders, the frigid zones are the regions where the sun may become visible at midnight during part of the year, and may not appear even at noon at the opposite time of the year. Extreme conditions occur at the poles themselves, where the sun shines continuously for six months and remains out of sight for the following six months.

The north and south **temperate zones** lie between the torrid and frigid zones. Here the sun never reaches the point overhead, nor does it ever fail to appear above the horizon at noon.

3.12 The midnight sun

The sun on 22 June is about as far north of the equator as it ever gets. The declination of its center is nearly 23°4N, so that its north polar distance is 66°5. How far north must we go on 22 June to see the sun circle around the pole without setting? According to our rule we must go beyond latitude 66°5. When we take into account the refraction effect and the size of the sun's disk, however, we can expect to see the sun at midnight on this date as far south as 65°75N.

The **midnight sun** (Fig. 3.15) is seen wherever the sun becomes circumpolar (Section 2.9). In far northern latitudes during the spring the sun enters the area of the heavens that is always above the horizon; it becomes circumpolar in far southern latitudes at the opposite time of the year.

Figure 3.15 *The sun is seen here as a circumpolar star. The multiple exposures were made over midnight at which time the sun is referred to as the midnight sun.*

Calendars

Calendars have been in use since the beginnings of civilization. They have tried to combine natural measures of time, the solar day, the lunar month, and the year of the seasons, in the most convenient ways and have encountered difficulties because these measures do not fit evenly one into another. Calendars have been of three types: the lunar, the lunisolar, and the solar calendar governed by the earth's revolution.

3.13 Lunar calendar

The **lunar calendar** is the simplest of the three types, and it was the earliest to be used by almost all civilizations. Each month began originally with the "new moon," the first appearance for the month of the crescent moon after sunset. This calendar, long controlled only by observation of the crescent, was eventually operated by fixed rules. In the fixed lunar calendar the 12 months of the common lunar year is 354 days in all. It is made up of months alternately of 30 and 29 days or a close average of the lunar month (new moon to new moon) of 29.53 days. The Mohammedan calendar is a survivor of this type.

3.14 Lunisolar calendar

The **lunisolar calendar** tries to keep in step with both the moon's phases and the seasons, and is the most complex of the three types. It began by occasionally adding a thirteenth month to the short lunar year to round out the year of the seasons. The extra month was later inserted at fixed intervals. The Hebrew calendar is the principal survivor of the lunisolar type.

3.15 Solar calendar

The solar calendar makes the year conform as nearly as possible to the year of the seasons, and neglects the moon's phases; its 12 months are generally longer than the lunar month. Only a few early nations, notably the Chinese, Egyptians, and eventually the Romans, adopted this type of calendar. The early Mayan calendar was based partially on the period of Venus but evolved into a highly developed solar calendar. This calendar was passed along to the Aztecs (Fig. 3.16) with its novel plan of leap years with extra days to keep it in step with the seasons.

Figure 3.16 *An Aztec calendar for the year 1790. (Photograph courtesy of the OAS.)*

The early Roman calendar formally begins with the founding of Rome in 753 B.C. It was originally a lunar calendar of a sort, beginning in the spring and having 10 months. The names of the months, if we use mainly our own style instead of the Latin, were March, April, May, June, Quintilis, Sextilis, September, October, November, and December. The years for many centuries thereafter were counted from 753 B.C. and were designated A.U.C. ("in the year of the founding of the City"). Two months, January and February, were added later and were eventually placed at the beginning, so that the number months have appeared in the calendar out of their proper order ever since that time.

In its 12-month form the Roman calendar was of the lunisolar type. Unlike most early people, Romans began the day at midnight instead of at sunset or sunrise. An occasional extra month was added to keep the calendar in step with the seasons. The calendar was managed so unwisely, however, that it fell into confusion; its dates drifted back into different seasons from the ones they were supposed to represent.

When Julius Caesar became the ruler of Rome, he was disturbed by the bad condition of the calendar and took steps to correct it. He particularly wished to discard the lunisolar form with its troublesome extra months. Caesar was impressed with the simplicity of the solar calendar the Egyptians were using, and he knew of their discovery that the length of the tropical year is very nearly 365¼ days. He accordingly formulated his reform with the advice of the astronomer Sosigenes of Alexandria. In preparation for the new calendar the "year of confusion," 46 B.C., was made 445 days long in order to correct the accumulated error of the old one. The date of the vernal equinox was thereby brought to 25 March. The Julian calendar began on 1 January 45 B.C.

The Julian calendar was a **solar calendar,** and so neglected the moon's phases. Its chief feature was the adoption of 365¼ days as the average length of the calendar year. This was accomplished conveniently by the plan of leap years. Three common years of 365 days are followed by a fourth year containing 366 days; this **leap year** in our era is divisible by 4.

In lengthening the calendar year from the 355 days of the old lunisolar plan to the common year of 365 days, Caesar distributed the additional 10 days among the months. With further changes made in the reign of Augustus, the months assumed their present lengths. After Caesar's death in 44 B.C., the month Quintilis was renamed July in honor of the founder of the new calendar. The month Sextilis was later renamed August in honor of Augustus.

Because its average year of 365^d6^h was 11^m14^s longer than the tropical year, the Julian calendar fell behind with respect to the seasons about 3 days in 400 years. When the council of churchmen convened at Nicaea in A.D. 325, the vernal equinox had fallen back to about 21 March. It was at that meeting that previous confusion about the date of Easter was ended.

As the date of the vernal equinox fell back in the calendar, 21 March and Easter, which is reckoned from it, came later and later in the season. Toward the end of the sixteenth century the equinox had retreated to 11 March. Another reform of the calendar was proposed by Pope Gregory XIII.

Two rather obvious corrections were made in the Gregorian reform. First, 10 days were suppressed from the calendar of that year; the day following 4 October 1582, became 15 October for those who wished to adopt the new plan. The date of the vernal equinox was restored in this way to 21 March. The second correction made the average length of the calendar year more nearly equal to the tropical year, so that the calendar would not again get so quickly out of step with the seasons. Evidently the thing to do was to omit the 3 days in 400 years by which the Julian calendar year was too long. This was done conveniently by making common years of the century years having numbers not evenly divisible by 400. Thus the years 1700, 1800, and 1900 became common years of 365 days instead of leap years of 366 days, whereas the year 2000 remains a leap year as in the former calendar. The average year of the new calendar is still too long by 26 seconds, which is hardly enough to be troublesome for a long time to come.

The **Gregorian calendar** was gradually adopted, until it is now in use, at least for civil purposes, in practically all nations. England and its colonies including America made the change in 1752. The countries of eastern Europe were the last to make the change.

For many astronomical purposes the **Julian Day** is used. This is not a calendar in the proper use of the term, but is a continuous numbering of the days from an arbitrary date. It was devised in 1582 by Joseph Scaliger and named in honor of his father, Julius Scaliger.

3.16 Easter

Easter was originally celebrated by some early churches on whatever day the Passover began, and by others on the Sunday included in the Passover week. The Council of Nicaea decided in favor of the Sunday observance and left it to the church at Alexandria to formulate the rule.

Easter is the first Sunday after the fourteenth day of the moon (nearly the full moon) that occurs on or immediately after 21 March. Thus if the fourteenth day of the moon occurs on Sunday, Easter is observed one week later. Unlike Christmas, Easter is a movable feast because it depends on the moon's phases; its date can range from 22 March to 25 April (Table 3.1).

Table 3.1
Dates of Easter Sunday

1980	6 Apr	1990	15 Apr
1981	19 Apr	1991	31 Mar
1982	11 Apr	1992	19 Apr
1983	3 Apr	1993	11 Apr
1984	22 Apr	1994	3 Apr
1985	7 Apr	1995	16 Apr
1986	30 Mar	1996	7 Apr
1987	19 Apr	1997	30 Mar
1988	3 Apr	1998	12 Apr
1989	26 Mar	1999	4 Apr

Questions

1 Astronomers use the terms rotation and revolution in a very strict way. Define these terms.
2 Explain aberration. Why does the aberration of starlight confirm the Copernican theory?
3 Why is the eccentricity of the orbits in the text figures so exaggerated?
4 Define the sidereal year.
5 How did the Tropic of Cancer get its name?
6 Neglecting refraction, where would you have to be in order for astronomical twilight to endure through the night on 21 June?
7 How has the failure to accept precession affected astrology?
8 What will be a good north pole star in A.D. 23,000?
9 Why will Canopus never be a satisfactory south pole star (see Map 6)?
10 Give the definitions of the four types of years.

Other Readings

Malstrom, V. H., "A Reconstruction of the Chronology of Mesoamerican Calendrical Systems," *Journal of The History of Astronomy,* **9,** 105–116 (1978).

Williamson, R., "Native Americans Were Continent's First Astronomers," *Smithsonian,* **9,** 78–85 (1978).

chapter 4

Telescopes and Instrumentation

The chief feature of the optical telescope is its objective which receives the light of a celestial object and focuses the light to form an image of the object. The image may be formed either by refraction of the light by a lens or by reflection from a curved mirror. Optical telescopes are accordingly of two general types: refracting telescopes and reflecting telescopes. In radio telescopes the radio radiations from a celestial source are received directly by an antenna or are concentrated on the antenna by a reflecting paraboloid or other surface of large aperture. Special telescopes are designed to receive infrared and x-ray radiation. The various radiations collected by the telescopes are then fed to various instruments for study.

Optical Telescopes

The optical telescope was first trained on the heavens by Galileo Galilei in 1609. His discoveries profoundly affected the thinking of his day and the optical telescope subsequently dominated astronomy for the next 350 years. Although today the most rapidly advancing areas of astronomy are occurring in fields which do not rely on observations in the visible part of the spectrum, we will start this chapter with a discussion of optical telescopes since they operate on principles with which most people are familiar.

4.1 Light

When light is emitted by a star, it travels away from its source as a spherical wave front, much as the concentric waves in a pond move out from the point where a stone is dropped into the water. Unlike water waves though, light is not conducted by oscillations in a material medium. Rather, light consists of oscillating electric and magnetic fields which perpetuate one another through space. For this reason, light is referred to as **electromagnetic radiation.**

The distance between successive crests (or troughs) in a wave is defined as the **wavelength** λ (Fig. 4.1). In the case of light, this is equivalent to the distance between successive peaks in the intensity of the electric (or magnetic) field. Since the speed of light is a constant, a wavelength of radiation may be defined by its **frequency**—the number of waves emitted by the source in a second. Mathematically, the frequency ν equals the velocity of light c divided by the wavelength λ, or $\nu = c/\lambda$. For example, an electromagnetic wave which oscillates with a frequency of 4.3×10^{14} cycles per second has a wavelength of 7×10^{-7} meters if we use the velocity of light as 300,000 kilometers per second. The wavelength 7×10^{-7} meters will often be written as 0.7 micrometers (μm) or 7000 Angstrom units, where one Angstrom unit (written as Å or more recently A) equals 10^{-10} meters. Obviously, light waves oscillate very rapidly and have extremely short wavelengths.

If light of this wavelength were to strike the retina of our eye, it would evoke the sensation of red. This is because the colors which our

Figure 4.1 *Wavelength of a wave is the distance from the crest to crest or trough to trough. It is denoted by the Greek letter λ.*

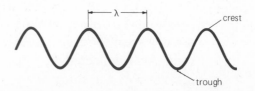

eyes perceive are dependent upon the wavelengths of the electromagnetic radiation being observed. However, our eyes are not sensitive to the entire range of wavelengths in the electromagnetic spectrum. In fact, visible light constitutes only a very small range of wavelengths between 3800 (deep violet light) and 7500 Angstroms (red light). Invisible to our eyes is a whole universe of sights which astronomers have only recently begun to observe.

The first evidence of radiation beyond the visible range was presented by the famous astronomer William Herschel in 1800. In his observations of the sun, Herschel was disturbed by the heat carried to his eyes by the sunlight. In order to determine which part of the spectrum was responsible for the sun's warmth, he placed thermometers in the dispersed light of a prism. Much to his surprise, the recorded temperatures increased towards the red end of the spectrum until reaching a maximum beyond the visible range of the light band. After further experimentation, Herschel concluded that this invisible light obeyed all the laws of visible light rays. He had discovered the infrared. Nearly a century later, Heinrich Hertz demonstrated that microwaves behave in the same way and it was soon realized that a continuous spectrum of electromagnetic radiation existed. The spectrum is shown in Fig. 4.2. It extends from gamma rays that have billions of waves per centimeter to radio waves that are many kilometers long.

Light from a star is not observed the instant that it is emitted. Its speed, although very fast, is finite. This means that when we look out at a distant star, we perceive it as it was at some time in the past. We are not only looking out in space, but back in time as well. And the further away the star is, the further back in time we see it. Because radiation emitted at the earliest times can still be viewed today, the early history of the universe can be observed! We will discuss this further in Section 19.5.

Figure 4.2 *The electromagnetic spectrum. Parts of the spectrum which are blocked out by the earth's atmosphere are shaded. The regions of the spectrum that penetrate to the earth's surface are known as windows. The scales are logarithmic.*

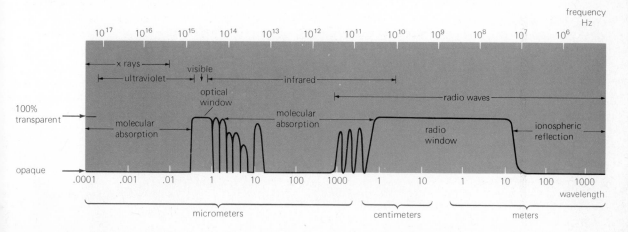

The **velocity of light** is about 299,792 kilometers per second; this is the speed of all electromagnetic radiation in a vacuum. For ease of calculation we may use 3×10^5 kilometers per second for the velocity of light. The speed is reduced in a medium, such as air or glass, by an amount depending on the density of the medium and the wavelength of light. This then is the reason for the refraction of light.

Although the wave nature of light suffices to explain the functions of telescopes and most other properties involving the propagation of light, there is an alternative explanation involving the particle nature of light. This concept was first introduced by Isaac Newton more than 300 years ago and has only recently been rejuvenated in this century in the form of quantum theory, which describes the interaction of radiation and matter. Light is therefore imbued with a dual wave–particle nature. Which behavior predominates depends on the situation under consideration. In the particle theory, light travels in little packets of energy called **photons.** When we consider the emission and absorption of light, its particle nature will become more evident. For now, the wave theory allows us to visualize more easily phenomena such as refraction and interference.

4.2 Refracting telescopes

Refraction of light is the change in the direction of a ray of light when it passes obliquely from one medium into another of different density, as from air into glass. A "ray" of light denotes the direction in which a narrow section of the wave system is moving.

When a ray of light obliquely enters a denser medium, where its speed is reduced (Fig. 4.3), one side is retarded before the other. The wave front is accordingly swung around, and the ray becomes more nearly perpendicular to the boundary between the two media. When the oblique

Figure 4.3 *Refraction of light. Light travels more slowly in glass than in air. Its direction is changed when it strikes their interface obliquely.*

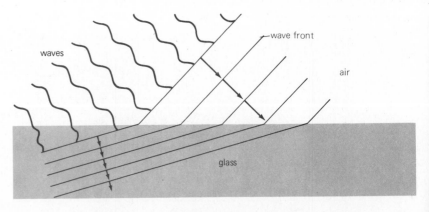

ray enters a rarer, thinner medium instead, it is refracted away from the perpendicular. The direction of the ray is not altered if it is originally perpendicular to the boundary.

Figure 4.4 shows how a double convex **lens** forms by refraction a real inverted image (that is, an image that can be projected on a screen) of an object that is farther from the lens than is the focal point *F*. This is the point where rays parallel to the axis of the lens are focused. Rays passing through the center of the lens are unchanged in direction. When the image formed by this lens is brought within the focal distance of a similar lens, the second lens serves as an eyepiece with which to view and magnify the image, leaving it inverted. A single lens, however, does not give a sharp image, mainly because it refracts shorter waves more than longer ones, violet light more than red, so that it does not bring them together at the same focus (Fig. 4.5). The failure of a refractive medium to refract all wavelengths equally is called **dispersion.** This is one of the major shortcomings of a refracting telescope and is called **chromatic aberration.** The confusion of colors is partly correctable by using a compound lens, that is, a lens made up of several lenses.

The **objective** of a refracting telescope, or **refractor,** is generally a combination of two lenses. Such an objective having two or more elements (or lenses) of various types of glass in order to minimize chromatic aberration is referred to as an **achromat.** Its aperture, or clear diameter, is given when denoting the size of the telescope. The **focal length,** or distance from the objective to its focus, is often about 15 times the aperture. The ratio of focal length to aperture is called the **focal ratio,** or *F*-number, of the objective. Thus a 60-centimeter telescope has an aperture of 60 centimeters. If it has a focal length of 9 meters then its focal ratio is $9/0.6 = 15$. For visual use the inverted image formed by the objective is viewed with an **eyepiece,** a magnifier constructed of small lenses set in a sliding tube. The objective of a visual refractor focuses together the yellow and adjacent colors of the spectrum, to which the eye is especially sensitive, but not the blue and violet light that most affects the ordinary photographic plate. Without a correcting device it does not serve well for general photography.

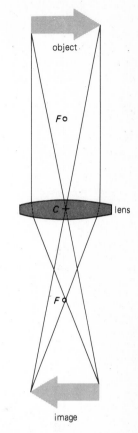

Figure 4.4   *A convex lens forms an inverted real image.*

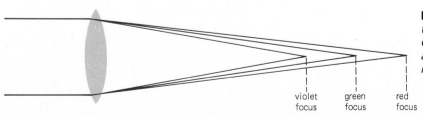

violet   green   red
focus   focus   focus

Figure 4.5   *Chromatic aberration occurs in a simple lens because it focuses the various colors at different distances from the lens.*

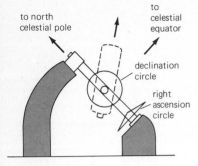

to north
celestial pole

to
celestial
equator

declination
circle

right
ascension
circle

When a refracting telescope intended for visual purposes is used as a camera, a plate holder replaces the eyepiece. A yellow filter and a yellow-sensitive photographic plate may be combined to utilize the light that is sharply focused, or a correcting lens may be introduced for a particular kind of photographic plate.

The largest refracting telescope is the 40-inch telescope of the Yerkes Observatory at Williams Bay, Wisconsin; its focal length is 21 meters. (Here we follow the common practice among astronomers of using the most common designation of a given telescope.) It has about the greatest permissible aperture for a refracting telescope because the objective can be supported only at the edge. A larger lens will sag seriously under its own weight. About 40 refracting telescopes have apertures of 50 centimeters or more (see the Appendix for a list of the larger ones).

The **equatorial mounting** (Fig. 4.6) is an example of the type generally used for the larger refracting telescopes. The **polar axis** is parallel to the earth's axis; around it the telescope is turned parallel to the celestial equator. The **declination axis** is supported by the polar axis; around it the telescope is turned along an hour circle, from one declination to another.

The polar axis carries a graduated circle showing the hour angle of the star toward which the telescope is pointing. There is also a dial on the pier, or on a console, which indicates the star's right ascension. A circle on the declination axis shows the declination of the star. An electronic system usually displays this on the console as well. By the use of these circles or displays, the telescope can be pointed toward a celestial object of known right ascension and declination; it is then kept pointing at the object by a driving mechanism in the pier. The dome is turned by a motor, so that the telescope may look out in any direction through the opened slit.

4.3 Reflecting telescopes

Light is **reflected** from a plane surface at the same angle at which it strikes it. This is familiar to everyone acquainted with mirrors. The light coming from a star is so far distant from its source that all the rays of light are essentially parallel. If this parallel light were to strike a perfectly spherical mirror, a clear image would not form because a common focal point would not exist for the individual rays reflected off the inner and outer parts of the mirror. This defect is called **spherical aberration.** In order to correct for this, a parabolic mirror is used rather than a spherical one (Fig. 4.7). A good image is then formed, which may be viewed with an eyepiece or photographed. Concave mirrors form inverted real images of celestial objects, just as convex lenses do, so that the field of view is again "upside down" (Fig. 4.8).

A reflecting telescope, or **reflector,** has a concave mirror as its objective

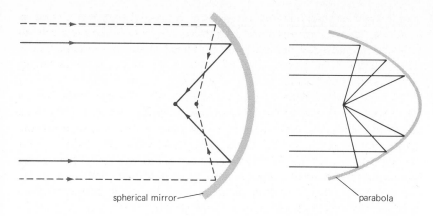

parabola

Figure 4.7 *Spherical aberration occurs because light rays from the outer zones of a spherical mirror are brought to a focus closer to the mirror than those from the inner zones. This can be overcome by using a parabolic mirror.*

at the end of a long tube. The mirror is usually composed of glass and is coated with a very thin film of highly reflective metal, such as aluminium. The glass serves only to give the desired shape to the metal surface. It need not have a high optical quality such as that required for a lens. Also, the entire back of the mirror may be supported without blocking the aperture. Consequently, a reflecting telescope can be much larger than a refracting telescope. Reflecting telescopes are achromatic because reflection does not disperse the light according to wavelength as refraction does. A reflecting telescope is also shorter than an ordinary refractor of the same diameter, so it is less costly to construct and to house. For these reasons, most modern telescopes are of the reflecting type.

The large mirror reflects the light of the celestial object to the **prime focus** in the middle of the tube near the upper end, where the image is accessible to the observer only in the very largest telescopes. In the **Newtonian** form a small plane mirror at a 45° angle near the top of the tube reflects the converging beam from the large mirror to a focus at the side of the tube. In the **Cassegrainian** form a small convex mirror replaces the plane mirror; it reflects the beam back through an opening in the large mirror to a focus below it. Where the large mirror has no opening, which is the case with the 2.5-meter Hooker telescope on Mt. Wilson, the returning beam is reflected to the **Nasmyth focus** at the side by a plane mirror in front of the large one. This and some other large telescopes provide another place of observation, the **coudé focus,** by reflection of the beam through the polar axis of the telescope to a laboratory below (Fig. 4.9). Here heavy equipment can be installed in fixed positions.

The world's largest reflecting telescope is the 6-meter reflector located at Zelenchukskaya in the Soviet Union. Construction of this great telescope began in 1948 and the first photographic plates were taken in 1976. The telescope is unusual and differs from all of the other great reflectors in that it is mounted in the horizon system of coordinates—altitude and

Figure 4.8 *A concave mirror forms an inverted real image.*

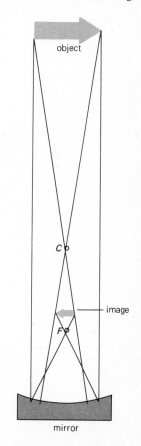

object

C

image

F

mirror

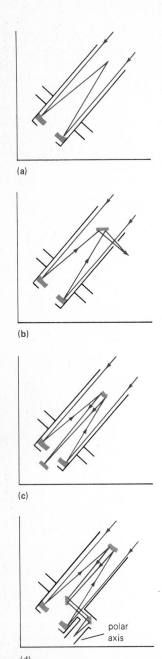

Figure 4.9 *Four possible forms for a reflecting telescope: (a) the prime focus, (b) the Newtonian focus, (c) the Cassegrainian focus, and (d) the coudé focus.*

azimuth. A computer takes the sidereal time and the equatorial coordinates of the object being observed and continuously computes the required altitude and azimuth. The prime focus of this telescope is 27 meters above the primary mirror.

The 5-meter Hale telescope of the Hale Observatories (Fig. 4.10), which was completed in 1948 after 16 years in construction, is one of the world's most famous optical telescopes. At the prime focus 16.8 meters above the aluminium coated mirror, the observer is carried in a cage that is 1.5 meters in diameter, obstructing less than 10 percent of the incoming light. At this focus the eye can detect a star 400,000 times fainter than the unaided eye can detect. When the 1-meter convex, secondary mirror is set in the converging beam it reflects the light to the Casse-grainian focus yielding an effective focal length of 81.4 meters. The observer is also carried in a cage at this focus as well. With the aid of a third mirror the light can be directed to the coudé focus in a laboratory below the observing floor yielding an effective focal length of 155.4 meters. This great telescope has been in the forefront of astronomical research for more than a third of a century.

The Hale telescope has been joined by a host of large reflectors during the past 10 years. Of special note are the twin 4-meter reflectors operated as national facilities by Kitt Peak National Observatory and the Cerro Tololo Interamerican Observatory. A list of the largest great reflectors is given in the Appendix.

The large reflecting and refracting telescopes are ideally suited for photographing small areas of the sky or studying individual objects with specialized instruments, such as photometers and spectrographs, soon to be described. Other telescopes commonly referred to as **astrographs** are more effective in recording large areas of the sky on photographic plates.

4.4 Astrographs

The large-field telescope plays a major role in present-day astronomy. The field size of these astrographs is often $10° \times 10°$, hence, hundreds and thousands of stars are recorded at one time (Fig. 4.11). Plates taken with various filters allow rapid, approximate star color and brightness estimates, and plates repeated after an appropriate interval allow studies of the motions and variability of stars. These plates also serve to locate objects of interest for the large telescopes.

The earliest telescopes of this type employed objectives made up of three or more lenses (Fig. 4.12). The purpose of the compound lens was to focus the large field on a reasonable size plate with as few aberrations as possible. Such telescopes often had focal ratios of, or somewhat less than, 8. Recently, it has been possible to combine lenses with mirrors to achieve the same purpose; very large apertures and short focal lengths

Figure 4.10 *A view of the 5-meter Hale telescope. Note the observing cage around the Cassegrainian focus. (Hale Observatories photograph.)*

make them very "fast" telescopes. The smaller the focal ratio the faster a telescope is said to be, just as with camera lenses. Focal ratios or F numbers of 1.5 are not uncommon. The Schmidt telescope is an example.

As its main element the **Schmidt telescope** has a spherical mirror, which is easy to make in an optical shop but is not by itself suitable for a telescope. Parallel rays reflected by the central part of such a mirror are focused farther away from the focus than those rays reflected from its outer zones. The appropriate correction is effected by a special type of thin lens called the **correcting plate,** at the center of curvature of the

Figure 4.11 *The region around omicron Persei taken with a wide-field Schmidt telescope. The field of view is about 6° in diameter.*

Figure 4.12 *A schematic drawing of the optics used for wide-field telescopes.*

astrograph

focal plane

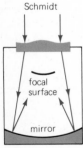

Schmidt

focal surface

mirror

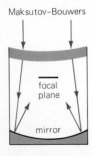

Maksutov–Bouwers

focal plane

mirror

mirror. The lens causes the outer parts of the entering beam to diverge slightly with respect to the middle, so that the entire beam is focused on a slightly curved surface. The photographic plate, suitably curved by springs in the plate holder, faces the mirror between the mirror and the correcting plate. The size of this type of telescope is denoted by the clear diameter of the correcting plate.

One of the largest telescopes of this kind is the 1.2-meter Schmidt of the Hale Observatories (Fig. 4.13) on Mt. Palomar. Its 1.83-meter primary mirror has a radius of curvature of 6.1 meters, which is about the length of the tube. The focal length is 3.05 meters giving a focal ratio of about 1.7. An important achievement of this telescope is the *National Geographic Society–Palomar Observatory Sky Survey*, a photographic atlas of the heavens north of declination 27°S. The atlas consists of 935 pairs of negative prints from blue- and red-sensitive plates; each print is 35.6 centimeters square and covers an area 6°6 on a side. The survey reaches stars of the twentieth visual magnitude and galaxies of the 19.5 magnitude.

Figure   4.13 *The   1.2-meter Schmidt telescope located on Mt. Palomar in California. (Hale Observatories photograph.)*

Figure 4.14 *The number of photons collected by a telescope is proportional to the square of the aperture. Note that in a steady rain the hole on the right, which is twice as large as the hole on the left, allows its glass to collect four times as much water. The function of a large telescope is to collect more photons.*

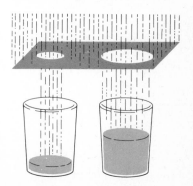

There are a host of specialized telescopes in use, for example, in solar observing, and we will describe a few of these in the appropriate sections.

4.5 Advantages of a large telescope

The **light-gathering power** of a telescope increases in direct proportion to the area of the objective, or the square of its diameter (Fig. 4.14). This is why large telescopes are so important. A particular star is 400 times brighter with the 5-meter Hale telescope than with a 25.4-centimeter telescope. Thus stars can be observed with the former telescope that are too faint to be detected with the latter.

The **magnifying power** of a given telescope increases as the focal length of the eyepiece decreases. When the telescope is employed visually, the magnifying power *(M)* equals the focal length *(F)* of the objective divided by the focal length *(f)* of the eyepiece being used, that is, $M = F/f$. Thus the 5-meter Hale telescope having a focal length of 81 meters at the Cassegrain focus has a magnifying power of 675 when an eyepiece with a focal length of 12 centimeters is used. It would appear that using shorter and shorter focal length eyepieces will give greater and greater magnification. However, there is a practical limit and magnifications greater than 1000 are rarely used at any telescope.

When the telescope is employed as a camera, the diameter of a celestial object having a noticeable size on a photograph equals the focal length of the objective times the angular diameter of the object in degrees, divided by 57°3. Thus the moon, having an angular diameter of about 0°5, appears 14.6 centimeters in diameter in a photograph at the prime focus of the 5-meter Hale telescope, and about 3.4 centimeters with a 25-centimeter refracting telescope of the usual type. In either case, the size can, of course, be increased by enlargement of the photograph. We should note however, that stars are point sources and their size on a photograph is the result of their apparent brightness, the aperture of the telescope, and the seeing and has nothing to do with their true dimensions.

The **resolving power** of a telescope is the angular distance between two stars that can be just separated visually with the telescope in the best conditions. This least distance *d,* in seconds of arc, is related to the wavelength λ and the aperture *a*. For visual wavelengths (0.55 μm) a handy formula is

$$d'' = \frac{13.8}{a},$$

where the aperture is expressed in centimeters. The formula comes from the more general formula

$$d = 1.22 \times 206{,}265 \times \frac{\lambda}{a}.$$

The value for the 5-meter Hale telescope is 0″023, and 0″47 for a 25-centimeter telescope. However, the theoretical high resolving power of a large telescope is never realized because of the disrupting effect of the earth's atmosphere (Section 1.12). We can see from this formula that the much longer wavelengths employed by radio telescopes make them much less effective in resolving fine detail when just a single telescope is used (Section 4.7).

Radio Astronomy

Radio astronomy is the study of the heavens by reception of cosmic radiations at radio wavelengths. The radio waves that can come through the atmosphere to the ground range in wavelength from somewhat less than 1 centimeter to about 30 meters. Waves much shorter than 1 centimeter are absorbed by molecules of our atmosphere, and those longer than 30 meters cannot ordinarily penetrate the electrified ionosphere. Selected parts of this wavelength range are known as **windows** because these specific wavelengths can penetrate the atmosphere more easily than others (Fig. 4.2).

There are two forms of cosmic radiation which are most easily identified at radio wavelengths. The first type is **thermal radiation,** which would be expected only as a result of the temperature of the source. A hot glowing star is a thermal source. It may also come from an emission nebula, where the gas has been highly heated and ionized by a blue star in the vicinity. This kind of radiation is stronger in the centimeter than in the meter wavelengths. The second type is **nonthermal radiation,** which may be the sort of radiation emitted by fast-moving ionized particles revolving in the synchrotron accelerators in particle physics laboratories. It is much stronger in the meter wavelengths than in the shorter ones.

Although radio reception from the heavens was achieved as early as 1931 by Karl Jansky, extensive activity in radio telescope building and operation did not begin until 1946, after the end of World War II.

4.6 Radio telescopes

The **radio telescope** is analogous to the optical telescope in principle and purpose. The most common type of telescope has a large paraboloidal reflector that concentrates the radiation at the focus of the concave surface, just as light rays are gathered by a mirror in an optical telescope. The focus is referred to as the **feed** because the first such telescopes were wartime radar reflectors that were "fed" radar pulses from the focus. Located at the focus is an antenna or a waveguide that receives the signal and sends it to an amplifier and from there to a recording device.

Another type of so-called telescope is the multielement array where series of antennas are interconnected in various ways to achieve a given purpose. Such an array may be a combination of simple dipole antennas, helixes, or even paraboloids.

The radio telescope is effective during the day as well as at night; the long-wave radiations it receives from the heavens are also not seriously obstructed by the clouds of our atmosphere nor by the interstellar dust that conceals much of the universe from the optical view. In addition to its reception from outside sources, the radio telescope is employed for recording radio pulses emitted from a station on the earth and reflected back by the nearer celestial bodies; thus far all of the major objects from the sun out to Saturn have been observed by **radar astronomy** techniques.

The National Radio Astronomy Observatory located at Green Bank, West Virginia has a 91.4-meter transit telescope which is a fine example of a very large paraboloid (Fig. 4.15). It is called a transit telescope because, like its optical counterpart, it can look only along its meridian. This and many other telescopes including a 42.7-meter steerable telescope are operated by this national observatory.

The University of Manchester in England has a 76.2-meter telescope at the Jodrell Bank Experimental Station. The Radiophysics Laboratory in Australia operates a 64-meter telescope at Parkes. Stanford University in California has a 45.7-meter telescope. A very large fixed reflector near Arecibo, Puerto Rico was designed by Cornell University and is operated as a national facility. This reflector is a 304.8-meter spherical dish placed in a natural depression in the mountains (Fig. 4.16). The shift in its pointing direction to as much as 20° from the zenith and the correction for spherical aberration are accomplished at the feed supported high above the dish.

The largest radio telescope is under construction at Soccoro, New Mexico by the National Radio Astronomy Observatory. It is a multielement instrument and is composed of 27 individual telescopes arranged with nine telescopes along each 21-kilometer long leg of a large-Y configuration. Because of its size it is referred to as the Very Large Array, or VLA (Fig. 4.17). Each telescope in the array is a paraboloid with an aperture of 28 meters. The telescopes of each leg can be moved to different positions along the leg. This makes the whole array highly versatile in overcoming the resolution problem discussed in Section 4.7.

Figure 4.15 *The large 91.4 meter telescope in Greenbank, West Virginia. This telescope points up and down along the meridian. Its surface is composed of a wire mesh. (National Radio Astronomy Observatory photograph.)*

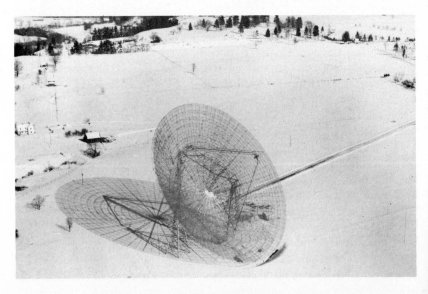

Figure 4.16 *The great 305-meter Arecibo zenith telescope. The dish surface is perforated aluminum sheets, enabling it to collect short-wavelength radiation. (Cornell University photograph.)*

Figure 4.17 *(a) Artist's conception of the very large array in Soccoro, New Mexico. This great instrument composed of 27 individual telescopes became fully operational in 1980. (b) See Figure 19.25. (National Radio Astronomy Observatory photograph.)*

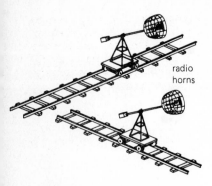

radio
horns

Figure 4.18 *A schematic representation of an aperture synthesis telescope. The effective aperture is a circle around the outside of the tracks. By moving the telescopes to various positions the entire pattern can be sampled and a radio picture constructed.*

4.7 Resolving power

The resolving power of a single radio telescope is a measure of the fineness of detail it can record. Calculated by the same formula mentioned in Section 4.5, the resolving power is the smallest angular distance between two radio point sources that can be detected with a particular telescope. Because it operates with the longer wavelengths, a paraboloidal radio telescope is less effective in separating fine detail than is an optical telescope of the same aperture. Thus the critical separation for a 21-meter radio dish at the wavelength of 21 centimeters is 35′ as compared with a theoretical least separation of only 0″023 with the 5-meter Hale telescope at visual wavelengths.

A partial remedy for the deficiency in resolving power of radio telescopes is the use of larger paraboloids or interference methods promoted by multielement antennas, or both.

Two large paraboloids properly connected together (Fig. 4.18) but separated by a known distance can be made to yield the resolution of a single telescope having a diameter equal to the separation of the two telescopes. This is called **long-baseline interferometry.** Thus, by our formula, two such telescopes separated by 2 kilometers and observing at a wavelength of 21 centimeters will have a resolution of 22″. With the introduction of very accurate atomic clocks (Section 2.16) there is no longer a need for the telescopes to be physically connected; their individual signals can be recorded along with the time on magnetic tapes and then the magnetic tapes can be brought together. This allows the telescopes to be separated by intercontinental distances. This is called **very-long-baseline interferometry.** Very high resolution is achieved by this technique. For example, suppose the two telescopes are 2000 kilometers apart and again working at a wavelength of 21 centimeters; the resolving power is now 0″002.

A second form of a high-resolution telescope is the aperture synthesis telescope. The interferometer mentioned earlier can be made to synthesize a large telescope if one of the paraboloids is movable. If the baseline is not east–west the rotation of the earth allows the "filling in" of the aperture to a certain extent. If both telescopes are movable, one east–west and the other north–south this filling in takes place more rapidly. It is even more efficient if there are numerous paraboloids on a given baseline, one or two of which are movable. This is called a supersynthesis telescope and uses the rotation of the earth for filling in as well. An instrument of this type is located at Westerbork in Holland (Fig. 4.19). Another instrument of this type is the VLA noted earlier.

Other Types of Telescopes

In recent years many new areas of astronomy have opened up because of technological advances in electronics and space flight. The capacity

Figure 4.19 *The great supersynthesis telescope at Westerbork, Holland. The telescope samples the entire pattern by being rotated by the earth instead of moving the telescopes around.*

to place telescopes in orbit above the earth's atmosphere has given birth to **x-ray, gamma-ray,** and **ultraviolet astronomy** while the invention of sensitive detection devices has enabled astronomers to take advantage of the infrared radiation which reaches the earth's surface. These fields have produced discoveries as unexpected as those derived from the pioneering of radio astronomy.

4.8 Infrared telescopes

Infrared radiation lies between radio waves and visible light on the electromagnetic spectrum. It is important in astronomy because it provides information about many sources that emit it which can not be obtained at any other wavelengths. This is because the wavelength of light being emitted by a source increases as its temperature decreases (Fig. 4.20). A relatively cool celestial object such as the moon or a new star just starting to heat up only emits radiation at longer wavelengths. In fact, at temperatures below 1200 K an object will radiate primarily in the infrared. Infrared

Figure 4.20 *As the temperature of a thermal source increases the peak of its radiation curve moves to shorter wavelengths. A relatively cool body has its peak in the infrared region of the spectrum. Only the hottest stars have their radiation peaks in the ultraviolet.*

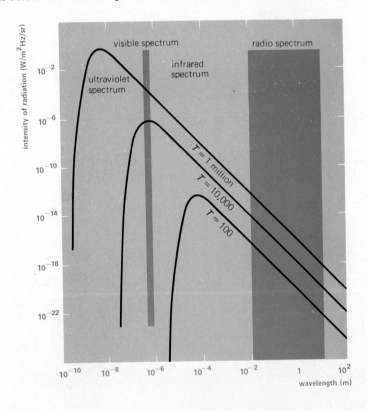

telescopes are therefore necessary if we are to probe the universe for these cooler objects.

The infrared region of the spectrum is difficult to observe for a number of reasons. First of all, water vapor in the atmosphere absorbs most of the infrared radiation that the earth receives before it reaches the ground. This problem is overcome by viewing the sky at selective wavelengths, known as windows, at which the absorption is a minimum (Fig. 4.2). The second difficulty is that the earth, as well as the air and even the telescope itself is a source of infrared radiation. In order to minimize these unwanted extraneous sources the telescope must be used at high altitudes in extremely arid regions. Thus these telescopes are located at observing sites such as Mount Lemmon in Arizona, the volcanic peak of Mauna Kea on the island of Hawaii, and the island of Tenerife in the Canary archipelago. Aircraft (Fig. 4.21), balloons, and rockets are also used to get telescopes above the opaque miasma of the atmosphere.

Reflecting telescopes must be used for infrared research because the glass lenses in a refracting telescope absorb the infrared radiation when it passes through them. Once the image is focused it can be photographed in the near infrared (up to 1 μm) using special photographic emulsions. However, in the longer-wavelength (and far more interesting) infrared regions photometric techniques must be employed. This depends on the use of a detection device much like the photographic exposure meters used by many people as a camera accessory. It operates on the principle that light falling on photosensitive material causes a change in its electrical resistance. An electric current flowing through the detector will therefore fluctuate according to the amount of radiant energy being received and this can be recorded with a meter.

Figure 4.21 *The C-141 airborne observatory. The aircraft carries a 95-centimeter telescope used mostly for infrared observations to altitudes of 15 kilometers. (National Aeronautics and Space Administration photograph.)*

4.9 X-ray astronomy

X rays lie on the high-frequency end of the spectrum. It is convenient to think of this short-wavelength radiation as photons, or particles of radiation. Ordinary mirrors can not be used to collect x rays for detection because photons of this energy would pass right through the reflecting surface. A special mirror must be designed which intercepts the x rays at a shallow angle. Highly polished surfaces on the inside of a tapered cylinder can selectively reflect x rays which strike them at a grazing incidence. This type of conical reflector is given a paraboloidal shape so that the light can be focused on a point. Its exact shape resembles a bullet with the tip removed.

Because the area of the mirror which reflects the x rays entering the telescope from the direction it is pointing is very small, several grazing incidence reflectors (grazing incidence is when the incoming radiation strikes the reflecting surface at a very shallow angle) nested inside one another are used to increase the effective collecting surface of the telescope. A metal disk with a pinhole in it can be placed at the focal plane of the reflector so that only radiation from the line of sight of the telescope reaches the detector (Fig. 4.22). The area of the sky which is observed depends on the diameter of the pinhole used. A field of view two minutes of arc across (2′) is possible with the smallest aperatures.

Although photographic emulsions are sensitive at x-ray wavelengths, they are highly inefficient. In order to determine the intensity of an x-ray source an electronic counter must be used. A counter consists of a gas-filled tube having a negative charge on its inner surface and a positively charged wire running down the middle. When a photon of sufficient energy enters the tube it breaks the gas up into positive and negative particles which allows a pulse of current to flow from the surface to the wire and out to a counter. This method of detection has enabled astronomers to measure the intensity and map the distribution of x-ray sources across the entire sky. However, in order to record an image photographic plates must be used. X-ray photographs have been particularly useful in studying the sun.

Figure 4.22 *A schematic drawing of a grazing incidence x-ray telescope.*

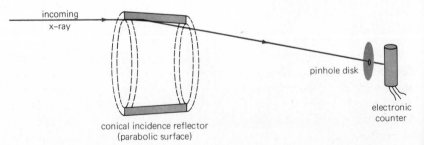

incoming
x-ray

pinhole disk

electronic
counter

conical incidence reflector
(parabolic surface)

Since the upper atmosphere absorbs x rays, radiation of this wavelength can only be observed at altitudes greater than 100 kilometers. Many rockets and satellites have been deployed for this purpose with surprising results. For example, some starlike sources in our galaxy have been found which are barely discernable in the visible part of the spectrum but which radiate 10,000 times more power at x-ray wavelengths than the sun does at all wavelengths combined! Thus high-energy processes occurring in our galaxy have been revealed which were unexpected from observations with optical telescopes.

4.10 Ultraviolet astronomy

The ultraviolet region of the spectrum lies between x rays and visible light. Shorter-wavelength ultraviolet [extreme-ultraviolet (XUV)] radiation is absorbed by ozone in the atmosphere. This impeded the progress of ultraviolet astronomy until the dawn of the space program. There are now many satellites equipped with ultraviolet telescopes.

Ultraviolet light can be focused with regular reflecting telescopes. Nevertheless, the stringent requirements imposed by space have necessitated some modifications. The mirror of the telescope is often made of berylium instead of glass because of its lighter weight and gold is used for the reflecting surface (rather than aluminum) to enhance the light-gathering ability of the smaller mirrors. It is usually too costly to return pictures from a satellite so photoelectric cells are employed instead. The measurements can then be relayed back to earth by radio. However, photographic plates were used on Skylab (the large instrumented space station launched in 1973) and returned with the astronauts after their mission was completed.

Ultraviolet astronomy has greatly furthered our knowledge of the composition of interstellar dust, comets, and the atmospheres of the planets. Studies of the ultraviolet and x-ray spectrum of the sun have also been helpful in elucidating the temperature and energy distribution in the gaseous outer layers of the sun. As with all the other new astronomies, this is a fast developing field and has promise of many more discoveries in the future.

4.11 Telescopes in space

We have already seen how telescopes above the earth's atmosphere facilitate the observation of wavelengths which are inaccessible at the earth's surface. Telescopes operated in space have a further advantage over ground-based telescopes in that their resolving power is much higher. In viewing through the atmosphere, turbulence refracts the starlight out of perfect parallelism so that all the rays of light do not pass through the focal point of the telescope. As a result, the image is blurred and resolution decreases. This limits the resolving power of even the largest earth-bound telescopes to about 0.5 seconds of arc under the very best conditions. A telescope above the atmosphere, however, is limited only by the quality of the objective. It is said to be **diffraction limited.** An object at the detection limit of a 3-meter telescope in space would therefore be ten times as distant as an object just barely detectable with a 5-meter telescope on earth. This obviously would mean a tremendous extension of our vision at the limits of observation. A telescope of these dimensions is planned and will be put into orbit by the space shuttle in the 1980s.

The Spectrograph

The previous sections have dealt with the methods of obtaining observations of the sky by light-gathering instruments. We must now discuss the means of analyzing the incoming radiation. A great deal can be learned about the universe by studying the spectral distribution of its radiation. In the case of stars, for example, by breaking their radiation up into individual wavelengths and measuring the intensity of each wavelength we can discover their temperature, the composition of the gases surrounding them, and their relative motions through space. All this is achieved with a spectrograph.

4.12 Dispersion of light

When a beam of light passes obliquely from one medium to another, as from air into glass, its direction is altered (Section 4.2). Because the amount of the change in direction increases progressively with decrease in wavelength of the light, from red to violet, the beam is dispersed by refraction into a spectrum, or an array of the components of light. Thus the rainbow is produced when sunlight is dispersed by raindrops. A spectrum is obtained by passing the visible light through a glass prism (Fig. 4.23). Radiation beyond the visible region extends into the ultraviolet in one direction and the infrared in the other.

Instead of a prism a **grating** is often used as the dispersing element. The ridges of the grating act to disperse the light into many distinct spectra called the orders of the grating. The chief advantage of a grating is that its spectra are linear with wavelength, which makes the subsequent analysis more simple. A grating, however, spreads the light among its many orders so it is very inefficient. In the early part of this century opticians learned to shape the ridges so that most of the light could be concentrated into a single order. Such a grating is called a **blazed grating** and looks very much like a saw tooth if sliced across the ridges. In most cases there

Figure 4.23 *Light is refracted by a prism (a), blue light more than red. This effect is the basic principle upon which the spectrograph was originally based. Light is diffracted by a grating (b) and most efficiently by a blazed grating (c). Modern spectrographs are based upon the properties of diffraction gratings.*

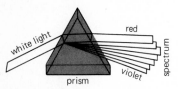

(a)

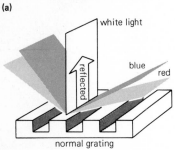

(b)

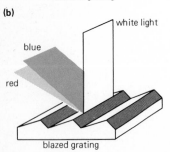

(c)

are at least 600 blazed faces to the millimeter and some of these gratings are as large as 25 centimeters on a side. Because of the very great efficiency of a blazed grating it has replaced the prism as the dispersing element in the spectrographs used by astronomers.

An instrument that makes use of the dispersion of light to record the spectrum is called a **spectrograph.** A photographic record or picture is a **spectrogram.** If the spectrograph is fitted with an eyepiece for direct viewing of the spectrum it is called a **spectroscope.** A familiar type of spectroscope consists of a blazed grating (or sometimes a prism) toward which a collimator and a telescopic camera are directed. The light enters the **collimator** (an inverted telescope) through a narrow slit formed by the sharpened and polished edges of two metal plates. The slit is at the focal point of the collimator. After passing through the collimator, the rays are accordingly parallel in all wavelengths as they fall upon the grating. The grating disperses this beam into a spectrum that is brought to focus by a small telescopic camera (often a small Schmidt telescope; Fig. 4.24). For the purposes of astronomy the spectroscope is attached to the telescope, which serves to concentrate the light of a celestial body and to focus it on the slit. Here, and in the laboratory, the eyepiece of a spectroscope is generally replaced by a film or a plate holder and hence it is referred to as a spectrograph.

4.13 Emission and absorption spectra

The **bright-line spectrum,** or **emission-line spectrum,** is an array of bright lines. The source of the light is a glowing gas at low pressure that radiates in a limited number of wavelengths. Each gaseous chemical element emits its characteristic selection of wavelengths and can therefore be identified by the pattern of lines of its spectrum (Fig. 4.25). The glowing gas of a neon tube, for example, produces a bright-line spectrum characteristic of and unique to neon. The 21-cm line of hydrogen is a bright emission line of great importance in radio astronomy.

The **continuous spectrum** is a continuous emission in all wavelengths. The source of the light is a luminous solid or liquid, or it may be a gas under conditions (great pressure) where it does not emit selectively. The glowing filament of a lamp and the sun's surface produce continuous spectra. There may also be emission and absorption continuums of limited extent in the spectra of certain gases.

The **dark-line spectrum,** or **absorption-line spectrum,** is an array of dark lines on a bright continuous background. Gas at low pressure intervenes between the observer and the source of light, which by itself may produce a bright-line spectrum. The gas abstracts from the light the pattern of wavelengths it emits in the same conditions, and thus reveals its own chemical composition. The spectrum of the sun's atmosphere is essentially

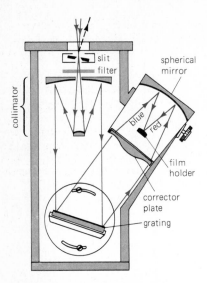

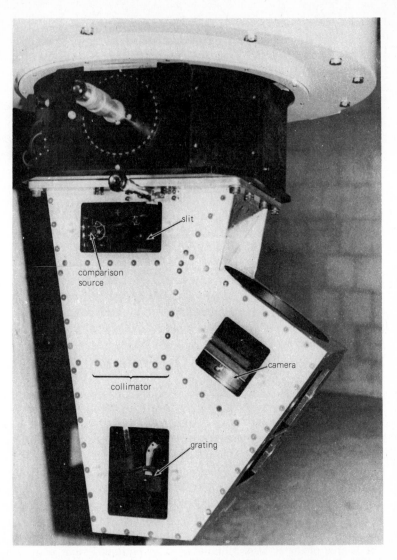

Figure 4.24   *(a) Schematic drawing of a grating spectrograph. (b) A grating spectrograph mounted on a 1-meter telescope. (Leander McCormick Observatory photograph.)*

a dark-line spectrum, because the sunlight has filtered through the atmospheres of the sun and earth before reaching us.

Where the gas consists of molecules, such as carbon dioxide or methane, the spectrum shows bands of bright or dark lines characteristic of these molecules.

4.14 The Doppler effect

When a source of sound waves, such as the whistle of a locomotive, is approaching the observer, the waves come to the observer crowded

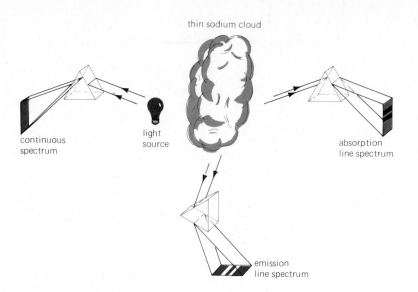

thin sodium cloud

continuous
spectrum

light
source

absorption
line spectrum

emission
line spectrum

Figure 4.25 *The formation of the basic types of spectra. A glowing filament yields a continuous spectrum as shown on the left. If the filament is viewed through a thin cool gas we see an absorption line (dark line) spectrum as shown on the right. If we view only the gas we see a bright line (emission line) spectrum as shown at the bottom. The bright line spectrum is the complement of the dark line spectrum.*

together, so that the pitch of the sound is raised (Fig. 4.26). When the source is receding, the pitch is lowered. In 1842, C. J. Doppler pointed out that a similar effect is required by the wave theory of light; thus a star should appear bluer when it is approaching the observer and redder when it is receding from us. In 1848, H. Fizeau explained that the perceptible effect would be a displacement of the absorption lines in the star's spectrum. The **Doppler effect**—often called the **Doppler–Fizeau effect** when referring to its occurrence in the spectrum of a star—can be stated as follows.

When a source of light is approaching or receding from the observer, the lines in its spectrum are displaced to shorter or longer wavelengths, respectively, by an amount that is directly proportional to the speed of approach or recession. Thus,

$$\Delta\lambda/\lambda = v/c \text{ or } v = c\,(\Delta\lambda/\lambda),$$

where $\Delta\lambda = \lambda' - \lambda$ (where λ is the laboratory wavelength of the light and λ' is the shifted wavelength). As is customary, v is the velocity of the source toward $(-)$ or away $(+)$ from the observer and c is the velocity of light. We can see that as the velocity of the source gets closer to the speed of light, the wavelength of the light is shifted more and more. If the source is approaching, the light is shifted to shorter wavelengths (blue shifted) and if the source is receding, the light is shifted to longer wavelengths (red shifted). Thus, by observing the amount that the lines in the spectrum of a star are shifted, its relative velocity can be determined.

Figure 4.26 *The Doppler effect. The radiation from a moving source is shifted to shorter wavelengths in the direction that the source is moving (observer C), and to longer wavelengths in the direction away from which it is moving (observer A). Note that the true wavelength is seen by observer B.*

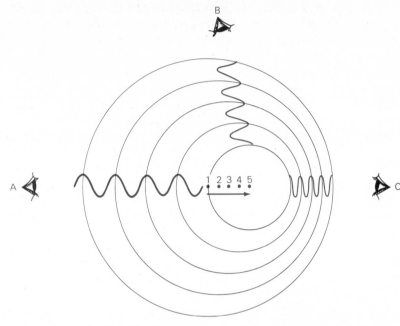

Figure 4.27 *The Doppler effect in the double star Mizar. The spectral lines of the approaching star are shifted to the blue and those of the receding star to the red thus showing two lines. When there is no motion along the line of sight, the lines blend and appear in their normal position.*

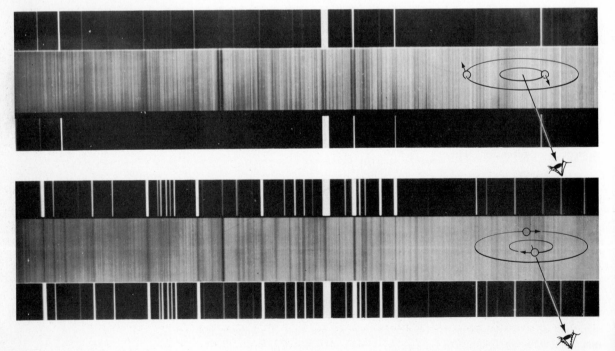

It is important to note that the Doppler effect applies only to the velocity of approach or recession, that is, the object's **radial velocity.** The velocity component along the line of sight is the only component affected. If the object's motion is entirely at right angles to the line of sight it will exhibit no Doppler effect because it has no radial component. Another important point is that the object viewed may be reflecting radiation falling upon it and the reflected radiation will exhibit the Doppler effect appropriate to the radial component of the reflecting object's motion.

The Doppler effect is so important in astronomy that we cite a few examples of where it can be applied. It has been used to establish the rotation periods of Venus and Mercury, the particle nature of the rings of Saturn, the rotation of the sun at various latitudes on the sun, the motion of stars towards or away from the sun (Fig. 4.27), and the motion of galaxies along the line of sight. Indeed the latter observation is the basis for the often stated fact that the Universe is expanding. More subtle effects are also observed. The broadening of absorption and emission lines due to the individual line-of-sight motions of the atoms or molecules are attributed to the Doppler effect on the atomic scale. This tells us about the temperature and pressure at the point where the lines are formed.

Photoelectric Instruments

Because of the inability of the human eye to accurately compare brightnesses and positions of stars, astronomers turned to photographic techniques for these purposes. Photographs have the advantage of recording numerous stars and objects on a single plate. For great accuracy in measuring brightness, however, the photographic plate has some drawbacks; in particular it does not store light linearly and it has a low efficiency. Because of this, astronomers have developed photoelectric devices.

4.15 Photoelectric photometer

The **photoelectric effect,** explained theoretically by Albert Einstein in 1905 for which he received the Nobel Prize, is the release of electrons from a surface when it is illuminated by light. This effect varies with intensity and surfaces can be made to have efficiencies 20 to 30 times greater than photographic emulsions. When such surfaces are placed in tubes they lend themselves to **photometry** (the measurement of brightness) at the telescope. The small current from the phototube is amplified and recorded by some convenient means (Fig. 4.28). Figure 4.29 shows a photometer on the telescope.

The trained human eye can generally estimate brightness in **magnitudes** (Section 12.16) accurate to $\frac{1}{10}$ of a magnitude. Using a photo-

Figure 4.28 *Schematic drawing of a photoelectric photometer. The background signal is limited by a diaphragm. The wavelength region is limited by a filter. The photomultiplier tube multiplies the signal manyfold after which it is amplified and then recorded.*

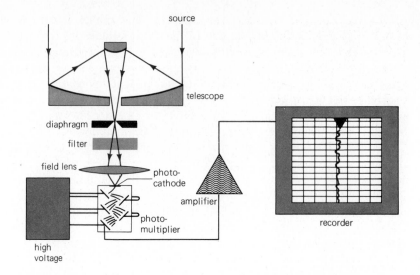

graphic plate magnitudes accurate to a few hundredths of a magnitude can be obtained. With the phototube we can measure magnitudes to a few thousandths. This accuracy explains the value of the photoelectric technique.

Photoelectric photometers are especially useful when attached to spectrographs. For such an application the spectrum is scanned across the photometer and a tracing of the spectrum results. Here as in the direct measurement of magnitudes, it is the extreme efficiency of the photoelectric effect as compared to the photographic emulsion that astronomers are taking advantage of.

4.16 Image tubes

Tubes using the photoelectric effect and forming images are now in common use on telescopes. The television pickup camera is such a tube. The **image tubes** that astronomers use are usually simpler than television tubes, however. In these simple tubes the photoelectrons are focused on a phosphor screen by means of a magnet and the glowing phosphor is then photographed (Fig. 4.30).

Image tubes are used in conjunction with the photographic plate and hence the system can record many stars, the entire field of view, at one time. The secret here is to take advantage of the efficient photocathode and then impart energy to the resulting photoelectrons. Two tubes in one unit (called a **cascade tube**) result in a considerable gain in time at the final photograph. A saving in exposure time of a factor of 10 or more is regularly achieved by such **image intensifiers** (Fig. 4.31).

Figure 4.29 *A two-channel photoelectric photometer mounted on a small telescope. (Leander McCormick Observatory photograph.)*

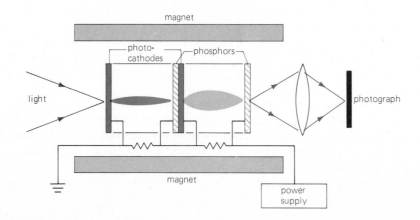

Figure 4.30 *A schematic drawing of a simple cascade-image-tube system. The light falls on the tube on the left. The light is converted to photoelectrons, amplification takes place in the center, and an image is formed which can be photographed.*

Figure 4.31 *A two-stage image intensifier as diagramed in Fig. 4.30. This tube is fabricated by RCA and is approximately 10 centimeters in diameter including the rubber casing. (Carnegie Institution of Washington, DTM photograph.)*

Another advantage of these tubes is that their cathodes can be made to respond to infrared or ultraviolet light. Such a tube still produces a blue image (because the phosphor gives off blue light) to be photographed, and is therefore known as an **image converter.** Image converters have gains of more than 1000 over comparable photographic techniques where the wavelength of the radiation falls in the near infrared. This large increase is chiefly because of the extreme slowness of infrared-sensitive films.

Astronomy except for planetary and solar physics, is in the main a passive science because the experiments are beyond the control of the scientist. These experiments, at least the interesting ones, always seem to occur at very great distances, hence, the amount of energy arriving at the earth per square centimeter is very slight. The astronomer must be able to collect and record as much light as possible to interpret these events. The search for ever more efficient techniques continues.

Questions

1 Define wavelength and frequency. How are the two related when we are dealing with light?
2 Distinguish between refraction and dispersion.
3 What is the purpose of an astrographic telescope?
4 Why do we build larger and larger telescopes?
5 Why do the large optical telescopes never reach their theoretical resolving power?
6 What are the two different types of radio radiation called?
7 How do we achieve high resolution with radio telescopes?
8 What is the purpose of a spectrograph?
9 List the three types of spectra and their causes?
10 Describe the Doppler effect.

Other Readings

Cameron, R. M., "NASA's 91-cm Airborne Telescope," *Sky & Telescope,* **52,** 327–331 (1976).

Cornell, J., "A New Breed of Optical Telescope—the Multiple Mirror Telescope—Brings Astronomical Mysteries Into Focus," *Smithsonian* **10,** 42–53 (1979).

Keene, G. J. and M. H. Sewall, "An Evaluation of 8 Films for Astrophotography," *Sky & Telescope,* **50,** 61–65 (1975).

Neary, J., "Huge New Radio Telescope Array Extends Man's Celestial Vision," *Smithsonian* **9,** 28–37 (1978).

chapter 5

The Moon in Its Phases

The moon is the closest object to the earth in space. It is also the only heavenly body ever to have been visited by man. We therefore know more about the moon than any other planetary body outside of the earth. In this chapter we consider the moon's motions relative to the sun and the earth, the tides it raises in the earth's oceans, and eclipses. We also review what has been learned about the moon from the Apollo space program because it is our first real step toward understanding the origin and evolution of the solar system.

Motions of the Moon

The moon is the earth's natural satellite. Together the earth and the moon revolve around the sun and at the same time mutually revolve around a point between their centers. This regular motion is the basis for the month in our calendars. It also affects us in more subtle ways. Many animals, even Homo sapiens, have reproductive cycles that correlate with the moon's period. Higher forms of life might not even have evolved on the earth if it were not for the moon. Other planets experience dramatic climate changes because their rotational axis wobbles erratically. But the earth has enjoyed a rather steady climatic history because the moon tends to stabilize the tilt of its rotational axis. The moon's motions are therefore integral to our very existence.

5.1 The moon's distance, size, and orbit

When you view your thumb at arm's length, first with one eye and then the other, it appears to jump back and forth against the background. This difference in the direction of an object as seen from two different positions is called **parallax.** Parallax provides us with a means of determining distance. We use it all the time. When viewing an object we cross our eyes a little so that they can focus on the same point. The closer the object is, the more our two lines of sight must converge in order to see it, and our brain interprets this as a measure of distance. If we want to measure the distance to the moon by parallax, we must view it from two points much further apart than the distance between our eyes. Observed from New York and San Francisco at the same time, the moon's position among the stars differs by half a degree (the full breadth of the moon), which is easily determined (Fig. 5.1). Then, since we know the distance between New York and San Francisco, the distance to the moon can be readily calculated.

Ordinarily, when we speak of the moon's parallax, it is as though one observer were at the center of the earth and the other observer at the equator with the moon on the horizon. It is this **equatorial horizontal parallax** that is given for sake of uniformity. The moon's parallax at its average distance from the earth is nearly 1°. The corresponding distance of the moon from the center of the earth is 384,400 kilometers, which is

Figure 5.1 *The moon's distance can be obtained by measuring the apparent angular shift (parallax) against a distant background. This shift amounts to almost ½ degree, or the diameter of the moon itself, when viewed from New York and San Francisco.*

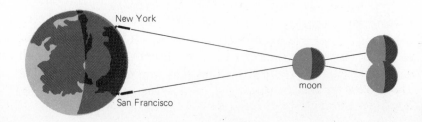

about 60 times the radius of the earth. This is a comparatively small distance in the vastness of interplanetary space. The sun, for example, is nearly 400 times farther away, even though it appears equally large in the sky. To the astronauts aboard the Apollo spacecraft, the distance to the moon represented a three day journey.

In 129 B.C. a total eclipse of the sun occurred in Greece, while in Egypt only $\frac{4}{5}$ of the sun was obscured. Using this difference in the direction of the moon (relative to the background sun) and knowing the distance between Egypt and Greece the Greek astronomer Hipparchus was able to calculate the distance to the moon.

Today we can measure the distance to the moon by various techniques. One is to bounce radio pulses off the moon and measure the length of time it takes them to return to earth. The results agree closely with the distance determined by parallax. Another technique uses special reflectors left on the moon by the astronauts. The two-way travel time of laser light reflected off these boxlike mirrors, called **retroflectors,** gives the distance to the moon to within an accuracy of a few centimeters. As a result, the motions of the moon can now be carefully monitored and its orbital characteristics closely determined.

Although the moon is not the largest of the satellites in the solar system, it is comparable with its primary in size and mass. For this reason the earth and the moon are sometimes referred to as the double planet. The moon's diameter of 3476 kilometers is more than one-quarter of the earth's diameter. However, the moon's mass is only $\frac{1}{81}$ that of the earth's. Consequently, its surface gravity is only $\frac{1}{6}$ as strong as the earth's. A 90-kilogram person on earth would only weigh 15 kilograms on the moon. Also, the moon's mean density is less than that of the earth's, being only 3.34 g/cm³.

Imagine the centers of the earth and moon joined by a stout rod. The point of support at which the two bodies would balance is the **center of mass** of the earth–moon system. It is this point around which the earth and the moon mutually revolve monthly, and it is the elliptical path of this point around the sun that we have hitherto called the earth's orbit (Fig. 5.2). The center of mass of the earth–moon system is only 4667 kilometers from the earth's center and is therefore within the earth. Thus the moon revolves around the earth, although not around the earth's center. In the descriptions that follow it is convenient to consider the moon's revolution relative to the earth's center.

The moon's orbit relative to the earth is an ellipse of small eccentricity ($e = 0.0549$), having the earth at one focus (Fig. 5.3). At **perigee,** where the moon is nearest to us, the distance between the centers of the moon and earth may be as small as 356,800 kilometers. At **apogee,** where the moon is farthest from us, the distance may be as great as 406,400 kilometers. The resulting variation of more than 10% in the moon's apparent diameter is still not enough to be conspicuous to the unaided eye.

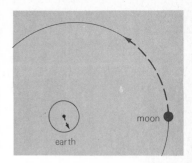

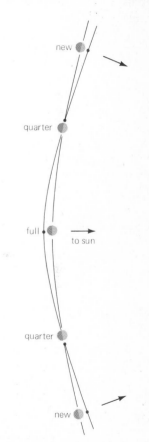

Figure 5.2 *The orbits of the earth and moon around the sun. (a) The moon and earth revolve about their common center of mass (not to scale) while the two of them (b) orbit the sun. It is not apparent from the drawing, but the orbit of the moon around the sun is actually everywhere concave to the sun.*

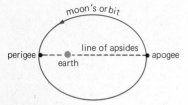

Figure 5.3 *The moon's orbit rel-ative to the earth is an ellipse. The eccentricity is greatly exaggerated in this figure.*

The moon's orbit around us is changing continually, mainly because of unequal effects of the sun's attraction for the moon and earth. For example, the major axis of the orbit rotates eastward in a period of about 9 years. This is one of the many variations that make the determination of the moon's motion an intricate problem, but a problem now being so well solved that the moon's course in the heavens is predictable with great accuracy.

5.2 The phases of the moon

The moon's phases are the different apparent shapes it shows (Fig. 5.4). Records of the phases of the moon have been found dating from 13,000 B.C. The moon is a dark globe like the earth; half is in the sunlight, while the other half turned away from the sun is in the darkness of night. The phases are the varying amounts of the moon's sunlit hemisphere that are turned toward us successively in the course of the month.

It is the **new moon** that passes the sun; the dark hemisphere is toward us. The moon is invisible at this phase unless it happens to pass directly across the sun's disk, causing an eclipse of the sun. On the second evening after the new phase the thin **crescent** moon is likely to be seen in the west after sundown. The crescent grows thicker night after night, until the sunrise line, or **terminator,** runs straight across the disk at the **first quarter.** Then comes the **gibbous** phase as the bulging sunrise line gives the moon a lopsided appearance. Finally, a round **full moon** is seen rising at about nightfall. The phases are repeated thereafter in reverse order as the sunset line advances over the disk; they are gibbous, **last quarter,** crescent, and new again.

The horns, or **cusps,** of the crescent moon point away from the sun's place, and nearly in the direction of the moon's path among the stars.

Figure 5.4 *The phases of the moon. The outer figures show the phases as seen from earth.*

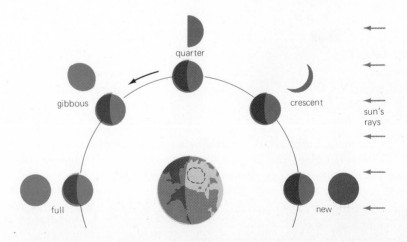

They are more nearly vertical in the evenings of spring, and more nearly horizontal in the autumn. Their direction in the sky follows the changing direction of the ecliptic with respect to the horizon during the year.

Often when the moon is in the crescent phase we see the rest of the moon dimly illuminated. This appearance has been called "the old moon in the new moon's arms," for the bright crescent seems to be wrapped around the faintly lighted part (Fig. 5.5). The thin crescent is in the sunlight. The rest of the moon's disk is made visible by sunlight reflected from the earth. Just as the moon tempers the darkness of night for us, so the earth shines on the moon.

Earthlight is plainest when the moon is a thin crescent, for the earth is then near its full phase in the lunar sky. It is a bluer light than that of the sunlit moon, because it is sunlight reflected by our atmosphere. Figure 5.5 shows the earthlight on the moon at crescent phase in the morning sky.

As we know from our television observations if anyone lived on the earthward side of the moon long enough, he would see the earth up among the stars in the sky (Fig. 1.1) going through all the phases that the moon shows to us. They are, of course, complementary; full earth occurs at new moon. The full earth would look 4 times as great in diameter as the full moon appears to us, and something like 60 times brighter. The earth is not only a larger mirror to reflect the sunshine but, owing to the atmosphere, it is a better reflector as well; it returns $\frac{1}{3}$ of the light it receives from the sun, although this varies somewhat depending upon the presence or absence of clouds.

The month of the phases from new moon to new again, averages slightly more than 29.5 days, and varies in length more than half a day. It is termed the **synodic month,** and is the lunar month of the calendars. Because it is shorter than our calendar months, with a single exception, the dates of the different phases are generally earlier in successive months.

The length of the **sidereal month** averages 27.3 days. This is the true period of the moon's revolution around the earth. At the end of this interval the moon has returned to nearly the same place among the stars. In the meantime the sun has been moving eastward as well, so that more than two days elapse after the end of the sidereal month before the synodic month is completed. In Fig. 5.6 between positions 1 and 2 the moon has completed one sidereal month. It does not complete the synodic month until it reaches position 3.

5.3 The moon's path among the stars

Two apparent motions of the moon are observed by everyone. First, the moon rises and sets daily; it circles westward around us along with the rest of the celestial scenery because the earth is rotating from west

Figure 5.5 *Earthlight on the moon at crescent phase in the morning sky. The earthlit moon can be seen in Fig. 5.7 also. (Yerkes Observatory photograph.)*

Figure 5.6 *The month of the phases is longer than the sidereal month. The moon completes a full orbit (360°) moving from 1 to 2. However, since the earth has moved with respect to the sun and stars, the moon must move to position 3 before the new moon again occurs.*

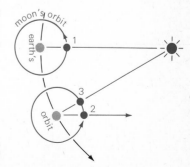

Figure 5.7 *Jupiter and its satellites emerge from occultation by the moon. Such pictures are striking evidence of the lack of an atmosphere on the moon. (Griffith Observatory photograph.)*

to east. Second, the moon moves eastward against the turning background of the stars, because it is revolving in this direction around the earth. In the course of a day the moon revolves 360°/27.3, or about 13°, moving slightly more than its own diameter in an hour. This motion is readily observed.

The moon's apparent path among the stars is nearly a great circle of the celestial sphere, which is inclined about 5° to the ecliptic. It therefore crosses the ecliptic at two opposite nodes. The ascending node is the point where the moon's center crosses the ecliptic going north; the descending node is the point where it crosses going south. The nodes regress, or slide westward along the ecliptic. Regression of the nodes goes on at a much faster rate than does precession of the equinoxes (Section 3.6). In only 18.6 years the nodes shift completely around the ecliptic. The moon's path through the constellations of the zodiac is considerably different from month to month. All of this, including the fact that the moon's orbit is not circular, was known at the time of Hipparchus.

The moon's revolution around the earth provides an important means of timekeeping that is independent of the earth's variable rotation. There is a practical observational problem, however. One must be able to determine precisely the universal time when the center of the moon's large disk arrives at a particular place among the stars. The corresponding uniform ephemeris time (Section 2.15) when the moon has this position can be determined by the theory of the moon's motion. Frequent comparisons of this kind permit accurate corrections from the predicted ephemeris times of celestial events to the universal times for ordinary use.

Occasions when the moon passes over (**occults**) a star have been employed for a similar purpose. The star disappears almost instantly at the moon's eastern edge and reappears as abruptly at the western edge. Occultations are interesting to watch with the telescope or with the unaided eye when the objects are bright enough. The photograph of Jupiter and its satellites emerging from behind the moon (Fig. 5.7) illustrates a spectacular example.

Occultations of stars by the edge of the moon and their reappearance have two scientific uses: (1) they provide a direct measure of the contour of the edge of the moon, and (2) they provide a means for measuring the angular diameter, or size, of the star being occulted. The latter measurements are extremely valuable and can make use of the diffraction effect because the moon has essentially no atmosphere. These measurements require ultrafast response electronics and recording techniques since the entire event takes place in less than 0.5 second.

5.4 Daily, monthly, and annual motions

We have seen that the solar day is about 4 minutes longer than the sidereal day because the sun moves eastward relative to the stars. The moon moves eastward still faster than the sun, and therefore the "lunar day" is longer than the solar day. The interval from upper transit of the moon to its next upper transit averages 24^h50^m of solar time, varying as much as 15 minutes either way.

The moon's crossing of the meridian as well as its rising and setting are delayed an average of 50 minutes from day to day. The time variation of the moonrise is even more marked, and we notice it particularly in the rising of the moon near its full phase.

The **harvest moon** is the full moon that occurs nearest the time of the autumnal equinox. The moon is then near its full phase and, as observed in our northern latitudes, it rises from night to night with the least delay. In the latitude of New York the least delay is shorter by an hour than the greatest delay. Thus the harvest moon lingers longer in our early evening skies than does the nearly full moon of other seasons, giving more light after sundown for harvesting. It can be shown that the least delay in the moonrise on successive nights occurs when the moon's path among the stars is least inclined to the east horizon, as it is at sunset at the time of the autumnal equinox. The full moon following the harvest moon is known as the **hunter's moon** for similar reasons.

The moon moves north and south of the celestial equator during the month, just as the sun does during the year and for the same reason. The moon's path among the stars is nearly the same as the ecliptic and is, therefore, similarly inclined to the celestial equator.

Consider the full moon. Opposite the sun at this phase, the full moon is farthest north when the sun is farthest south of the equator. The full moon near the time of the winter solstice is above the horizon for a longer time than are the full moons of other seasons. In summer it is the other way around. The full moon of June rises in the southeast, transits low in the south, and soon sets in the southwest, like the winter sun.

In some years the moon ranges farther north and south than in other years (Fig. 5.8). The greatest range in its movement in declination occurs at intervals of 18.6 years. This was the case in 1969, when the moon was going fully 5° farther than the sun both north and south from the celestial equator. Many people remarked on it at the time, probably because of the great general interest in the moon due to the impending manned lunar landing. The moon reached its highest point on 25 March 1969 when its declination was +28°43′. It reached only 18°36′ north at full moon on 30 October 1979.

As often as it is revealed in the changing phases, the face of the "man in the moon" is always toward us. This means that the moon rotates on its axis once in a sidereal month, while it is revolving once around the earth. Although the statement is true in the long run, anyone who watches the moon carefully during the month can see that the same hemisphere is not turned precisely toward us at all times. Spots near the moon's edge are sometimes in view and at other times turned out of sight, as

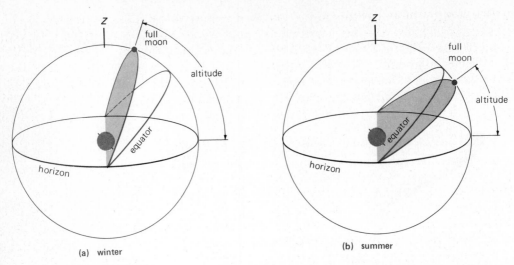

(a) winter

(b) summer

Figure 5.8 *Depending upon the orientation of the nodes of the lunar orbit, full moon may occur above the ecliptic or below the ecliptic. Hence, the moon can be far above (a) or below (b) the equator and hence nearer or farther from the zenith, respectively.*

Fig 5.9 clearly shows. The moon seems to rock as it goes around us. These apparent oscillations, or **librations,** arise chiefly from three causes:

1 Libration in latitude. The moon's equator is inclined about 6°.5 to the plane of its orbit. Thus its north pole is brought toward us at one time, and its south pole is toward us two weeks later, just as the earth's poles are presented alternately to the sun during the year.

2 Libration in longitude. The moon's revolution is not uniform. In its elliptical orbit around us, the law of equal areas (Section 6.2) applies: The nearer the moon to the earth the faster is its revolution. Meanwhile the rotation of the moon is practically uniform. Thus the two motions do not keep perfectly in step, although they come out together at the end of the month. The moon rocks in the east and west direction because of this, allowing us to see farther around it in longitude at each edge than we could otherwise.

3 Diurnal libration. An observer on earth can be almost 6000 kilometers east or west of the line joining the center of the earth with the center of the moon. This allows the observer to see "around" the edge of the moon by about 1° at moonrise and another 1° at moonset.

Fully 59% of the moon's surface has faced the earth when the lunar month is completed. The remaining 41% is never seen from the earth.

The Ocean Tides

The obvious correlation between the rise and fall of the sea with the motions of the moon was observed from the earliest times. But it

Figure 5.9   *The libration of the moon is clearly detected in these two photographs taken at the same phase but several years apart. The maria are much closer to the limb in the left-hand picture. (Lick Observatory photographs.)*

was not until the discovery of the Law of Gravity by Isaac Newton that it could be successfully explained. We now know that the tides are influenced not only by the moon but by the sun as well. In addition, the tides are affecting the rotation of the earth and the motions of the moon in previously unsuspected ways.

5.5   Lunar tides

The moon's gravitational attraction raises two tidal bulges in the water covering the earth's surface; one on the side of the earth facing toward the moon and the other on the side facing away from the moon. Since the earth rotates beneath these tidal bulges, an observer at a fixed position on its surface sees two high tides and two low tides each day.

The tidal bulges form because of the gravitational attraction of the moon. The side of the earth facing towards the moon experiences more of a gravitational attraction than the center of the earth, and the side of

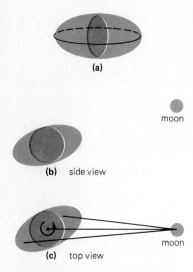

(a)

moon

(b)   side view

moon

(c)   top view

Figure 5.10   *The tidal figure is an ellipsoid of revolution called a prolate spheroid (a), whose long axis extends toward the moon (b). However, since the earth is rotating beneath the tidal figure, the tidal figure is carried slightly ahead of the line joining the earth–moon (c). Because the bulge closest to the moon exerts more gravitational attraction on the moon than the bulge on the opposite side of the earth, the moon is accelerated. As a result, the moon is slowly receding from the earth. Figures are not to scale.*

the earth facing away from the moon experiences a smaller attraction. Since water can flow in response to this difference in the gravitational attraction of the moon, a bulge is raised on the side of the earth facing the moon. On the opposite side of the earth, the difference in the moon's gravitational attraction between the surface and the center of the earth is just as great as the difference in the gravitational attraction from the center to the near side, so the tidal bulge on the far side of the earth is just as large as the bulge on the near side. In consequence, the water on the earth's surface is drawn out into a prolate sphereoid whose long axis points toward the moon (Fig. 5.10).

If the moon were fixed in position, the earth would rotate beneath the tidal figure in exactly 24 hours. Then the two high tides would occur exactly 12 hours apart at the same time each day and the low tides would occur halfway between them. But as the earth rotates, the moon advances in its orbit about the earth. This advance is about 13° per day. Thus, in one-half day, the earth must turn an extra $6°.5$ to rotate under the next high tide and this takes about 25 minutes more than half a day. As a result, each high or low tide occurs about 25 minutes later than the last one.

In the open ocean the difference in water level between high and low tide is scarcely ¾ meter. However, the exact times and heights of the tides on the seacoast depend upon the declination of the moon and the local geography. High tide on the middle Atlantic coast of North America occurs essentially at the time predicted by the complete tidal theory and reaches heights of about 2 meters. At the Bay of Fundy in Newfoundland, high tides occur well after the theoretical time and reach heights in excess of 18 meters.

5.6 Spring and neap tides

The sun also causes tides. The sun, however, is less than half as effective as the moon in raising tides.

The lunar and solar tides may be thought of as acting independently. Thus, when the moon is in **syzygy,** that is, at full moon or new moon, the two tides reinforce each other and we have the highest high tide, referred to as **spring tide.** When the moon is in quadrature, that is when the moon is at first quarter and last quarter, the two tides are out of phase and the high tides are due to the moon alone. These tides are called the **neap tides.**

So far we have discussed only the ocean tides. Since the earth is not an absolutely rigid body the moon and sun raise tides in the earth itself. Of course these **earth tides** are considerably smaller. The tidal bulge of the earth amounts to only a few centimeters.

5.7 Tidal effects

The continual wash of the sea on the coastline and shallow ocean floors causes a drag on the earth's rotation. The effect of this friction on the earth has been a gradual slowing of its rotation and a corresponding increase in the length of the day. The change is imperceptibly small; the earth's rotational period increases by only one-hundred-millionth of a second each day. But when considering the billions of years since the earth has formed, the effect is significant. Three-hundred-million years ago the day was only 22 hours long in terms of our present hours and the year contained 400 days.

The earth and the moon comprise an interrelated system of rotating and revolving bodies. Any change in the earth's rotation must therefore affect the moon as well. This is because rotation and revolution are both carrying angular momentum, a physical quantity that is always conserved. Angular momentum is a function of the mass of a body, the rate at which it rotates or revolves, and the distance of the rotating or revolving body from its center of rotation or revolution. Since the rotation of the earth is decreasing, then its angular momentum is decreasing. To compensate and keep the total angular momentum constant the moon must move farther from the earth. Compounding this effect is the fact that the earth drags the tides forward from the line joining the earth and the moon. Thus the tidal ellipsoid actually points ahead of the moon and the normal high tides occur after the moon has transited the meridian. The tidal bulge nearest the moon exerts a greater gravitational attraction on the moon than the bulge on the opposite side of the earth. As a result the moon is accelerated and it moves farther away from the earth (Fig. 5.10). The net result is that the moon is presently receding from the earth at a rate of 2 centimeters per year.

The Features of the Moon

The exploration of the moon really began almost 400 years ago when Galileo first trained his telescope on the moon. Our knowledge of the moon's features grew steadily after this great leap forward in our ability to perceive other worlds. Then, in the 1960s, the United States undertook a program of manned exploration of the moon. In order to accomplish this monumental task, scientists needed to acquire a thorough understanding of where the safest landing areas on the moon were and what the astronauts might expect to find once they got there. Towards this end, lunar research from ground-based telescopes was intensified and a program of unmanned spacecraft exploration was embarked upon. The result was an intimate familiarity with the lunar surface.

5.8 Lunar lands and seas

In 1609 Galileo heard of the discovery by a Dutch spectacle maker, Hans Lippershey, that two lenses held at a suitable distance before the eye gave a clearer view of the landscape. Galileo fitted two small lenses into a tube and went out to view the heavens. Naturally, he directed his telescope toward the moon. One of his first observations was that the lunar surface consists of lighter and darker areas (Fig. 5.11) which are actually readily distinguished by the unaided eye. The darker areas were smooth and roughly circular in appearance. Galileo thought that these regions might be covered with water and so he named them **maria,** the Latin word for "seas." The brighter areas around the maria were noticeably more rugged, with mountains and valleys. Galileo called these areas **terrae** or "lands." We now know that the moon is entirely waterless. What Galileo had mistaken for oceans are immense, gently rolling plains which were formed by volcanic lava flows billions of years ago. Nevertheless, the name

Figure 5.11 *A photograph of full moon (left). The maria and manned landing sites are identified in the drawing (right). The numbers refer to the Apollo program numbers. (Yerkes Observatory photograph.)*

maria has survived. The terrae are often referred to as the lunar highlands. It is the lunar maria and highlands that form the familiar "man in the moon."

As telescopes improved, better and better maps of the moon's surface were made. From Hevelius' map of 1647 the names of mountain chains such as the Alps, Apennines, and Carpathians came to be applied to the moon as well. A map of the moon made by Riccioli was published in 1651 that also left its legacy in terms we are now familiar with. Examples are Mare Tranquillitatis (Sea of Tranquillity), Oceanus Procellarium (Ocean of Storms), Mare Imbrium (Sea of Rains), and Mare Serenitatis (Sea of Serenity). Riccioli also initiated the practice of naming lunar craters after philosophers, scientists, and historical figures (Plato, Tycho, Copernicus, Julius Caesar, etc.). These conventions have now been formalized by the **International Astronomical Union** (IAU).

No matter how closely astronomers looked, though, all that could be seen was the front side of the moon. Since the moon always keeps

the same face toward the earth, the back side of the moon was always hidden from view. Spacecraft provided the first glimpse of the back side. What was revealed could hardly have been expected. The far side of the moon has almost no maria at all. Except for a few small dark patches, it is entirely covered by highlands. Two of the largest dark regions on the far side are Mare Moscoviense (Sea of Moscow) and the inside of the crater Tsiolkovsky. Since the far side of the moon was first photographed by the Soviet space probe Luna 3, several of its features were given Russian names.

The highlands are characterized by extremely mountainous regions. Galileo was able to estimate the height of these mountains by observing the moon near the edge of the sunlit part of the disk referred to as the terminator. Just before sunrise on the moon the peaks in the unlit portion jut up out of the shadow and reflect the sun's light. The very top of the peaks are the first to catch the sun's rays. They appear like little dots in the darkness (see Fig. 5.12). By measuring the distance from the sunrise

Figure 5.12   *The lunar surface in Mare Imbrium. Note how the crater rims and ridge tops beyond the terminator jut up into the sunlight. The low sun angle makes small features visible because of the long shadows they cast. Lava flows extending from the lower left are 10 to 25 kilometers in width and about 35 meters high. Wrinkle ridges can be seen at the top of the picture. (National Aeronautics and Space Administration photograph.)*

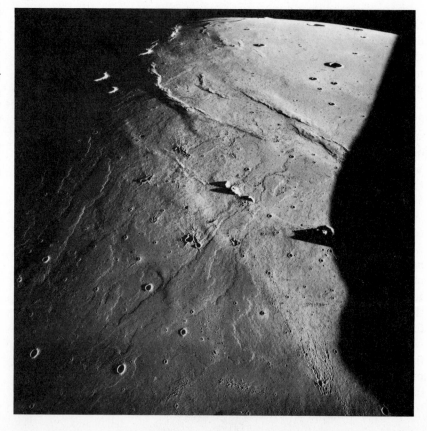

line to the position of first appearance of the moon's peaks, Galileo concluded that the mountains on the moon were about the same size as the mountains of the earth.

When the sun is low in the sky, the shadows that are cast are very long. At these low sun angles, small features become visible and a striking amount of detail can be observed. For this reason, the moon is best observed near the terminator where features are exaggerated (Fig. 5.12).

The Apollo spacecraft carried a laser altimeter which could measure the elevation of the lunar surface to an accuracy of 1 meter. Since the spacecraft remained in a nearly circular orbit around the moon, the absolute height of the lunar features could be determined. The lunar maria were found to lie considerably below the average level of the lunar surface. Some maria are as much as 5 kilometers below the average elevation. These large depressions in the lunar surface are called the maria basins. On the other hand, most of the far side of the moon was found to lie above the average level of the lunar surface. The greatest difference in height between the mountains and the maria is about 8 kilometers, which is nearly the elevation of Mt. Everest above sea level on the earth.

The maria appear darker than the highlands because the material that covers them has a lower **albedo,** or reflectivity. The general albedo of the moon is about 0.07. This means that only 7% of the sunlight that strikes the moon is reflected. The rest of the light is absorbed. The maria have albedos between 0.05 and 0.08 while the highlands have albedos between 0.09 and 0.12 (Fig. 5.13). This difference in albedo is due to a difference in the type of rocks forming the maria and highlands. The rocks in the maria are mostly basalts which contain dark-colored minerals such as ilmenite. The highlands, on the other hand, are composed mostly of plagioclase (probably derived from anorthosites), a light-colored mineral. This difference in mineralogy is also reflected by a difference in chemistry, the maria being richer in the elements iron, titanium, and magnesium and the highlands being richer in calcium and aluminum.

Before the space age, astronomers carefully studied the properties of the light reflected from the moon's surface in an attempt to determine what the surface of the moon was like. One useful property was the **photometric function** of the moon. This is a measure of the variation of brightness of the lunar surface as the angle at which sunlight strikes the moon changes. It was found that sunlight is reflected back toward the earth even when the sunlight strikes the moon at a very low angle. This means that the full moon is almost equally bright from its center to its edge. If the moon were a perfectly smooth surface, like a mirror, only the center of the full moon would be bright. In contrast, the material at the moon's surface must be rough on a small scale. By comparing the photometric function of the moon with various materials in the laboratory, it was concluded that the moon must be covered with gravel or coarse sand.

Figure 5.13 *The Apennine Mountains along the southeast rim of Imbrium basin. Hadley rille is in the center of the photograph. Apollo 15 landed where the rille turns abruptly towards the lower left. The rille is over 100 kilometers long. A linear rille parallel to Hadley rille is at the bottom of the picture. (National Aeronautics and Space Administration photograph.)*

Radar was also used to study the moon. By bouncing radar pulses off the moon, astronomers could construct a picture of the moon at radar wavelengths. Instead of being equally illuminated over the entire disk, the moon's radar reflectivity became dimmer toward the edge of the disk. Therefore, at radar wavelengths (centimeters to meters) the moon must be relatively smooth. This is because the dimensions of an object must be larger than a wavelength of the radiation in order for that object to reflect the radiation. The depth to which radiation penetrates a surface is also approximately equal to its wavelength. Consequently, radar enabled astronomers to examine the first few meters of the moon's surface. The conclusion was that the surface of the moon is covered with a thick porous layer of material.

The combined picture of the lunar surface obtained from earth-based observations was that on a scale of millimeters the surface possessed considerable roughness, but at a scale of meters it is a smooth undulating surface with only occassional objects larger than a few centimeters. The lunar soil was thought to consist of fine-grained sand and dust interspersed with gravel. The total depth of this rubble was believed to be a few meters. When the astronauts arrived on the moon, these predictions were largely confirmed. What they saw was a smooth boulder-strewn terrain pockmarked with craters.

5.9 Lunar craters

Craters cover the entire surface of the moon. Ranging in size from microscopic pits to basins hundreds of kilometers in diameter, they are the most pervasive feature of the lunar surface. But craters are not unique to the moon. Pictures of Mars and Mercury taken by the Mariner space probes reveal cratered surfaces on these planets as well. The moons of Mars and Jupiter also are cratered. Radar mapping shows that Venus has craters as well. Even the earth has craters, the most familiar example being Meteor Crater in Arizona. It therefore appears as though craters are a feature common to all of the solid bodies in the solar system.

Craters are produced by the impact of objects moving at high speed. The size, mass, and velocity of the projectile determine how large a crater is formed. Cosmic dust, or micrometeorites, form tiny glass-lined pits in the dust and rocks on the moon's surface while larger craters a few meters in diameter are formed by the impact of meteorites the size of a softball. The largest craters result from the impact of asteroids or the nuclei of comets. All gradations in size occur between these extremes, but as a rule, the impact of a body becomes increasingly more unusual as its size increases. Millions of micrometeorites strike the lunar surface every day and about 100 meteorites larger than a softball strike the moon every year, but bodies the size of asteroids (kilometers in diameter) only collide with moon about once in a million years.

The impact origin of craters was contested for a long time because almost all the craters on the moon are nearly circular, whereas a meteorite striking a surface obliquely might be expected to gouge out an oblong depression. What the early lunar investigators failed to appreciate was the tremendous amount of energy that would be released by a fast-moving object striking the moon. Asteroids in our vicinity, for example, fly past us with a velocity of 10 to 20 kilometers per second. If an average-sized asteroid, about 2 kilometers in diameter, were to strike the moon at this speed, the amount of energy released would be equivalent to a million one-megaton hydrogen bombs. The resulting explosion would form a crater 90 kilometers in diameter (the size of the crater Copernicus). The violent nature of an impact is therefore capable of producing a circular depression many times larger than the impacting projectile. In general, the size of a crater is about 50 times larger than the meteorite which produced it and the volume of material excavated is at least several hundred times the volume of the impacting body.

The material thrown out during the cratering event is distributed as a layer of debris referred to as **ejecta.** The ejecta blankets the area around a crater for 2 to 4 times its diameter. Longer radial extensions of the ejecta are called **rays.** Rays form some of the most prominent features on the moon. The bright rays of the craters Tycho and Copernicus are particularly noticeable at full moon (Fig. 5.11). The rays from Tycho reach fully halfway around the moon.

The most common rock type collected by the Apollo astronauts were **breccias.** Breccias are composed of thousands of tiny rock fragments of all kinds welded together by the heat generated during a meteorite impact. The highlands are composed almost entirely of breccias. Meteorite bombardment was so intense there early in the moon's history that a layer of rubble many kilometers in thickness was produced.

Because of the earth's atmosphere, only large meteorites can survive the fall to the earth's surface. Even when a meteorite does strike the earth, the crater it produces is destroyed rather rapidly by the erosion effects of wind and water. However, on the airless and waterless moon, any meteorite striking its surface will produce a crater that will be preserved for billions of years. The lunar surface therefore serves as a record of the flux of cosmic debris in the vicinity of the earth's orbit throughout the history of the solar system.

If it is assumed that the meteorite impacts occur continually and with equal probability over the entire surface of the moon, the more craters there are per unit area of the moon's surface, the older that area must be. By carefully counting the number of craters in different regions of the moon, scientists have worked out an age sequence for the different lunar features. The highlands comprise the oldest terrain on the moon. Here, craters are so densely packed together that no part of the surface has been unaffected by impacts. The highlands are saturated with craters. The maria are far less cratered, indicating a younger age for the maria. The youngest features on the moon are the rayed craters, such as Tycho and Copernicus.

Impact cratering is the agent of erosion on the moon. Slowly chipping away at the rocks on the surface, the impact of micrometeorites has gradually produced the layer of sand and dust that makes up the lunar soil. Occasionally a larger impact will blast through the soil and excavate large blocks of rock, scattering them around the rim of the crater. Then the smaller impacts will begin relentlessly wearing down the freshly exposed rocks again. Although this process is extremely slow, it has been going on for billions of years and in places the lunar soil, or **regolith,** extends to a depth of 60 meters. The result is the gently rolling appearance of the lunar surface. Even the highest mountains on the moon have well rounded peaks (Fig. 5.14) in sharp contrast to the craggy summits of high mountains on earth. Meteorite erosion also wears down the impact craters themselves, rounding their rims and filling in their depressions until some completely disappear. The age of a crater can therefore be determined by the sharpness of its features.

Craters do not all have the same shape. Aside from the gradual modification they undergo with time, craters also differ in form with increasing size. Small craters are generally bowl shaped with sharp raised rims. Larger craters have flat floors. Above 20 kilometers in diameter a central peak frequently occurs in the crater and the walls are terraced rather than smooth. An astronaut standing in the center of these large craters would not see the crater walls. This is because the radius of the moon is relatively small and the wall is well beyond the local horizon. Above 200 kilometers in diameter, craters are characterized by several rings around the central peaks.

The largest impact structures on the moon, called multi-ringed basins,

Figure 5.14 *The Apennine Front rises behind the Apollo 15 lunar module. The mountains are well rounded. (See Fig. 5.13 also.) (National Aeronautics and Space Administration photograph.)*

have up to five concentric rings. They are the oldest features recognizable on the lunar surface. Such features are seen on Mercury and on Callisto, a satellite of Jupiter, as well.

On the near side of the moon the ringed basins form the depressions occupied by the maria. Orientale basin is the youngest and hence best preserved. It has three concentric rings, the outer one having a diameter of 900 kilometers (Fig. 5.15). The Apennines are part of a ring formed around the Imbrium basin. Altogether, 31 ringed basins have been recognized on the moon. They are believed to have formed in the final stages of the moon's formation when large protoplanetary bodies in the vicinity of the earth's orbit were still being swept up. Presumably, each ringed basin represents the impact of one of these bodies. The earth probably was covered with ringed basins early in its history, but all traces of these structures were subsequently erased by the processes of erosion and mountain building occurring on earth, processes which do not occur on the moon.

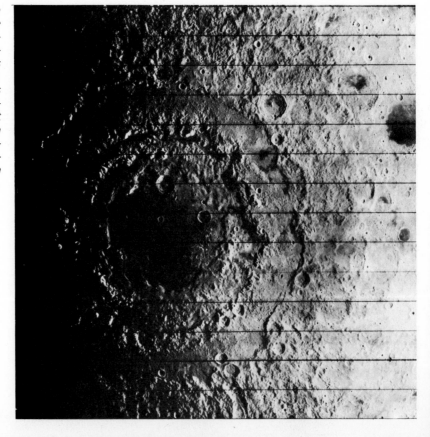

Figure 5.15 *Orientale basin showing its three rings. Two large craters can be seen within the inner ring. The northern (top) crater, 55 kilometers in diameter, is Maunder and is an impact crater. The eastern crater, 45 kilometers in diameter, is Kopff and is volcanic in origin. The ejecta blanket of Orientale basin can be seen around the outer ring. Note its radial pattern. (National Aeronautics and Space Administration photograph.)*

5.10 Lunar volcanism and tectonism

The maria were formed by prolonged volcanic eruptions of basaltic lava early in the moon's history. Spreading out in long, thin sheets, the molten lava flowed downhill and filled up the great ringed basins on the near side of the moon. The resulting lava lakes which cover one-fifth of the moon's surface are recognized as the smooth, dark plains on the moon today.

The evidence for a period of volcanic activity on the moon is abundant, the most convincing proof being the volcanic rocks brought back by the astronauts. High-resolution pictures of the maria reveal lobe-shaped lava flows bounded by sinuous scarps (Fig. 5.12). Small mounds about 100 meters high and up to a few kilometers across, called **domes,** look like scaled-down versions of the volcanoes found on earth. **Chain craters** are sets of craters along almost straight lines that may have formed by outgassing along cracks in the lunar surface. Alternatively, they could have formed by slumping into lava tubes that carried molten lava beneath the surface.

There is also evidence of **tectonic activity** (motions of the crust) on the moon, although on a much more limited scale than occurs on earth. Faults on earth are dislocations in the crust along which vertical or horizontal movement occurs. Some faults, like the San Andreas fault in California, record horizontal displacements of hundreds of kilometers. The best known fault on the moon is the Straight Wall, a linear scarp in Mare Nubium over 100 kilometers long and 300 meters high. This fault appears to have undergone a vertical displacement of a few hundred meters, which is miniscule compared with the great upheavels associated with continental drift on earth. The moon is therefore a tectonically inactive body in comparison to the earth.

A number of lunar features can be interpreted as either volcanic or tectonic in origin. Mare ridges, or **wrinkle ridges,** are sinuous irregular elevations up to 30 kilometers in width and 100 meters in height (Fig. 5.12). They could be the result of lava that was squeezed up along cracks in the solid crust of the still molten lava lake that once covered the maria. Alternatively, they may be faults along which the lunar crust was buckled and uplifted.

Rilles are long trenches in the lunar surface up to 5 kilometers wide, but with floors rarely more than 400 meters deep. They are of two types. Linear rilles are generally flat-floored, steep-walled troughs tens to hundreds of kilometers long. They frequently cross over craters and go from maria to highlands without changing their direction or depth. This disregard for topography suggests that they are depressed segments between two long parallel faults, analogous to a graben on earth. Sinuous rilles, on the other hand, look like meandering rivers and often start in craters or depressions. They probably result from the collapse of lava tubes through which the maria lavas flowed on their way to the surface (Fig. 5.13).

Some craters on the moon are also attributed to volcanic activity. One example is the crater Kopff in Mare Orientale (Fig. 5.15). This crater lacks the ejecta blanket and terraced walls typical of other craters of its size. Instead, it is surrounded by a smooth, undisturbed terrain that may have formed from falling volcanic ash. Its floor has an irregular pattern of fractures that look like cracks formed in cooling lava.

The Physics and Origin of the Moon

Despite all that was learned in preparation for the manned Apollo landings on the moon, this knowledge was completely overshadowed by what has been learned since men have actually been there. On 20 July 1969, man visited another heavenly body for the first time. He was to return another five times in the course of the next two years. While on the moon, the astronauts collected 400 kilograms of rock and soil samples, took hundreds of pictures, and left instruments to monitor the moon after they left. As a result, we now have a much better understanding of the history and internal constitution of the moon.

5.11 The atmosphere of the moon

It has been known for centuries that the moon lacks an appreciable atmosphere. This was concluded from observations of the occultation of stars by the moon. Instead of gradually dimming as the stars pass behind the moon, they stay bright until the last instant before the moon blocks out their light (Fig. 5.7). Nonetheless, the region over the moon's surface is not totally empty space. Extremely sensitive instruments left by the Apollo program were able to detect very slight traces of gas.

The total pressure of the lunar atmosphere is less than one-ten-trillionth of the earth's. This still amounts to several million atoms per cubic centimeter. Most of this gas consists of hydrogen and helium atoms that are continuously supplied by the solar wind emanating from the sun. In addition, argon-40 was detected, which probably forms inside the moon from the decay of radioactive potassium in the rocks and then slowly leaks to the surface.

The reason for the lack of a lunar atmosphere is the small escape velocity of the moon. Because of the moon's weak gravity, the escape velocity is only 2.4 kilometers per second at the surface. The kinetic energy of all particles in a gas is essentially the same, therefore the light particles must move faster than the heavy particles since the energy of a particle is one-half the mass times the square of its velocity ($E = \frac{1}{2}mv^2$). The rule is that if the mean velocity of the gas particles is less than one-fifth the escape velocity the planet will retain its atmosphere for more than a billion years. If the mean velocity is much less, the atmosphere will endure for a very much longer time. If the mean velocity is even only slightly greater than one-fifth the escape velocity, the atmosphere rapidly "boils" away.

For the moon, with its high sun-side temperature and low surface gravity, the only trace of a primitive atmosphere that could exist would consist of heavy inert gases such as xenon and krypton. On the earth only hydrogen and helium escape in significant amounts, thus our oxygen–nitrogen atmosphere.

The moon must have had an atmosphere at one time because bubbles found in the lunar lavas contained gases that would have been released when they erupted on the surface. However, it would require only 3×10^8 years for half of this atmosphere to escape and essentially all of it would have disappeared within a billion years. Since the last volcanic eruptions occured on the moon over 3 billion years ago, the moon has lacked an atmosphere for at least the last 2 billion years.

This lack of an atmosphere makes the moon a very inhospitable place. On earth we are mostly protected from cosmic rays, x rays, and ultraviolet light by the atmosphere. On the moon this deadly radiation strikes the surface directly. Without an atmosphere to act as an insulator the surface temperature varies from 134°C during the day to −170°C during the night. And without an atmosphere to act as a cushion, meteorites strike the moon with their full force. The material at the lunar surface is therefore subjected to all the ravages of space.

5.12 Lunar geology

Before the moon was explored, some scientists speculated that the moon was a primitive undifferentiated body composed of the primordial material from which the solar system formed. By sampling its surface, they hoped to find this material in an unaltered state. This hope was not realized; the rocks on the moon turned out to be much like the rocks on earth. This similarity implies that the moon is a highly evolved body, like the earth, which melted early in its history and subsequently became radially stratified into concentric shells of differing compositions, that is, it became **differentiated,** the heavier materials sinking and the lighter materials floating on top.

Seismographs left on the moon have enabled scientists to decipher the nature of the lunar interior. The moon has a distinct crust, mantle, and core. The moon's crust ranges in thickness from 60 kilometers on the near side to 100 kilometers on the far side. The lunar mantle extends from the crust to a depth of 1200 kilometers. Like the earth's mantle, it is composed of silicate minerals rich in iron and magnesium, but unlike the earth's mantle, which is hot and tends to flow plastically under pressure, the moon's mantle is cold and rigid. It is this thick and rigid mantle that prohibits the occurrence of volcanic eruptions or continental drift on the moon such as occurs on the earth. The lunar core is almost molten, and it is probably composed of iron. However, the moon's core, which is only 500 kilometers in diameter, comprises less than two percent of its volume. The small size of the lunar core explains why the moon's density (3.34 g/cm³) is so much lower than the earth's (5.5 g/cm³).

The seismographs on the moon recorded two kinds of **moonquakes;** those coming from within the moon and those originating on the moon's surface from the impact of meteorites. The internal moonquakes all originate from 800 to 1100 kilometers below the surface. Since the quakes occur in a regular monthly cycle, they are attributed to stress caused by the tidal effect of the earth on the moon's interior. The strength of moonquakes is very low, usually less than 2 on the Richter scale. A person standing on the moon during a quake of this magnitude would not even feel the vibrations. Moonquakes are also less frequent than earthquakes. Compared with the earth, the moon is obviously a geologically quiet body.

Radioactive age dating of lunar rock samples has produced an absolute time scale for the events which occurred early in the moon's history. Shortly after the moon was formed 4.6 billion years ago, its outer layer was melted by heat generated from meteorite bombardment and a tidal interaction with the earth. By 4.2 billion years ago the highland crust had formed by solidification of the molten outer layer of the moon and the extrusion of highland basalts. Starting around this time and extending to 3.9 billion years ago, intense bombardment of the lunar surface produced the great ringed basins and the crater-saturated highlands that we see today. In the meantime, the decay of radioactive elements was heating up the interior of the moon. Then, around 3.8 billion years ago, vast amounts of lava began flooding the near side of the moon. This impressive volcanic episode lasted for 600 million years, until the maria basins were filled to their present level with basalt. Except for the appearance of more and more craters, its surface features were formed over three billion years ago and are essentially the same today.

Significant differences in the chemistry of the lunar rock as compared with terrestrial rocks were revealed when the lunar samples were returned to laboratories on earth. Most notable is the complete absence of water in the moon rocks. Terrestrial rocks commonly contain about 1% water in their mineral structures. In fact, it was the gradual release of this trapped water component from rocks in the earth's mantle that gave rise to the oceans on our planet. Lunar rocks were also found to have only about one-tenth as much sodium and potassium as occur in terrestrial lavas. These elements and many others which were depleted in abundance relative to the earth have one thing in common; they are volatile substances. Volatiles are easily boiled out of rocks when they are heated to high temperatures. It therefore appears as though the moon was either derived from a region of much higher temperature than the earth or it was heated to high temperatures shortly after it formed.

The moon does not have an appreciable magnetic field. This is to be expected from the small size of its core and the slow rotation of the moon. However, frozen into the rocks on the moon is the record of a once strong magnetic field in the vicinity of the moon 4.2 to 3.2 billion years ago. This finding is one of the greatest mysteries to have come out of the Apollo program. Conflicting theories have been advanced to explain it, but none are satisfactory.

One of the most surprising results of the lunar exploration program was the discovery of large mass concentrations beneath the lunar maria, known as **mascons.** They were detected from slight changes in the orbital paths of spacecraft circling the moon. Analysis of the accelerations and decelerations show that the mascons are attributable to the higher density of the basaltic lavas filling the maria. Since the maria were formed over 3 billion years ago, the moon's mantle must have been very rigid. Otherwise, the extra mass would have sunk and adjusted with its surroundings. The existence of the mascons is the most compelling evidence for the view that the moon has been a cold and geologically lifeless body throughout most of its history.

5.13 The origin of the moon

The earth–moon system is unique in the solar system, at least in so far as the earthlike planets are concerned. The moon is a relatively large satellite, almost as large as the planet Mercury, and yet it revolves around a relatively small planet, the earth. The two objects are of similar size, only Pluto and its moon duplicate this but on a much smaller scale. Because of the similar sizes of the earth and moon, some scientists thought that the moon and earth might have formed as a double planet out of the dust cloud from which the solar system condensed. In this view, the earth and the moon originated as separate bodies at approximately the same orbital distance from the sun. But the earth is much denser than the moon. This is because the earth contains more iron, which is concentrated in its core. Explanations of how this difference in composition could have arisen between two bodies formed in the same region of the solar system are so cumbersome that the double-planet hypothesis has fallen into disfavor.

Another theory, known as the fission hypothesis, was based on the observation that the moon is slowly receding from the earth and that the earth's rotation rate is slowing down due to tidal friction. Extrapolating back in time, the moon would have been much closer to the earth and the earth would have rotated in only a few hours. G. H. Darwin (Charles Darwin's son) seized upon this fact to suggest that the moon had split off from the earth early in its history. He envisioned a fast rotating molten earth being pulled in two by centrifugal force. Since the moon only formed from the outer layers of the earth, the earth's iron core would have been left behind. The fact that the density of the moon is about equal to the earth's mantle lends support to this idea. But the fission hypothesis has problems too. For one thing, the moon's orbit is only inclined 5° to the ecliptic, whereas the earth's equator is inclined 23°5. If the moon split off from a rapidly spinning earth its orbit should be around the earth's equator where the centrifugal effect is greatest. More troublesome is the difference in the chemical composition of the moon and the earth's outer layers. As mentioned earlier, the moon is heavily depleted in volatile substances like water and the alkali metals (lithium, sodium, potassium, rubidium). This difference in chemistry is difficult to explain if the moon simply spun off the earth.

Another set of theories holds that the moon was captured from elsewhere in the solar system. A. G. W. Cameron has suggested that the moon initially formed inside the orbit of Mercury and was later slingshotted to its present position by a close encounter with that planet. This could explain the rather large eccentricity of Mercury's orbit as well as the fact that the moon's orbit is closer to the ecliptic than the plane of the earth's equator. It also explains the low volatile content of the moon because the moon would have acquired less volatile constituents if it had accreted at the higher temperatures of the primitive solar nebula closer to the sun. But a problem arises when the mechanics of capture are considered.

Such a capture is extremely improbable because it requires an almost perfect coincidence of positions and velocities. It is much more likely that the moon would have whizzed right past the earth or even crashed into it. But even if the moon was captured by the earth, the tremendous amount of energy carried by the moon would have had to be absorbed by the earth. This would have raised the temperature of the earth above 3000 K and all of the earth's volatile components would have boiled away. Since the earth has an ocean and an atmosphere, this obviously did not happen. Some astronomers have tried to alleviate this problem by proposing that the moon was initially captured in a **retrograde** orbit (opposite to the direction it now orbits). Tidal interactions with the earth would have then caused the moon to move away from the earth and the retrograde orbit would have been converted to the **prograde** orbit observed today. However, this only reduces the energy of capture by a few percent.

Thus, in spite of all that we have learned about the moon, its origin remains a mystery. There is no general consensus among the scientific community as to which of these theories is the correct one, if any one is. They are all tinged with improbability. It was hoped that the exploration of the moon would settle the question of the origin of the moon once and for all. But the only conclusion to have been reached so far is that it was by a most remarkable event that the moon came to exist at all.

Eclipses of the Moon

When the earth is between the moon and the sun and the earth's shadow falls on the moon an eclipse occurs. Such a lunar eclipse can be viewed by everyone with a clear sky and located on the night side of the earth.

5.14 Conditions of a lunar eclipse

Like any other opaque object in the sunshine, the earth casts a shadow in the direction away from the sun. The **umbra** of the shadow is the part from which the sunlight would be completely excluded if some light were not scattered into it by the earth's atmosphere (Fig. 5.16). The umbra is a long, thin cone reaching an average of 1,382,500 kilometers into space before it tapers to a point. This darker part of the shadow is often meant when we speak of the shadow. It is surrounded by the **penumbra,** from which the direct sunlight is only partly excluded. The umbra of the earth's shadow has a diameter of 9170 kilometers at the moon's distance, or nearly 3 times the moon's diameter of 3476 kilometers.

The earth's shadow points directly away from the sun and it sweeps eastward around the ecliptic once in a year as the earth revolves. At

Figure 5.16 *The moon in the earth's shadow. Light is scattered into the umbra by the earth's atmosphere so that the moon generally can be seen even in totality.*

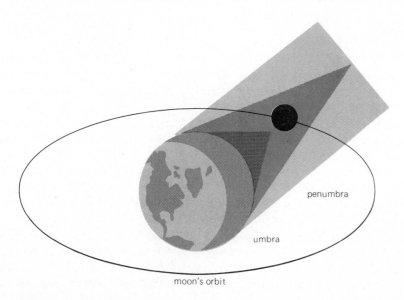

penumbra

umbra

moon's orbit

intervals of a synodic month, the faster moving moon overtakes the shadow and sometimes passes through it. This occurs when the nodes of the moon's orbit are near the earth–sun line, since the moon's orbit is tilted about 5°8′ with respect to the ecliptic. The geometry of an eclipse is very simple and worth working through as an exercise. Since the line of the nodes must be near the earth–sun line, it is possible that no umbral eclipses will occur in certain years, as in 1976, but by the same token as many as three lunar eclipses may occur, as in 1982.

Umbral eclipses occur when the moon passes through the umbra of the earth's shadow. The longest eclipses, when the moon goes centrally through the shadow, last about 3 hours and 40 minutes; this duration is counted from the first contact with the umbra until the moon leaves the umbra completely. Total eclipse can last as long as 1 hour and 40 minutes, preceded and followed by partial phases, each of about 1-hour duration. Usually the eclipse is not central (Fig. 5.17), so that its duration is shorter. Often the moon passes so far from the center of the shadow that it is never completely immersed in the umbra, and the eclipse is partial throughout.

Penumbral eclipses occur when the moon passes through the penumbra of the earth's shadow without entering the umbra. The weakened light of the part of the moon that is in the penumbra is visible to the eye when the least distance of the edge of the moon from the umbra does not exceed one-third of the moon's diameter. It is detected by photography when the shortest distance does not exceed two-thirds of the moon's diameter and by photometric means when the distance is still greater.

5.15 Observing a lunar eclipse

A lunar eclipse is visible wherever the moon is above the horizon during its occurrence, that is, over more than half the earth, counting

Figure 5.17 *Circumstances for the total lunar eclipse of 30 December 1982. The entire umbral phases are visible from eastern North America because it occurs in deep winter. All times are Universal Time. (Details courtesy of R. Schmidt, U.S. Naval Observatory.*

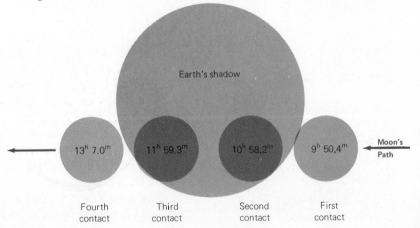

| Fourth contact | Third contact | Second contact | First contact |

Earth's shadow

13ʰ 7.0ᵐ   11ʰ 59.3ᵐ   10ʰ 58.2ᵐ   9ʰ 50.4ᵐ

Moon's Path

the region that is rotated into view of the moon while the eclipse is in progress. The ephemeris times of the circumstances of all umbral eclipses and of penumbral eclipses when they occur are published in advance in the *American Ephemeris and Nautical Almanac.*

The first conspicuous effect of the lunar eclipse is seen soon after the moon enters the umbra. A dark notch appears at the eastern edge of the moon and slowly spreads over the disk. The shadow is so dark in comparison with the unshaded part (Fig. 5.18) that the moon might be

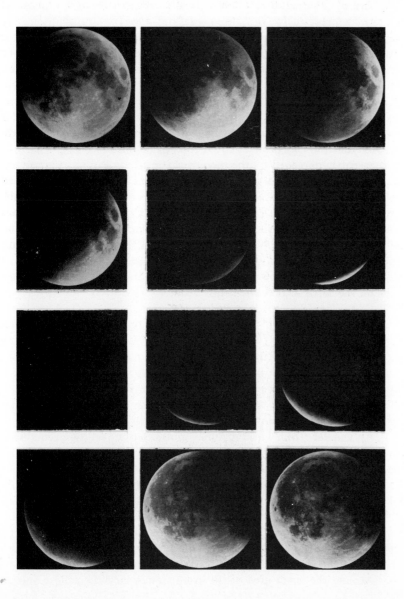

Figure 5.18   *A total lunar eclipse sequence. (Griffith Observatory photograph.)*

expected to disappear in total eclipse. As totality comes on, however, the moon usually becomes plainly visible. The explanation for this is as follows: The moon in total eclipse is still illuminated by sunlight that filters through the earth's atmosphere around the base of the shadow. The light is diffused by the air into the shadow and onto the moon. It is redder than ordinary sunlight for the same reason that the sunset is red, so that the totally eclipsed moon has an unfamiliar hue. The brightness of the moon then depends on the transparency of the atmosphere around the base of the shadow. Enough light usually sifts through to show the surface features clearly. On a dozen occasions since 1601 the totally eclipsed moon became very dim or even invisible, usually because of volcanic dust in the atmosphere. The last such case was the eclipse of December 1963.

Eclipses of the Sun

An eclipse of the sun occurs when the moon passes directly between the sun and the earth, blocking part or all of the sun's disk. The earth is then partly darkened by the moon's shadow.

The average length of the umbra, or conical shadow, of the moon is 373,000 kilometers, which is 4800 kilometers less than the average distance of the moon from the earth's surface. Generally, the umbra will not reach the earth's surface; however, it often does reach the earth because of the eccentricity of the moon's orbit around the earth and the earth's orbit around the sun. When the moon is nearest the earth and the earth is also farthest from the sun, the conical umbra of the moon's shadow falls on the earth 29,000 kilometers inside its apex.

5.16 Total and annular eclipses

A **total eclipse** of the sun occurs when the umbra of the moon's shadow extends to the earth's surface (Fig. 5.19). The area encompassed by the umbra rarely exceeds 240 kilometers in diameter when the sun is overhead. The observer within the area sees the dark circle of the moon completely hiding the sun's disk.

An **annular** (from Latin annulus, meaning ring) **eclipse** occurs when the umbra is directed toward the earth but is too short to reach it. Within a small area of the earth's surface the moon is seen almost centrally projected upon the disk of the sun, but the moon appears slightly the smaller of the two, so that a ring of the sun's disk remains uncovered. Annular eclipses are 20% more frequent than total eclipses. Occasionally the umbra is long enough to reach the earth at the middle of its path but at the beginning and end fails to extend to the surface. In this event the eclipse is total around the middle of the day and is otherwise annular.

Around the small area of the earth from which the eclipse appears

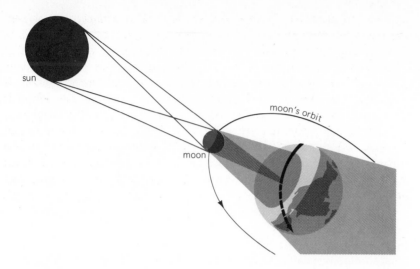

Figure 5.19 *Path of a total solar eclipse. The moon's shadow moves in an easterly direction over the earth's surface. The eclipse is total for an observer in the umbra and is partial in the larger area of the penumbra.*

total or annular, is the larger area, some 3200 to 4800 kilometers in radius, which is covered by the **penumbra** of the shadow. Here a **partial eclipse** is visible, and the fraction of the sun that is hidden decreases as the observer's distance from the center of the eclipse increases. Eclipses are entirely partial where the axis of the shadow is directed slightly to one side of the earth. All total and annular eclipses are preceded and followed by partial phases.

The revolution of the moon causes its shadow to sweep eastward at the average rate of 3380 kilometers an hour. Since the earth is rotating toward the east at the rate of 1670 kilometers an hour at the equator, the speed of the shadow over the surface is 1700 kilometers an hour at the equator when the sun is overhead. The effective speed becomes greater with increasing distance from the equator, where the rotation is slower, and may become as fast as 8000 kilometers an hour when the sun is near the horizon.

Considering its high speed and small area, we see that the umbra can darken any part of its path for only a short time. The greatest possible duration of a total solar eclipse at a particular place can scarcely exceed 7.5 minutes, and that of an annular eclipse can be only a little greater. The partial phase accompanying either type of eclipse may have a duration of more than 4 hours from beginning to end, but it is usually much less. Astronomers have been able to observe solar eclipses from jet airplanes, thus extending the times of total eclipses.

5.17 Eclipse seasons and predictions of eclipses

In order to eclipse the sun, the moon must be almost directly between the sun and the earth. This condition is not fulfilled every time the moon arrives at its new phase, because the moon's path around the heavens is inclined 5° to the ecliptic. Thus the new moon is more likely to pass north or south of the sun.

Eclipses occur during two eclipse seasons, about the times the sun is passing the two opposite nodes of the moon's path (Fig. 5.20). Owing to the rapid westward shifting of the nodes along the ecliptic (Section 5.3), these seasons come about half a month earlier in the calendar from year to year; in 1979 they fell in February and August. The length of the eclipse year is 346.62 days, which is the interval between two successive returns of the sun to the same node. Thus if the first season is early in the year, another season around the same node may begin before the end of the year.

Each solar eclipse season lasts a little more than a month, or somewhat more than the month of the moon's phases. During each interval the moon becomes new at least once, and may do so twice. Two eclipses are accordingly inevitable each year, one near each node. Five may occur, two near each node and an additional eclipse if the sun comes around again to the first node before the year ends. Similarly, it can be shown that three umbral eclipses of the moon are possible in the course of the year, although a whole year may pass without a single one.

Accurate predictions of solar eclipses, when they will occur, and where they will be visible are published in various astronomical almanacs a year or two in advance. Tracks of total solar eclipses for several years in advance are published in the U.S. Naval Observatory *Circulars*. A current compilation of solar eclipses covering the interval 1898 through 2510 is given in *Canon of Solar Eclipses* by J. Meeus, C. Grosjean, and W. Vanderleen.

The predictions of eclipses are made possible by knowledge of the motions of the earth and moon. They are facilitated by a relation between the occurrences of eclipses, which has been known from very early times— at least since the ancient Babylonians. The saros is the interval of 18

Figure 5.20 *Eclipse seasons. Since the moon's orbit is inclined about 5° to the plane of the earth's orbit, eclipses can occur only at two opposite seasons when the sun is near the line of the nodes of the moon's orbit (b). At other times in the year the moon does not pass between the earth and the sun or into the earth's shadow.*

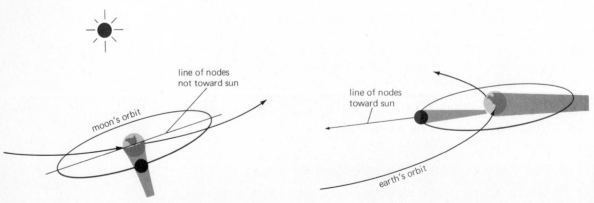

years 11.3 days (or a day less or more, depending on the number of leap years included) in which eclipses of the same **family** are repeated. It is equal to 223 synodic months containing 6585.32 days, and is nearly the same as 19 eclipse years having 6585.78 days. After this interval the relative positions of the sun, moon, and node are nearly the same as before. The sun is about one diameter west of its former position relative to the node; accordingly the paths of the eclipses of a family are displaced progressively in latitude, being shifted gradually from pole to pole until the shadow fails to touch the earth and the particular series is completed.

The third of a day in the saros period causes the path of each eclipse to be displaced in longitude a third of the way around the earth with respect to its predecessor. After 3 intervals, about 54 years and 1 month, the path returns to about the same region as before (Fig. 5.21) although shifted in latitude as mentioned. Occasionally a family that starts as total eclipses becomes annular eclipses as the series progresses (Fig. 5.22). Many families are occurring at the same time as can be deduced from Figs. 5.21 and 5.22.

The dates, durations at noon, and land areas of the principal current total and annular eclipses are given in Table 5.1. The tracks are shown in Fig. 5.23.

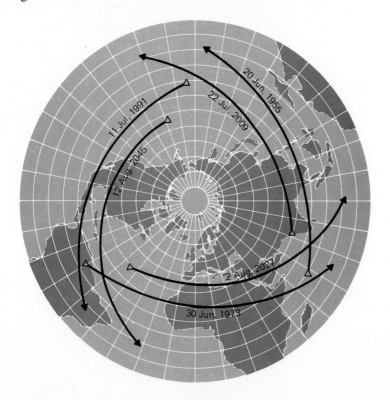

Figure 5.21 *Paths of a family of solar eclipses through the year 2045. This family is slowly moving northward.*

Figure 5.22 *Families of eclipses. A family is just beginning at the south pole (left). Another family moving southward is becoming annular (right).*

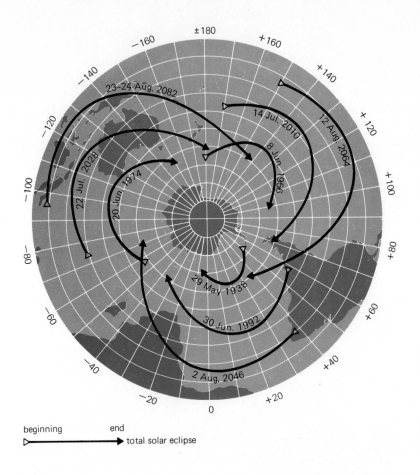

beginning	end	
▷	———————▶	total solar eclipse

Table 5.1
Solar Eclipses 1980–2000[a]

Total Eclipses

16	February	1980	Atlantic to China
31	July	1981	Turkey, Central Asia to Pacific
11	June	1983	Indian Ocean to South Pacific
22	November	1984	Indonesia through South Pacific
12	November	1985	Antarctica
18	March	1988	Indian Ocean to Alaska
22	July	1990	Europe, Siberia to Pacific
11	July	1991	Pacific to South America
30	June	1992	South America to Indian Ocean
3	November	1994	South Pacific to Indian Ocean
24	October	1995	Arabia to Pacific
9	March	1997	East Asia
26	February	1998	Pacific to Atlantic
11	August	1999	Atlantic, Europe, India

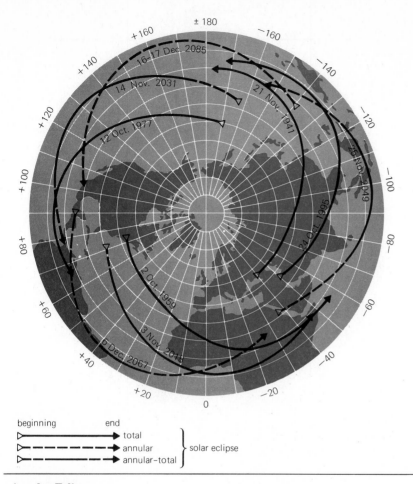

beginning   end

▷ ————————→ total

▷ – – – – – – → annular   } solar eclipse

▷ –·–·–·–·–·–·→ annular–total

Annular Eclipses

10	August	1980	Pacific to South America
4	February	1981	South Pacific
4	December	1983	Atlantic, Africa
30	May	1984	(part total) Pacific to Africa
3	October	1986	(part total) North Atlantic, Ireland
29	March	1987	(part total) South America to Africa
23	September	1987	Asia to South Pacific
11	September	1988	South Indian Ocean
26	January	1990	Antarctica
15	January	1991	Australia, South Pacific
4	January	1992	Pacific
10	May	1994	Pacific, North America, Africa
29	April	1995	South Pacific, South America
22	August	1998	Indonesia to South Pacific
16	February	1999	South Atlantic to Australia

[a] There are 18 partial eclipses in this interval. All solar eclipses in the year 2000 are partial.

Figure 5.23 *Eclipse tracks for total solar eclipses through the year 1988.*

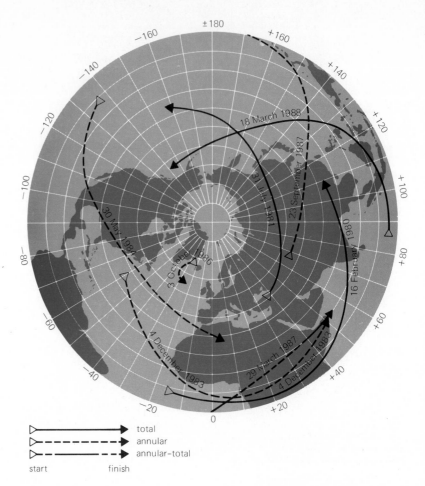

5.18 Total solar eclipse

The sun in total eclipse is an impressive sight to be remembered always. The beginning of eclipse, called first contact, is manifested by the appearance of a dark notch at the sun's western edge. Gradually the sun's disk is hidden by the moon. When only a thin crescent remains uncovered, the sky and landscape have assumed a pale and unfamiliar aspect, because the light from the sun's rim is redder than ordinary sunlight.

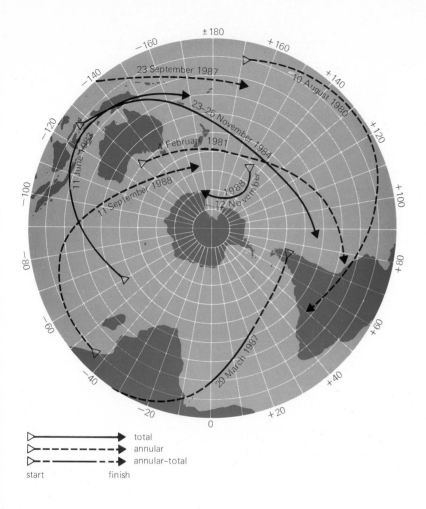

The light fades rapidly as total eclipse approaches. There is a chill in the air, birds seem bewildered, and some flowers begin to close. Just before the last sliver of the sun breaks into brilliant beads and disappears the filmy corona bursts into view. When the sun disappears, called second contact, red prominences are often seen close to the edge of the eclipsing moon; some planets and bright stars may appear. Totality ends at third contact as abruptly as it began; the sunlight returns and the corona vanishes. Finally the moon moves off of the disk of the sun at fourth contact.

Questions

1 Before 50 B.C. the Romans used polished stones, such as rubies, as aids to vision. Suppose they had arranged two such lenses and observed the moon; how do you think the course of astronomy would have changed?

2 What is the phase of the earth as viewed from the moon when the moon is new as viewed from the earth?

3 What is the definition of albedo?

4 Why is earthlight bluer than sunlight?

5 Explain the harvest moon phenomenon.

6 Why are there two high tides each day?

7 What are earth tides?

8 What are the general rules for naming features on the moon?

9 Why is there no significant atmosphere around the moon?

10 Why is the moon red during lunar eclipses?

Other Readings

Bruin, F., "The First Visibility of the Lunar Crescent," *Vistas of Astronomy,* **21,** 331–358 (1977).

Burgess, E., and J. E. Oberg, "Science on the Moon," *Astronomy,* **4,** 684–690 (1976).

French, B. M., "What's New on the Moon?," *Sky & Telescope,* **53,** 164–169 (1977).

Haas, W. H., "The Partial Lunar Eclipse of April 4, 1977," *Strolling Astronomer,* **27,** 77–79 (1978).

Hammond, A. L., "Lunar Science: Analyzing the Apollo Legacy," *Science,* **179,** 1313–1315 (1973).

chapter 6

The Paths of The Planets

Seven bright celestial bodies move about among the "fixed stars" that form the constellations. They are the sun, the moon, and the five planets, Mercury, Venus, Mars, Jupiter, and Saturn, which have the appearance of stars to the unaided eye. These were the planets, or "wanderers," of the ancients. These seven bodies are among our nearest neighbors in space. In the foreground of the starry scene they are conspicuous in our skies. Their brightness and their complex movements against the background of the stars have made them objects of special interest through the ages.

Motions of the Planets

Anyone who has viewed the spectacle of the heavens that is displayed in one of the larger planetariums knows how the planets swing back and forth in their courses among the stars. Celestial movements of a year can be represented in a short time in the sky of the planetarium. The looped paths of the planets are shown very clearly there.

For the most part, the planets move eastward through the constellations. This is their **direct** motion, for it is in this direction that they revolve around the sun. At intervals, which are different for the different planets, they seem to turn and move backward, toward the west, that is, they **retrograde** for a while before resuming the eastward motion. They are said to be **stationary** at the turns. Thus the planets seem to march and countermarch among the stars, progressing toward the east around the heavens in series of loops.

These apparent movements of the planets are readily observed in the sky itself. Watch the red planet Mars from week to week, for example, beginning as soon as it rises at a convenient hour of the night. Notice the planet's position among the stars on each occasion, and mark the place and date on a star map. The line of dots will show presently, as it does in Fig. 6.1, that Mars steers a devious course.

Just how do the planets move around the earth so as to proceed in loops among the constellations? By what combinations of uniform circular motions centered in the earth can their observed movements be represented? This was the problem the early scholars wished to solve. Intuitively they felt that celestial motions ought to be uniform and in circles; then, too, the circle is an easy figure for calculations.

The globe of the earth was stationary at the center of their universe. The sphere of the stars turned around it daily. Within that sphere the

Figure 6.1 *The predicted path of Mars during its 1982 opposition. The stars in Virgo, visible to the unaided eye, are indicated. (Based on data furnished by R. Schmidt, U.S. Naval Observatory.)*

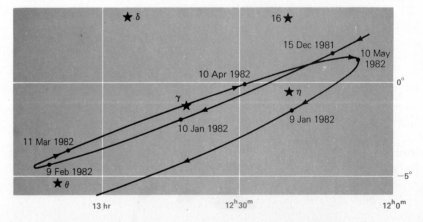

seven wanderers shared its daily turning, and also moved eastward around the earth at various distances from it.

6.1 The geocentric system

The most enduring early plan for solving the problem of the planetary motions was developed by Ptolemy at Alexandria in the second century and is, accordingly, known as the **Ptolemaic system.** It was a plan of epicycles. In the simplest form of the system (Fig. 6.2) each planet was supposed to move on the circumference of a circle, the **epicycle,** while the center of that circle revolved around the earth on a second circle, the **deferent.** By such combinations the attempt was made to represent the observed movements of the planets. The Ptolemaic plan became more complex as time went on. During the many centuries that intervened between the decline of Greek culture and the revival of learning in Europe, Arabian astronomers undertook to improve the system so that it would more nearly represent the planetary movements. Among other devices they tried to get a better fit to the observations by adding more epicycles. Each planet was eventually provided with from 40 to 60 epicycles turning one upon another. It was then that King Alphonso of Castile remarked that had he been present at the Creation he might have given excellent advice. The theory of the central earth had begun to seem unreasonable.

From the times of the early Greek scholars there was an undercurrent of opinion that the earth is not stationary. The followers of Pythagoras, who taught that the earth is a globe, supposed that it is moving. Aristarchus of Samos, in the third century B.C., is said to have been convinced that the earth rotates daily on its axis and revolves yearly around the sun. There were others who caught glimpses of the truth, such as Hipparchus who measured the moon's orbit and found it was oval rather than circular (Section 5.1). Such ideas, however, then seemed unbelievable to almost everyone and received little attention, simply because a turning globe would throw off all its occupants just as a spinning wheel does. Besides, the geocentric model seemed reasonable and consistent with the observations.

By the time Columbus sailed west on his famous voyage, there was growing dissatisfaction with the theory of the central, stationary earth. Before the companions of Magellan had returned from the first trip around the world, another European scholar had independently reached the conclusion that the earth is in motion. His name, as we say it, was Nicolas Copernicus. His theory of the moving earth was published in 1543.

The **Copernican system** set the sun in the center instead of the earth, which now took its rightful place as one of the planets revolving around the sun. It is therefore referred to as a **heliocentric system.** His theory retained the original idea that the planets moved uniformly in circles, and accordingly retained a system of epicycles. Copernicus could offer

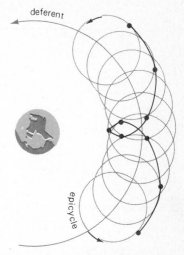

deferent

epicycle

Figure 6.2 *The epicyclic motion of a planet in the Ptolemaic system to explain retrograde motion.*

no convincing proof of either the earth's revolution or daily rotation, which he also advocated.

6.2 Observations and interpretations

Tycho Brahe greatly improved the instruments and methods of his time for observing the positions of the celestial bodies. He saw clearly that an improvement in the theory of the motions of the planets required more reliable data on their apparent movements among the stars. Born in 1546, the most fruitful years of his life were spent in his observatory on the formerly Danish island of Hven, 30 kilometers northeast of Copenhagen. He died in Prague in 1601.

Tycho's observations were made before the invention of the telescope; his chief instruments were large quadrants and sextants having plain sights. With these, he and his assistants observed the planets night after night and determined their right ascensions and declinations with a degree of accuracy never before attained. He gave special attention to the planet Mars, a fortunate choice because its orbit is more elliptical than the orbits of some of the other bright planets.

Tycho seemed to have carried out his programs in an effort to prove Copernicus wrong. This led him to uncover some problems with the Ptolemaic model, namely, that the orbits of Venus and Mercury would overlap the moon's orbit. He therefore proposed a new model, the **Tychonic model,** which had Mercury closer to the sun than Venus and both orbiting the sun on circles. His model was, in a sense, a compromise between those of Ptolemy and Copernicus.

Johannes Kepler was Tycho's assistant in his last year in Prague. He inherited the records of the positions of the planets, which his mentor had kept for many years, after an interesting conflict with Tycho's survivors. Kepler studied the records patiently in the hope of determining the actual motions of the planets. In 1609 he announced two important conclusions, and in 1618 he discovered the third. They are known to us as **Kepler's laws.**

1 **The planets move around the sun in ellipses having the sun at one of the foci.**

2 **Each planet revolves in such a way that the line joining it to the sun sweeps over equal areas in equal intervals of time.** This is called the **law of equal areas** (Fig. 6.3). The nearer the planet comes to the sun, the faster it moves, as we have already noticed (Section 2.13) in the case of the earth.

3 **The squares of the periods of revolution of any two planets are in the same ratio as the cubes of their mean distances from the sun.** This useful relation is called the **harmonic law.**

Figure 6.3 *The planet orbits the sun on an ellipse with the sun at one focus (a). This illustrates Kepler's first law. The areas swept out by a line joining the planet to the sun are equal for equal intervals of time (b). This illustrates Kepler's second law.*

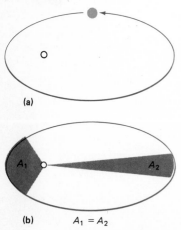

(a)

(b) $A_1 = A_2$

Note that Kepler's third law as stated can be written symbolically as

$$\left(\frac{P_1}{P_2}\right)^2 = \left(\frac{a_1}{a_2}\right)^3,$$

where the subscripts refer to planets 1 and 2, which can be the earth and Jupiter or any other planet. This relation can be rewritten in the following way:

$$\frac{a_2{}^3}{P_2{}^2} = \frac{a_1{}^3}{P_1{}^2}$$

and if we use years for the periods and astronomical units for the distances the ratio is unity. To check this statement we can use Jupiter, where $P = 11.86$ years and $a = 5.203$ astronomical units. Then $P^2 = 140.659$, $a^3 = 140.851$, and the ratio is 1.0014, which is very nearly unity. The reason that it is not is that we are using approximate values and Kepler's relation is itself an approximation, a more precise form being given by Newton (Section 6.5).

Here ended the attempts, even those of the Copernican model, to represent the movements of the planets by uniform circular motions centered around the earth. Kepler's laws predicted the positions of the planets with a precision far exceeding other methods. There was still no evidence, however, that the earth itself revolves around the sun.

The Law of Gravitation

While Kepler was deriving his laws that describe the movement of the planets around the sun, Galileo Galilei, his contemporary in Italy, was laying the foundations of an area of physics called mechanics. Galileo questioned the traditional ideas about the motions of things and set out to determine for himself how they really move. It remained for Issac Newton in England to formulate clearly the new laws of motion and to show that they apply not merely to objects immediately around us but to the celestial bodies as well. The principal feature of the new mechanics was the concept of an attractive force that operates under the same rules everywhere in the universe.

6.3 The laws of motion

Before the time of Galileo, an undisturbed body was supposed to remain at rest. Hence, it seemed appropriate that the earth should be

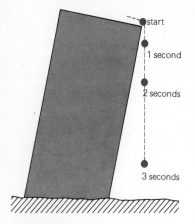

start

1 second

2 seconds

3 seconds

Figure 6.4   *A ball falling from a building accelerates rapidly.*

stationary. Anyone who asserted that the earth is moving might well be asked to explain by what process it is kept in motion.

Galileo's experiments led him to the new idea that uniform motion in a straight line is the natural state. An object will go on forever in the same direction with the same speed unless it is disturbed. Rest is the special case where the initial speed happens to be zero. Uniform motion in a straight line, therefore, demands no explanation. It is only when the motion is changing either in direction or in speed that an accounting is required. We say then that a force is acting on the body, and inquire where the force originates.

The strength of the **force** F is measured by its effect on the body on which it acts; it equals the **mass** m of the body multiplied by its **acceleration** a, or the rate of change of its velocity (directed speed)

$$F = ma.$$

The acceleration may appear as increasing or diminishing speed, or changing direction, or both. A stone falling vertically moves faster and faster; it is accelerating. Galileo was first to state this (Fig. 6.4). An object moving in a circle with constant speed is accelerated. The direction of its motion is constantly changing. In both cases a force is acting.

The laws of motion formulated by Newton in his *Principia* (1687) are substantially as follows.

1   **Every body persists in its state of rest or of uniform motion in a straight line unless it is compelled to change that state by a force impressed upon it.**

Where a force is applied:

2   **The acceleration is directly proportional to the force and inversely proportional to the mass of the body, and it takes place in the direction of the straight line in which the force acts.**

3   **To every action there is always an equal and contrary reaction.**

The first law states that there is no acceleration where no force is acting; the motion of the body remains unchanged. The second law defines force in the usual way. The third law asserts that the force between two bodies is the same in each direction. A baseball bat exerts no greater force on the ball than the ball exerts on the bat; but the lighter ball experiences a greater acceleration than the heavier bat and batter combined.

Armed with these laws of motion, Newton succeeded in reducing Kepler's three laws of the planetary movements to a single universal law. It is said that the fall of an apple one day as Newton sat in his garden started the great mathematician to thinking of this problem. Does the attractive force that brings down the apple also control the moon's revolution around the earth? Does a similar force directed toward the sun cause the planets to revolve around it?

A force is continuously acting on the planets, because their courses around the sun are always curving. It is an attractive force directed toward the sun; this fact can be deduced from Kepler's law of equal areas. From further studies of Kepler's laws, Newton discovered the law of the sun's attraction. He found that the force between the sun and a planet is directly proportional to the product of their masses, and inversely proportional to the square of the distance between their centers. Newton next calculated the law of the earth's attraction. An apple falls 4.9 meters in the first second. The moon, averaging 60 times as far from the earth's center, is drawn in from a straight-line course 0.13 centimeter per second, which is about 4.9 meters divided by the square or 60. Using more exact values than these, he showed that the force of the earth's attraction for objects around it is inversely proportional to the squares of their distances from its center. Although his studies could not extend beyond the planetary system, Newton concluded that he had discovered a universal law and so announced it in his **law of gravitation:**

Every particle of matter in the universe attracts every other particle with a force that varies directly as the product of their masses and inversely as the square of the distance between them:

$$F \propto \frac{m_1 m_2}{r^2}.$$

This may be written

$$F = G \frac{m_1 m_2}{r^2},$$

where G is the universal constant of gravity. The value of G is given in the Appendix.

6.4 Revolution of the planets

Using the earth as our example, let us try to understand its revolution following Newton's law. By Newton's law there is an attractive force between the earth and the sun, which is the same in the two directions. Started from rest they would eventually come together. The earth is moving, however, nearly at right angles to the sun's direction at the rate of 29.8 kilometers per second, and in 1 second it is attracted less than 0.3 centimeter toward the sun. It is this deviation from a straight-line course

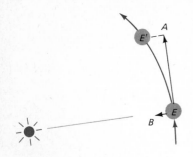

Figure 6.5 *The earth's motion as explained by Newton's laws of motion. The earth (E) is attracted to the sun along line EB. At the same time the earth moves along line EA due to its orbital velocity. The net effect is that as the earth falls toward the sun it moves along a curved path (EE').*

Figure 6.6 *Various possible orbits from circular to hyperbolic. The tangential velocity and distance from the sun at perihelion determine the form of the orbit.*

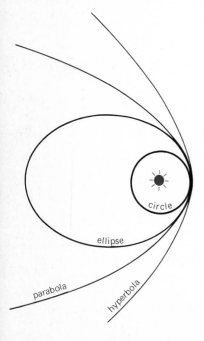

second after second through the year that causes the earth to revolve around the sun (Fig. 6.5). At the position E, the earth, if undisturbed, would continue on to A, by the first law of motion. It arrives at E' instead, having been attracted in the meantime toward the sun the distance EB.

Properly speaking, the earth and sun mutually revolve around a point between their centers. The ancient problem of whether the sun or the earth revolves was not well stated. Both revolve. If the earth and sun were equally massive, the point around which they wheel yearly would be halfway between their centers. Because the sun is a third of a million times as massive as the earth, this center of mass is only 450 kilometers from the sun's center; it is not far from the center for all other planets excepting Jupiter and Saturn.

Thus the planets revolve around the sun, although not precisely around its center. The first law of motion explains their continued progress, and the force of gravitation causes them to revolve around the sun instead of going away into space.

The orbits of the planets relative to the sun are ellipses having the sun's center at one focus. These **relative orbits** are the ones we employ for the planets generally, and similarly for the satellites revolving around their planets. They are the same in form as the actual orbits and differ from them only slightly in size.

Newton showed from his law of gravitation that the orbits of revolving bodies in general may be any one of the three conic sections. These are the ellipse (Section 3.3), parabola, and hyperbola (Fig. 6.6). The ellipse includes the circle, where the eccentricity is zero. The parabola, eccentricity 1, is open at one end, and the hyperbola is open wider. Evidently the permanent members of the sun's family have closed, elliptical orbits.

If the earth, now revolving in nearly a circle at the rate of 29.8 kilometers per second, could be speeded up, its orbit would become larger and more eccentric. At the speed of 42 kilometers per second (which is $\sqrt{2}$ times the circular velocity) the orbit would become a parabola, and the earth would depart from the sun. This is the velocity of escape from the sun at the earth's distance.

6.5 The masses of the planets

The law of gravitation views the physical universe as a scheme of masses and distances. It is, therefore, of considerable interest to inquire how the masses are measured. The **masses** of some planets (i.e., the quantities of material they contain) can be found by Kepler's harmonic law as restated more precisely by Newton: The **squares of the periods** of any **two pairs** of mutually revolving celestial bodies, **each multiplied by the**

combined mass of the pair, are the same proportion as the cubes of the mean distances that separate the pairs. This is shown symbolically by

$$\frac{P_1{}^2 m_1}{P_2{}^2 m_2} = \frac{a_1{}^3}{a_2{}^3},$$

where m_1 represents the total mass of the first pair and m_2 that of the second pair.

Suppose that we wish to find the mass of the planet Saturn. We write this proportion, taking Saturn and one of its satellites as one pair and the earth and sun as the second pair. Let the unit of mass be the combined mass of the earth and the sun, the unit of distance the mean distance between the earth and sun, and the unit of time the period of the earth's revolution around the sun. The relation becomes simply the mass of Saturn and its satellite (the mass of the satellite may be neglected), equals the cube of the mean distance of the satellite from Saturn divided by the square of its period of revolution around Saturn.

The masses of planets having satellites have been found in this way. It is more difficult to weigh planets such as Mercury and Venus which have no satellites; their masses are determined by their disturbing effects on the motions of neighboring bodies, such as orbiting or passing spacecraft or, as in former times, other planets.

6.6 Courses and forces

Early astronomers tried to represent the planetary movements by combinations of circular motions centered in the earth. Kepler discovered that the planets revolve around the sun in ellipses instead of circles and epicycles. So far the interest was confined to the courses themselves.

Newton's law of gravitation directed that attention to mighty forces controlling the courses of the planets. This law has made possible the present accurate predictions of the planetary movements. It has promoted the discoveries of celestial bodies previously unknown, from their effects on the motions of known bodies. It applies equally well to mutually revolving stars.

For most purposes astronomers make their calculations on the basis of Newton's law of gravitation. However the advance of Mercury's perihelion around the sun at a faster rate than is predicted by the law of gravitation raised a problem and required a better theory for its solution.

In 1905, Albert Einstein propounded the **principle of relativity,** which states as axiomatic that the laws of physics are the same as determined by one observer as by another observer. According to this principle the velocity of light is a constant, measured to have the same numerical value for all experimenters, independent of time, direction, velocity, or position. A basic difference between Newtonian mechanics and relativistic mechanics appears in the concept of time. In Newtonian mechanics time is absolute, in relativistic mechanics time is relative—relative to the observer. As a result no single observer is favored or, to make a positive statement, all observers are equally favored.

Although the mathematics are beyond the limits of this text, certain general statements can be made. At low speeds (or low velocities) relativistic mechanics and Newtonian mechanics are identical. At high speeds they yield quite different results (Section 18.15). For example, for two masses at rest with respect to each other we can define the rest mass of one as m_0. The same mass moving at some high velocity (v) will have an **apparent mass** m which is greater than m_0 by the amount

$$m = \frac{m_0}{[1 - (v/c)^2]^{1/2}}.$$

This simple relation tells us that as v gets very close to c (the velocity of light), m becomes very, very large.

The same thing happens for time. The moving clock of any sort has a period given by

$$t = \frac{t_0}{[1 - (v/c)^2]^{1/2}},$$

Figure 6.7   *The advance of perihelion of Mercury's orbit. The eccentricity of the orbit and the amount of the advance are greatly exaggerated in the drawing.*

where t_0 is the clock period of an identical stationary clock. Near the velocity of light the period becomes very, very long in terms of the period t_0. This is known as **time dilation.** An example of this occurs in experimental cosmic-ray physics: when a cosmic ray strikes an atom in the upper atmosphere it may destroy the nucleus releasing particles called **mesons.** Independent mesons live only a brief period of time before they spontaneously (automatically) decay into either electrons or **positrons** (positive electrons). This decay time is well measured in the laboratory, and it is too short for the meson to make the journey from the upper atmosphere to the surface of the earth, whatever the meson's velocity. However, mesons as by-products of cosmic rays are observed at the surface of the earth and the only explanation is that they are traveling at velocities such that their time with respect to ours is dilated.

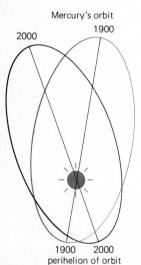

Mercury's orbit

2000   1900

1900   2000
perihelion of orbit

The examples of time and mass above are derived from the **special theory of relativity.** The earlier example of the excessive advance of the perhelion of Mercury (Fig. 6.7) and the apparent outward displacement of stars near the sun's limb (Fig. 15.9) are consequences of Einstein's

general theory of relativity in which the principle of relativity is extended to systems accelerating with respect to one another.

According to Newtonian theory (and relativistic theory) the major axis of the orbit of Mercury will **precess** (rotate) at a rate of about 532 seconds of arc per century resulting from effects of the sun and other planets, whereas it was well known from observations that it was rotating at a rate of about 574 seconds of arc per century. According to general relativity an accelerated body in the presence of a large mass should show an additional precessional advance. For Mercury this additional advance amounts to 43 seconds of arc per century and this accounts for the difference. In rounding off the observed values we are leaving the impression that there is a disagreement of 1 second of arc per century. The situation is much better than that, the actual difference being only 0.08 second of arc per century.

The Planetary System

The meaning of the word **planet** as a body revolving around the sun began with the acceptance of the Copernican system, which added the earth to the list of planets, subtracted the sun from the original list, and reduced the moon to its proper place as a satellite of the earth.

The known membership of the planetary system has increased greatly since Copernicus' time. Knowledge of satellites attending other planets began in 1610 with Galileo's discovery of the four bright satellites of Jupiter. The planet Uranus, barely visible to the naked eye, was discovered in 1781. Neptune, which is always too faint to be seen without the telescope, was found in 1846. The discovery of the still fainter and more remote Pluto in 1930 completed the list of the nine known **principal planets.** In addition to the planets there are smaller bodies, the **asteroids** or **minor planets,** which also orbit the sun. The first and largest of the asteroids to be discovered was Ceres in 1801.

The earth is one of the principal planets. The moon is one of the satellites that accompany seven of these planets. Thousands of asteroids and great numbers of comets and meteoroid swarms are also members of this large family that, including the sun itself, is known as the solar system.

6.7 Aspects and phases of the planets

The names of the planets in order of mean distance from the sun are:

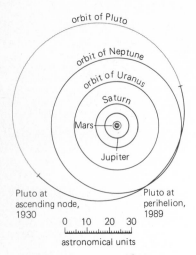

Pluto at
ascending node,
1930

Pluto at
perihelion,
1989

0   10   20   30

astronomical units

Figure 6.8 *The orbits of the superior planets. On this scale the
orbits of the earth and inferior
planets cannot be drawn. The
part of Pluto's eccentric orbit
south of the ecliptic plane is indicated by the lightly tinted section.*

Figure 6.9 *Aspects and phases
of an inferior planet. The elongations are limited and the phases
are like those of the moon.*

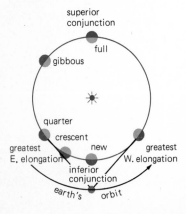

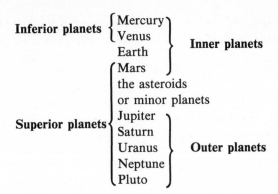

They are classified as inferior and superior planets, and also as inner and
outer planets according to the scheme just presented. The four inner planets
are sometimes known as the **terrestrial planets,** because they are small
and dense as compared with the **giant planets,** Jupiter, Saturn, Uranus,
and Neptune. The giant planets are often called **Jovian** or **gaseous planets.**

Not all the asteroids are confined to the main zone between Mars
and Jupiter; some of them invade the regions of the principal planets. A
part of the projected orbit of Pluto (Fig. 6.8) is nearer the sun than is
part of Neptune's orbit.

Certain terms are used to describe the position of a planet in its
orbit as viewed from the earth.

The **elongation** of a planet at a particular time is its angular distance
from the sun as seen from the earth. Certain positions of the planet relative
to the sun's place in the sky have distinctive names and are known as
the **aspects** of the planet. The planet is in **conjunction** with the sun when
the two bodies have the same celestial longitude, so that the planet's elongation is not far from 0°. It is in **quadrature** when the elongation is 90°,
and is in **opposition** when its celestial longitude differs by 180° from the
sun's.

The inferior planets, Mercury and Venus, have limited elongations;
they appear to us to oscillate to the east and west of the sun's place.
From superior conjunction, beyond the sun, they move out to greatest
eastern elongation, which does not exceed 28° for Mercury and 48° for
Venus. Here they turn westward relative to the sun, pass between the
sun and the earth at **inferior conjunction** when they may transit (or cross)
the face of the sun, then move out to **greatest western elongation,** and
finally return toward the east behind the sun. As Fig. 6.9 shows, the
inferior planets go through the complete cycle of phases, just as the moon
does.

The superior planets, such as Mars and Jupiter, revolve around the
sun in periods longer than a year. They accordingly move eastward through
the constellations more slowly than the sun appears to do. With respect

to the sun's place in the sky, they seem to move westward (clockwise in Fig. 6.10), and attain all values of elongation in that direction from 0° to 180°. At conjunction they pass behind the sun to subsequently appear in the east before sunrise. At **western quadrature** they are near the celestial meridian at sunrise. At **opposition** they rise around the time of sunset, and at **eastern quadrature** they are near the meridian at sunset.

From Fig. 6.10 we also see that the superior planets show the full or nearly full phase at all times to the earth. Mars near its quadrature appears conspicuously gibbous, because it is the nearest of these planets to the earth, so that its hemisphere turned toward the sun and earth are considerably different.

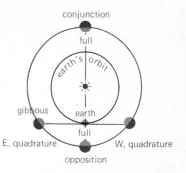

Figure 6.10 *Aspects and phases of a superior planet. The aspects are the same as those of the moon. The phases are either full or gibbous.*

6.8 Retrograde motions of the planets

Consider again the loops in the planets' movements among the stars, which mystified the early astronomers and promoted the complex machinery of the Ptolemaic system. At intervals the planets seem to interrupt their motions toward the east, and retrograde, or move back toward the west, for a while. The retrograde motions occur because we are observing from a planet that is revolving at a different rate from the others.

A superior planet, such as Mars, retrogrades near the time of its opposition. The earth, which is moving faster, then overtakes it and leaves it behind. On the other hand, Mars has its fastest direct motion near its conjunction with the sun, where its own notion and its displacement by ours are in the same direction. The inferior planets retrograde near inferior conjunction. In general, a planet retrogrades when it is nearest the earth.

6.9 Mean planetary motions of the planets

The distances of the planets from the sun are given in Table 6.1, where other information about the planets and their orbits appears as well. These are mean distances; the distances vary because the orbits are ellipses with the sun at one focus. To find the greatest distance that a planet departs from the mean, multiply the mean distance by the numerical value of the eccentricity of the planet's orbit. For example, the mean distance of Mercury from the sun is 57.9×10^6 kilometers, and the eccentricity of its orbit is 0.206; the greatest variation from the mean is therefore 11.9×10^6 kilometers. Mercury is 46×10^6 kilometers from the sun at perihelion and 69.8×10^6 kilometers at aphelion.

A relation known as the **Titius-Bode relation,** often called **Bode's law,** is an easy way to remember the relative distances from the sun of all except the most remote planets. Write in a line the numbers: 0, 3, 6, 12, and so on, doubling the number each time to obtain the next one. Add 4 to each number, and divide the sums by 10. The resulting series

Table 6.1
The Planets

	Name	Symbol	Mean Distance from Sun: Astron. Units	Mean Distance from Sun: Kilometers $\times 10^6$	Period of Revolution: Sidereal	Period of Revolution: Synodic	Eccentricity of Orbit	Orbital Inclination to Ecliptic
					days	days		
Inner	Mercury	☿	0.3871	57.91	87.969	115.88	0.206	7° 0′
	Venus	♀	0.7233	108.20	224.701	583.92	0.007	3 24
	Earth	⊕	1.0000	149.60	365.256	. . .	0.017	0 0
	Mars	♂	1.5237	227.94	686.980	779.94	0.093	1 51
					years			
	Ceres	①	2.7673	413.98	4.604	466.60	0.077	10 37
Outer	Jupiter	♃	5.2028	778.33	11.862	398.88	0.048	1 18
	Saturn	♄	9.5388	1426.99	29.458	378.09	0.056	2 29
	Uranus	⛢	19.1819	2869.57	84.013	369.66	0.047	0 46
	Neptune	♆	30.0579	4496.60	164.794	367.49	0.009	1 46
	Pluto	♇	39.439	5900.00	247.686	366.74	0.250	17 10

Name		Equatorial Diameter (Kilometers)	Mass ⊕ = 1	Density Water = 1	Period of Rotation in Days	Inclination of Equator to Orbit	Oblateness	Stellar Magnitude at Greatest Brilliancy	Albedo
Sun	⊙	1,392,000	332,960	1.41	25.38	7°15′	0	−26.8	. . .
Moon	☾	3,476	0.012	3.34	27.322	6 41	0	−12.6	0.07
Mercury		4,868	0.05	5.44	58.65	0°	0	−1.9	0.06
Venus		12,112	0.82	5.26	−243	2°36′	0	−4.4	0.76
Earth		12,756	1.00	5.52	0.997	23 27	0.003	. . .	0.36
Mars		6,787	0.11	3.94	1.026	25 12	0.009	−2.8	0.16
Jupiter		142,800	317.9	1.36	0.413	3 4	0.063	−2.5	0.73
Saturn		120,000	95.12	0.704	0.426	26 44	0.108	−0.4	0.79
Uranus		50,800	14.6	1.31	0.67	97 53	0.058	+5.7	0.90
Neptune		48,500	17.2	1.77	0.77	28 48	0.026	+7.6	0.84
Pluto		3000[a]	0.002[a]	0.7[a]	6.39	?	?	+14.9	0.6[a]

[a] Uncertain values.

of numbers, 0.4, 0.7, 1.0, 1.6, 2.8, . . . , represents the mean distances *(D)* of the planets expressed in astronomical units. (The **astronomical unit, AU,** is the earth's mean distance from the sun.)

Compare the distances found by this rule with the actual mean distances in astronomical units given in Table 6.1. The agreement is quite close except for Neptune.

When the distance of one planet from the sun is given, the distance of the others can be calculated from their periods of revolution by Kepler's harmonic law. The earth's mean distance from the sun is taken as the yardstick that sets the scale for the distances of planets from the sun, of satellites from their planets, and of stars as well. This is the reason for calling the yardstick the astronomical unit and for determining its value as accurately as possible.

We have seen in Chapter 5 that the moon's distance can be found by observing its parallax from two stations on the earth. The distance of the sun would be measured less reliably in this way, because the sun's parallax is much smaller and also because the stars are less available as reference points in the daytime. More dependable values of the astronomical unit have been derived by observing the orbits of the nearer planets at their closest approaches to the earth. The value of the solar parallax adopted since 1896 in the astronomical almanacs by international agreement is 8.″80. This is the difference in the direction to the sun's center as it would be viewed on the equator from the center and edge of the earth when the earth is at its average distance from the sun. The corresponding mean distance of the earth from the sun is about 149.6×10^6 kilometers.

The timing of many radar echoes from the planet Venus near its inferior conjunction beginning in 1961, as reported by several radio observatories, gave remarkably consistent results among themselves. The calculated average value of the solar parallax is 8.″794, which gives a value for the astronomical unit of about 149,597,896 kilometers. This value is slightly less than the traditional value. In any case, it is readily appreciated that the radar values are the more accurate and the averages of these values are now used to calculate tables of positions of the planets, stars, etc.

It should strike the reader as interesting that we really do not measure the parallax of the sun, but derive it. The methods used, whether optical or radar, are identical to those of Kepler. The relative orbits of the earth and Venus (in the radar case) are known very accurately in terms of the astronomical unit. Determining the distance in some measuring units (meters, kilometers, etc.) when the two planets are close together allows us to scale the orbits in our measuring units and hence determine the astronomical unit. Knowing the radius of the earth we can then derive the parallax of the sun.

The astronomical unit is often referred to as unit distance which is often given in terms of light travel time. The light travel time is 499.012 seconds. In later discussions we use two light travel time units: the distance light travels in one year, called the **light year,** and 206,265 times the astronomical unit distance, called the **parsec** (3.26 ly) (Section 12.1).

6.10   Regularities of the revolving planets

The revolutions of the planets around the sun and of the satellites around their planets exhibit some striking regularities, which apply more generally to the larger bodies. These regularities and the exceptions in the cases of the less massive members may provide important clues concerning the origin of the solar system.

1 **All the planets revolve around the sun from west to east, as the earth does.** This includes all the asteroids. Most satellites have direct revolutions around their planets, and this is also the favored direction of all rotations in the system excepting Venus and a few satellites and asteroids.

2 **The orbits of the planets and satellites are nearly circles.** However, the orbits of the smallest principal planets, Mercury and Pluto, and of some asteroids have greater eccentricities.

3 **The orbits of most planets and satellites lie nearly in the same plane.** With the exception of Pluto's orbit, the orbits of the principal planets are inclined less than 8° to the ecliptic plane, so that these planets are observed always near the ecliptic, and mostly within the boundaries of the zodiac. As will be presented in Chapter 9, we will find that these regularities do not apply to many comets and meteoroid swarms, especially to those having the longer periods of revolution.

The true periods of revolution of the planets increase with distance from the sun in accordance with Kepler's harmonic law, from 88 days for Mercury to nearly 250 years for Pluto. The synodic periods are also given in Table 6.1. They are the intervals between two successive conjunctions of the planet with the sun, as seen from the earth; for the inferior planets the conjunctions must both be either inferior or superior. In other words, the synodic period is the interval in which the inferior planets that are moving faster gain a lap on the earth, or in which the earth gains a lap on the slower superior planet. Mars and Venus have the longest synodic periods because they are nearest the earth and run it the closest race around the sun.

6.11 Transits of Mercury and Venus

The inferior planets occasionally **transit,** or cross directly in front of the sun at inferior conjunction. They then appear as dark dots against the sun's disk.

About 13 transits of Mercury occur in the course of a century; they are possible only within 3 days before or after 8 May, and also within 5 days of 10 November, when the sun passes the nodes of the planet's path. Transits are scheduled for the remainder of the century on 13 November 1986, 6 November 1993, and 15 November 1999, which will be a grazing transit. Transits of Mercury are not visible without the telescope.

These transits, which can be timed rather accurately, have been useful for improving our knowledge of the planet's motions. Since Mercury's

Figure 6.11 *The transit of Mercury on 10 November 1973 is shown above. (Photograph by H. Caulk and R. W. Hobbs.) The diagram (bottom) shows the transits of Mercury from 1973 to the year 2000. Accurate measurements of sunspot diameters can be obtained on such occasions.*

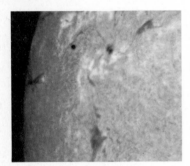

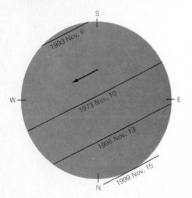

Table 6.2
Satellites of the Principle Planets

Name	Year of Discovery	Mean Distance from Primary	Mean Sidereal Period of Revolution	Diameter	Magnitude at Mean Opposition
Satellite of Earth					
Moon	—	348,397	27^{d}322	3476km	−12.6
Satellites of Mars					
Phobos	1877	9,379	0.319	22	+11.6
Deimos	1877	23,459	1.262	13	12.7
Satellites of Jupiter					
1979J3	1979	128,000	0.29	40[a]	16[a]
1979J1	1979	128,500	0.30	30[a]	16.0
5th, Amalthea	1892	181,500	0.498	193	14.1
I Io	1610	422,000	1.769	3640	5.0
II Europa	1610	671,400	3.551	3130	5.2
III Ganymede	1610	1,071,000	7.155	5270	4.6
IV Calisto	1610	1,884,000	16.689	4850	5.6
13th	1974	11,094,000	238.7	15[a]	20[a]
6th	1904	11,487,000	250.58	190[a]	14.7
7th	1905	11,747,000	259.65	80	16.6
10th	1938	11,861,000	263.55	20	18.6
12th	1951	21,250,000	−631	19[a]	18.8
11th	1938	22,540,000	−692	19[a]	18.1
8th	1908	23,510,000	−739	19[a]	18.8
9th	1919	23,670,000	−758	19[a]	18.3
14th	1975	?	?	20[a]	20[a]
Satellites of Saturn					
—	1977	151,800	0.73	200[a]	14.0[a]
Janus	1967	168,700	0.815	300	14.0
Mimus	1789	185,500	0.942	400	12.9
Enceladus	1789	238,000	1.370	600	11.8
Tethys	1684	294,700	1.888	1040	10.3
Dione	1684	377,400	2.737	825	10.4
Rhea	1672	527,000	4.517	1575	9.8
Titan	1655	1,222,000	15.945	5825	8.4
Hyperion	1848	1,484,000	21.276	225	14.2
Iapetus	1671	3,562,000	79.331	1600	11.0
Phoebe	1898	12,960,000	−550.45	250[a]	15[a]
Satellites of Uranus					
Miranda	1948	129,800	1.413	300[a]	16.5
Ariel	1851	190,900	2.520	800[a]	14.4
Umbriel	1851	266,000	4.144	650[a]	15.3
Titania	1787	436,000	8.706	1130[a]	14.0
Oberon	1787	583,400	13.463	960[a]	14.2
Satellites of Neptune					
Triton	1846	355,550	−5.876	4400	13.5
Nereid	1949	5,567,000	359.875	320[a]	18.7
Satellite of Pluto					
Charon	1978	18,000	6.387	1600	17[a]

[a] Uncertain values

diameter is well known it can also be used to accurately scale the size of sunspots and other solar features as it makes its transit (Fig. 6.11). The mathematician Leverrier discovered from records of many transits that the perhelion of Mercury's orbit is advancing faster than would be predicted by the law of gravitation. The observed excess in its turning has been explained by the theory of relativity as we have noted in Section 6.6.

Transits of Venus are less frequent; they are possible only when the planet arrives at inferior conjunction within about 2 days before or after 7 June or 9 December, the dates when the sun passes the nodes of the planet's path. They are now coming in pairs having a separation of 8 years. The latest transits occurred in 1874 and 1882; the next ones are scheduled for 8 June 2004 and 6 June 2012. Transits of Venus are visible without a telescope. However, the normal precautions for observing the sun must be taken.

Questions

1 What is meant by the term retrograde motion? Explain Mars' retrograde motion.
2 Rationalize a flat earth in the face of repeated observations that the top of the mast of a ship going to sea disappears from view last.
3 Why were Tycho's observations of Mars so valuable to Kepler?
4 If a planet had a 500-year period, what would be its average distance from the sun?
5 How can we determine the masses of planets?
6 Distinguish between inferior planets and inner planets.
7 Why are superior planets never seen in the crescent phase from earth?
8 If the Titius–Bode relation held what would be the distance to the next planet after Pluto?
9 What are minor planets?
10 List some regularities of the solar system.

Other Readings

Alexander, Jr., J. K., "New Vistas in Planetary Radio Astronomy," *Sky & Telescope,* **51,** 148–153 (1976).

Sagan, C., "The Solar System," *Scientific American,* **233**(3), 22–31 (1975).

chapter 7

The Terrestrial Planets and the Asteroids

The planets have been objects of detailed study since the application of the telescope to astronomy. First the emphasis was on their visible features, then on the physical features of their surfaces and atmospheres. More recently, space probes have been deployed to all of the terrestrial planets, providing us with direct measurements of their properties and close up photographs of their surfaces. In this chapter we will consider each of the inner planets in detail, with the goal of gaining some insight into the origin and evolution of the solar system.

Mercury

Mercury is the innermost planet in the solar system. Orbiting at an average distance of 0.39 astronomical units, it requires a mere 88 days to complete one revolution around the sun. Its orbit is so eccentric that at perihelion it is 40% closer to the sun than at aphelion. In addition, Mercury's orbital plane is inclined 7° to the ecliptic. Only Pluto has a more highly inclined and eccentric orbit than this. Mercury has the distinction of being the smallest of the inner planets. Its diameter (4868 kilometers) is more than one-third of the earth's but only slightly larger than the moon's. In fact, Mercury is smaller than some of the moons of Jupiter and Saturn.

7.1 Mercury observed from earth

Mercury is occasionally visible to the naked eye for a few days near the times of its greatest elongations (Fig. 7.1). It then appears in the twilight near the horizon as a bright star, sometimes even a little brighter than Sirius, and twinkles like a star because of its small disk and low altitude. Elongations occur about 22 days before and after inferior conjunction. Because the synodic period is only 116 days, several greatest elongations occur in the course of a year; they are, however, not equally favorable.

Because Mercury is an inferior planet, it swings from one side of the sun to the other in its orbit, appearing successively as a morning and an evening star. Mercury's altitude above the horizon at sunset and sunrise varies considerably on these occasions in our latitudes. The altitude

Figure 7.1 *Mercury near elongation. The phase effect is obvious but no surface detail can be seen. (Lowell Observatory photograph.)*

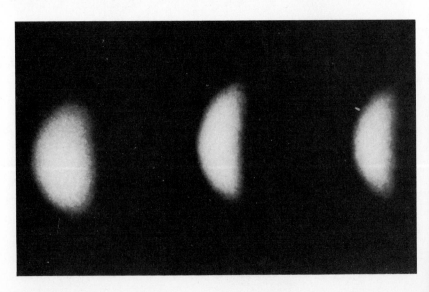

is greatest, and the planet is therefore most easily visible, when the ecliptic is most inclined to the horizon. The most favorable times to see Mercury as an evening star are at its greatest eastern elongations in the early spring, and as a morning star at its greatest western elongations in the early autumn. Such occasions are especially favorable at that time if the planet is also near aphelion. Mercury's distance from the sun's place in the sky at its greatest elongations varies from 28° at its aphelion to as little as 18° at its perihelion because of its eccentric orbit. (See Table 7.1.)

The best telescopic views of Mercury are obtained in the daytime when the planet is well above the horizon. In addition to the phases, some dark markings, reminiscent of the lunar maria, are glimpsed and are also recorded in photographs. The great increase in the planet's brightness from quarter to full phase, as the shadows become shorter, indicated that its surface is as mountainous as the surface of the moon. Also, its albedo of only 7% is similar to the moon. Like the moon, Mercury has no atmosphere, and would not be expected to have any because of its small size and low surface gravity.

In the 1960s, radar pulses were beamed towards Mercury with the Arecibo radio telescope and faint reflected signals were recorded. Radar signals bounced off a moving surface are Doppler shifted by an amount proportional to the surface's velocity; the faster the surface is moving towards the observer, the shorter the wavelength of the reflected signal. Conversely, the faster the surface is receding, the more the radar signal will be Doppler shifted to longer wavelengths. A radar signal which is bounced off a rotating planet will be spread out over a range of wavelengths by the planet's rotation, because the side of the planet rotating towards

Table 7.1
Elongations of Mercury

	Eastern		Western	
1980	19 Feb.	18°1	2 Apr.	27°8
	14 June	24°5	1 Aug.	19°5
	11 Oct.	25°2	19 Nov.	19°6
1981	2 Feb.	18°3	16 Mar.	27°6
	27 May	22°8	14 Jul.	20°8
	23 Sep.	26°3	3 Nov.	18°7
1982	16 Jan.	18°8	26 Feb.	26°8
	8 May	21°3	26 June	22°3
	6 Sep.	27°1	17 Oct.	18°2
	30 Dec.	19°6		
1983			8 Feb.	25°7
	21 Apr.	20°0	8 June	24°0
	19 Aug.	27°4	1 Oct.	17°9
	13 Dec.	20°6		
1984			22 Jan.	24°3
	3 Apr.	19°1	19 May	25°6
	31 July	27°2	14 Sep.	17°9
	25 Nov.	21°8		

the earth is moving with a greater velocity relative to the radar signal than the side of the planet rotating away from the earth. By measuring the range of frequencies over which the reflected radar signal is spread by the planet's rotation, astronomers accurately determined the rotation rate of Mercury to be 58.65 days. Visual observations had always shown the same dark markings on the planet's surface facing towards the sun because the period between elongations when observations could be made was almost exactly twice as long as the rotational period of the planet. Thus, every time the planet could be seen, it had the same face pointing towards the sun.

The rotation period is exactly ⅔ of the planet's orbital period. Instead of rotating once each time it orbits the sun, Mercury rotates three times for every two orbits it completes (Fig. 7.2). If Mercury's mass is unevenly distributed, this would be an extremely stable rotational period because of the planet's highly eccentric orbit. Each time the planet is at perihelion, where the sun's gravitational influence is the greatest, Mercury would have its oblong side facing either towards or away from the sun, so that gravity could not tug on the bulge, and alter its rotational period. Thus, Mercury's 3:2 **spin–orbital relationship** is not merely coincidental, but is the result of a **tidal lock** with the sun.

The combination of Mercury's slow rotation rate and its highly eccentric orbit makes the sun appear to follow a rather astonishing course through the Mercurian sky. Since the planet's orbital speed increases as it approaches perihelion, the effect of its orbital motion dominates over the effect of the planet's rotation in this part of its orbit. Consequently, the sun stops advancing from east to west across the sky when Mercury is at perihelion and for 8 days it moves in a retrograde direction from west to east, before resuming its normal course through the heavens. This remarkable reversal would be even more striking to an observer watching a Mercurian sunrise at perihelion; the sun would first rise above the horizon, hover there for a few days, and then head back in the direction it came,

Figure 7.2 *Mercury's 3:2 spin–orbit coupling. Mercury rotates three times on its axis during two complete orbits around the sun. Since Mercury rotates 1½ times between its closest approaches to the sun (0 to 6 and 6 to 12), its bulge alternately points towards and away from the sun at perihelion. Mercury's two orbits of the sun are drawn as a spiral to avoid confusion.*

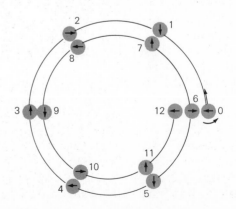

only to rise again about a week later. The opposite would occur on the evening terminator of the planet; after setting, the sun would rise above the horizon, and set again about 8 days later. The sun rises and sets six times in the course of a Mercurian day!

7.2 Mercury from space

The surface of Mercury was photographed at close range for the first time in 1974 when Mariner 10 flew past Mercury. For 6 days before its closest approach, Mariner 10 photographed the planet near the evening terminator, and for 4 days after closest approach it photographed the morning terminator. Mariner 10 also carried instruments to measure other characteristics of Mercury. In order for these other experiments to be carried out, the probe had to pass behind the dark side of the planet on 29 March 1974. Since pictures of the planet could not be taken on the nighttime side, images of Mercury were restricted to the areas near the terminator for the "incoming" and "outgoing" views (Fig. 7.3).

At first glance, Mercury looks very much like the moon. It is gray in color and heavily cratered. Some craters have rays and ejecta blankets around them, and with increasing size, the craters display central peaks and then rings of peaks, just as on the moon. However, when viewed more closely, Mercury displays systematic differences from the moon. Mercury's craters are shallower, they have narrower rims, and the ejecta around them do not extend as far from the crater's rim as they do on the moon. The larger craters have more secondary craters around them. All these differences are attributable to the stronger gravitational pull on Mercury's surface, which would prevent the material blasted out of the crater from traveling as far as ejecta excavated from a similar sized crater on the moon. When a meteorite strikes Mercury, much of the debris falls back into the resulting depression, filling the crater's floor and burying its rim. The larger blocks which fall to the ground around the crater strike with greater force, resulting in more secondary craters close to the rim of the primary crater.

On the terminator of the "outgoing" view, a large ringed basin was photographed which is reminiscent of the lunar basins. Known as **Caloris Basin** (Fig. 7.4), it is as large as Imbrium Basin on the moon. Caloris Basin was formed by the impact of a large protoplanetary body early in the planet's history. It is surrounded by mountains up to 2 kilometers in height and the floor of the basin is severely fractured. The entire basin is located on the side of Mercury that faces the sun on every other perihelion passage causing excessive heating of the area, hence the name.

Large smooth plains are found in and around the Caloris Basin. They are generally thought to have formed from volcanic eruptions early in the planet's history, just like the maria on the moon. The Mercurian

Figure 7.3 *The "incoming" (top) and "outgoing" (bottom) views of Mercury from Mariner 10. Mercury's surface is heavily cratered making it look like the moon. (National Aeronautics and Space Administration photograph.)*

Figure 7.4 *Caloris Basin. The half of the basin seen along the terminator is defined by a ring of mountains 1400 kilometers in diameter. The smooth plains are probably volcanic in origin. (National Aeronautics and Space Administration photograph.)*

plains do not have a significantly lower albedo than the more densely cratered regions around them, which gives Mercury a relatively dull appearance in comparison with the moon. The low albedo of the lunar maria is due to a high abundance of iron and titanium in the basaltic lavas from which they formed. Perhaps the lavas which formed Mercury's plains are not as rich in these dark-mineral-forming elements.

On the opposite side of Mercury from Caloris Basin is an unusual area which is unique to the surface of Mercury. The appearance of this region was so unfamiliar to scientists examining the photographs that they called it *weird terrain* (Fig. 7.5). The weird terrain has a rippled appearance which appears to have resulted from a violent jolt which jumbled up the surface. It has been suggested that the shock waves from the impact which formed Caloris Basin were transmitted through Mercury's core and focused at the antipodal point on the planet's surface, severely disrupting the crust of the planet there.

Mercury does not have a moon, so its mass can only be determined by the degree to which it perturbs objects that pass by it. By carefully tracking the path of Mariner 10 as it passed Mercury, the mass of Mercury was accurately determined to be $\frac{1}{18}$ of the earth's mass. Dividing Mercury's mass by its volume then gives a density of nearly 5.5 g/cm³, which is close to the density of both Venus and the earth. This high density indicates that Mercury must have a large iron core. It is estimated that Mercury's core is 3600 kilometers in diameter. Relative to its size, Mercury has a bigger core than any other planet.

Mariner 10 discovered a magnetic field around Mercury which is about 1% as strong as the earth's magnetic field. This came as a surprise to scientists because Mercury was not expected to have a magnetic field. Although Mercury has a large iron core, the planet is rotating much too slowly to be generating a magnetic field. It is thought that the magnetic field must be a remnant magnetic field which was frozen into Mercury's interior earlier in its history. At that time, Mercury would have been rotating much more rapidly, and a dipole magnetic field would have been generated parallel to the planet's spin axis. Although it is weak by terrestrial standards, Mercury's magnetic field is stronger than the magnetic field around either Venus or Mars.

Long steep cliffs which have a semicircular form are a pervasive feature of Mercury's surface. Some of these arcuate scarps are hundreds of kilometers in length and over a kilometer in height. Their global distribution suggests that they formed as a result of a decrease in the surface area of the planet. If Mercury's iron core was once molten and it cooled, it would have contracted, causing the crust of the planet to shrink around it. This decrease in surface area would be accommodated by faulting of the crust. A decrease of only 1 to 2 kilometers in the diameter of the core is sufficient to account for the abundance of scarps on Mercury's surface.

Mariner 10 was equipped to measure the temperature at the surface of the planet as the probe passed by. The noontime temperature at the equator was in excess of 700 K (hot enough to melt lead and tin), but it dropped quickly to 100 K as the surface of the planet rotated past the terminator into the dark hemisphere. Thus, at the equator, the surface temperature varies by more than 600 K. At the poles of Mercury, on the other hand, sunlight never strikes the floor of some of the steep walled craters, because Mercury's rotational axis is aligned nearly perpendicular to the planet's orbital plane. In these depressions, the temperature has

Figure 7.5 *The weird terrain opposite from Caloris Basin. Old craters have rippled floors while young craters do not thus dating the event and showing that the event occurred over a short period of time. (National Aeronautics and Space Administration photograph.)*

probably remained close to absolute zero for eons. These cold traps may serve as repositories for any volatiles that have been released from the interior of Mercury early in the history of the planet.

The ultraviolet spectrometer aboard Mariner 10 detected an atmosphere around Mercury, although its density was only a few billionths of the earth's atmospheric density. Some of the elements recognized were oxygen, nitrogen, carbon, argon, and xenon. Because of Mercury's high surface temperature on its daytime side, and its low gravitational attraction, these gases rapidly escape from Mercury's atmosphere. They must therefore be constantly replenished by the solar wind which streams past Mercury's surface. Helium was also detected in Mercury's atmosphere, which is particularly surprising, since helium is such a light element that it would require only a few hours to escape from Mercury's gravitational influence. Some of this helium is probably being released at Mercury's surface from the interior of the planet.

Thanks to an ingeniously devised orbital flight pattern, Mariner 10 was able to orbit the sun in 176 days, and come around to take another look at Mercury on 21 September 1974. On its second encounter, Mariner 10 flew over Mercury's south pole, and photographed this region of the planet. Mariner 10 passed over Mercury's north pole on 16 March 1975. On this third and final approach before its fuel supply was exhausted, Mariner 10 flew within 300 kilometers of the planet's surface (its closest approach), and obtained pictures with a resolution as small as 50 kilometers. However, since the orbital period of Mariner 10 was exactly twice as long as Mercury's orbital period, each time it returned, the same hemisphere of the planet was facing towards the sun.

Venus

Venus was once thought to be a planet much like the earth. Its size, mass, and density are nearly identical to the earth's, and it is the next closest planet to the sun in the solar system. These similarities led to the widespread belief during the nineteenth century that life thrived on the planet's surface. However, this notion was quickly dispelled when more detailed studies of its surface conditions revealed its true character. As we shall see, Venus is the most inhospitable of all the terrestrial planets.

7.3 Venus observed from earth

Besides the sun and the moon, Venus is the brightest object in the sky. It is the second planet from the sun and at inferior conjunction, it comes closer to the earth than any other planet. At this time it is only 44 million kilometers away or about ¼ of the distance from the sun to

Year	ELONGATIONS			CONJUNCTIONS	
	Eastern		Western	Superior	Inferior
1980	Apr. 5	45°9	Aug. 24 45°9		Jun. 15^d7^h
1981	Nov. 10	47°2			
1982			Apr. 1 46°5		Jan. 21
1983	Jun. 16	45°3	Nov. 4 46°6		Aug. 25

Table 7.2
Elongations and Conjunctions of Venus

the earth. Because it is an inferior planet, Venus is never very far from the sun in the sky. At greatest elongation it is about 47° from the sun. Venus is best viewed with a telescope at these times just after sunset, when it appears as an evening star, or just before sunrise, when it is a morning star. It is then high enough above the horizon that its image is not severely distorted by the earth's atmosphere. Elongations of Venus are listed in Table 7.2.

The first telescopic observations of Venus were made by Galileo. His most striking discovery was that Venus displays phases, just like the moon. At superior conjunction Venus is at full phase. At its greatest elongation, it is in quadrature, and at inferior conjunction it is in new phase (Fig. 7.6). As Venus moves from superior conjunction to its greatest elongation it moves closer to the earth and comes still closer as it approaches inferior conjunction. Its diameter in the sky therefore increases from 10 seconds of arc at full phase to 64 seconds of arc at new phase (Fig. 7.7). Galileo regarded these changes as convincing proof of the Copernican heliocentric theory.

Venus is at its brightest 36 days before and after inferior conjunction, when it is 39° from the sun in the sky (Fig. 7.8). This is because its brightness is a function of area. The planet gets larger as it approaches us, but the illuminated portion visible to us decreases. Around the time of its greatest brilliance, it is visible as a star in broad daylight.

Telescopic observations of Venus, even with the largest telescopes, reveal nothing about the planet's surface characteristics. This is because Venus is perpetually cloaked by a thick blanket of clouds that visible light cannot penetrate. The cloud cover gives Venus a very high albedo (0.76), which accounts for its brilliancy. In white light the clouds have a yellowish hue and show no detail. It is only in ultraviolet light that the cloud cover takes on an interesting pattern.

Studies of Venus in ultraviolet light showed vague markings which enabled the rotational period of Venus' atmosphere to be determined. The unexpected conclusion was reached that Venus' atmosphere rotates in a retrograde direction with a period of only 4 days. Radar observations made in 1961 showed that the surface of Venus rotates with a much longer period of 243 days, but still in a retrograde sense, which makes

Figure 7.6 *The phases of Venus sketched to show the changing position and size.*

Figure 7.7 *Photographs of the phases of Venus reprinted to the same relative scale. The complete halo at inferior conjunction proves that Venus has an atmosphere. (Lowell Observatory photograph.)*

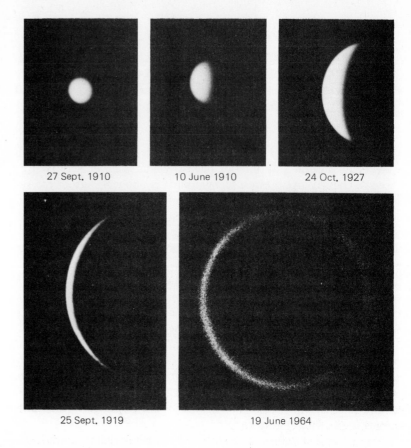

27 Sept. 1910   10 June 1910   24 Oct. 1927

25 Sept. 1919   19 June 1964

Venus unique among the planets in this regard. To an inhabitant of Venus, the sun would appear to rise in the west and set in the east! The fast apparent rotation of Venus is due to rapid circulation in Venus' upper atmosphere.

Radio observations were much more informative about the conditions at the surface of Venus because their longer wavelengths can penetrate Venus' atmosphere and investigate the properties of the planet's surface. The first observations of Venus in the mid-1950s revealed that its surface temperatures were startlingly high. Although the temperature at the top of the cloud cover was known to be 240 K, the temperature at the surface was determined by radio observations to be 700 K. This measurement was regarded as implausibly high by many scientists. It was not until space probes reached Venus that these incredibly high temperatures were

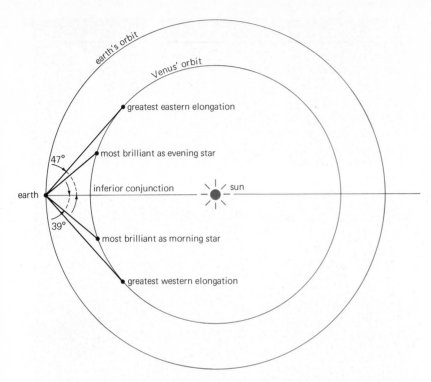

Figure 7.8 *Venus' greatest brilliancy occurs 36 days before and after inferior conjunction when it is 39° from the sun.*

universally accepted. Aside from the sun, Venus has the hottest surface temperature in the solar system.

Radar maps of Venus' surface have been constructed by sending radar signals from the earth and examining the time delay in the signals that are reflected off the planet and recorded back on earth. The point on the planet's surface closest to the earth is the first part to be reached by the radio pulse, and this is the first part of the signal to be reflected back towards the earth. As the radar wave passes the planet, it is reflected off successively larger concentric circles surrounding this point. The last part of the radar pulse to be reflected back towards earth is reflected off the planet's limb. Thus, a sharp pulse of radio waves sent from the earth becomes spread out when it is reflected by the planet's spherical surface. Since the returned signal is only spread out by a hundredth of a second, extremely precise detecting equipment is required to distinguish different segments of the reflected radar beam. Various areas in each circle on the planet are then distinguished on the basis of the Doppler shift in the signal resulting from the planet's rotation. In this way, each part of the planet's surface can be studied. Elevations or depressions are revealed by the early or late return of the signals from each area of the planet.

The composite picture of the surface which results displays irregularities in the topography as bright spots on the radar map (Fig. 7.9).

With this technique, a resolution of 20 kilometers or better can be obtained. Many craters have been mapped on Venus. The largest, which is 1000 kilometers in diameter, is as large as some of the ring basins on the moon. A huge peak has been discovered with a base over 700 kilometers in diameter, which is comparable in size with the island of Hawaii. Just north of this peak are 20 smaller peaks, which may be a cluster of volcanoes similar to volcanic fields found on earth. Along the equator an immense trough, 1500 kilometers in length, has been found which may be a tectonic graben like the Great Rift Valley that runs down the length of East Africa.

Venus revolves around the sun with a sidereal period of 225 days. Combined with a retrograde rotation period of 243 days, this results in a solar day on Venus which is 117 days in duration. In other words, the length of time that elapses between two successive "high noons" on the planet's surface (one Venusian day) is equivalent to 117 earth days.

Figure 7.9   *A radar "picture" of Venus. Elevations and depressions in the surface show up as bright spots. (Jet Propulsion Laboratories photograph.)*

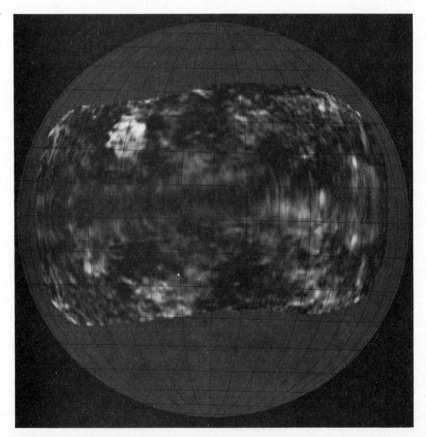

Because the earth is revolving around the sun in the same direction as Venus, but at a slower rate, Venus must make more than 1½ revolutions in order to "catch up" with the earth in its orbit. The synodic period of Venus is therefore 584 days. This is exactly 5 Venusian days. Consequently, at every inferior conjunction, the same side of Venus always faces towards the earth. Since radar maps of Venus can only be made when the planet is closest to the earth, all radar maps of Venus are made near inferior conjunction and they all show the same surface features. The far side of Venus is always hidden from view at inferior conjunction, just as the moon's back side is hidden from view by its synchronous rotation around the earth.

The moon has a synchronous orbit because it is shaped like an egg and its bulge has been drawn into a tidal lock with the earth. In order for Venus to be tidally locked to the earth at its great distance, a huge bulge on its surface would be required. Spacecraft missions to Venus have failed to detect any significant distortion of Venus' mass from spherical symmetry. It is therefore a mystery why Venus should show such a precise spin–orbital relationship with the earth.

7.4 Venus observed from space

Mariner 2 flew past Venus in 1962. Although Mariner 2 was equipped with a relatively meager instrument package by today's standards, it did reveal a few things about Venus which could not be determined from the earth. For one thing, it was discovered that Venus lacks an appreciable magnetic field. Radiation belts comparable to the Van Allen belts around earth cannot form around Venus. The solar wind therefore strikes the Venusian atmosphere directly. Consequently, Venus has a much weaker ionosphere than the earth.

Spectroscopic observations of Venus from earth had revealed the presence of carbon dioxide in the Venusian atmosphere. In addition, trace amounts of hydrogen chloride and hydrogen flouride had been detected, compounds that form highly corrosive acids when mixed with water. However, water had not been identified with certainty. In order to accurately determine the composition of Venus' atmosphere, a probe would have to sample it directly. In 1967 as Venera 4 parachuted through Venus' atmosphere, it measured the temperature, pressure, and composition. Venera 4 found that the atmosphere of Venus is composed almost entirely of carbon dioxide ($> 90\%$), while nitrogen, which was assumed, by analogy with the earth, to be a major constituent of the atmosphere, was relatively scarce. The amount of water vapor in the atmosphere was found to be less than a thousandth of that in the earth's atmosphere. At this low-water-vapor concentration, the surface of Venus would be drier than the driest desert on earth.

The path of Mariner 5, launched in 1967, was purposefully designed to take it behind Venus as seen from the earth. Radio signals beamed through the atmosphere of Venus by Mariner 5 as it flew behind the planet enabled scientists to deduce the structure of the atmosphere from the signal received on earth. The temperature was estimated to increase from 240 K (below the freezing point of water) at the top of the clouds to 760 K (above the melting point of lead) at the planet's surface. In addition, the atmospheric pressure was calculated to be more than 90 times the atmospheric pressure at sea level on earth. This is a greater pressure than is found at the bottom of the earth's deepest oceans. The pressure and temperature profile of Venus' atmosphere is shown in Fig. 7.10.

The extremely high surface temperature and pressure of Venus is due to the large carbon dioxide content of its atmosphere. Solar radiation which reaches the surface of Venus is absorbed and reemitted as infrared radiation. Since carbon dioxide absorbs infrared radiation, most of this heat energy is then trapped, and the temperature of the atmosphere rises

Figure 7.10   *The structure of Venus' atmosphere. The solid curve indicates the temperature (bottom scale) and pressure (right-hand scale) with height (left-hand scale). Note that the pressure scale is logarithmic.*

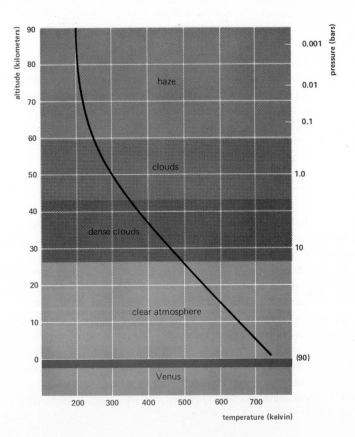

tremendously (the **greenhouse effect**). If the earth had an almost pure carbon dioxide atmosphere, its surface temperature would be correspondingly high as well.

The high atmospheric pressure on Venus is due in turn to the high surface temperature. On earth, carbon dioxide in the atmosphere is almost entirely dissolved by water and is deposited as carbonate rock (limestone, chalk, etc.). However, on Venus, the surface temperatures are too high for liquid water to exist at the surface, and all the carbon dioxide remains in the atmosphere. Even if carbonate rocks did exist on Venus, the carbon dioxide would be baked out of the rocks by the high surface temperature. Consequently, the surface of that planet is burdened by an atmosphere comparable in weight to all the carbonate rocks on earth.

The abundance of carbon dioxide on Venus and the earth is nearly the same, though it exists as a gas on Venus and is trapped as a solid in the rocks on earth. Nitrogen is also equally abundant on Venus and the earth, although on Venus it is so diluted by the incredibly dense carbon dioxide atmosphere that it only accounts for a few percent of the atmospheric composition. It is therefore a puzzle why water should be so scarce on Venus and plentiful on earth, when the other constituents appear to be equally abundant on the two planets. Perhaps, early in the history of the solar system, Venus had oceans like the earth? However, since Venus is closer to the sun, its surface temperature rose to the boiling point of water, and the oceans evaporated. Without oceans to remove the carbon dioxide from the atmosphere its concentration built up, and the greenhouse effect took over, increasing the temperature still more. In addition, ultraviolet light striking the cloud tops would dissociate all the water vapor into hydrogen, which escaped into space, and oxygen, which combined with rocks on the surface.

Other spacecraft showed that the surface temperatures (750 K) and pressures (90 atmospheres) were nearly identical on the day and night sides of the planet in spite of the long duration of the solar day on Venus. Apparently, Venus' dense atmosphere retains heat so well that the temperature is the same over the entire surface of the planet. Venus is unbearable everywhere all of the time!

The spacecraft Venera 8 analyzed the abundance of radioactive elements on the surface of Venus with a gamma-ray spectrometer. The abundances are typical of granitic rocks on earth, which only form through extensive melting of the outer layers of a planet. Venus is therefore believed to be a geologically active planet like the earth.

During its decent through the Venusian atmosphere, Venera 8 measured the wind velocity and found that it decreased from 50 meters/second at an altitude of 45 kilometers to 2 meters/second at an altitude of 10 kilometers. Thus the high-speed winds in Venus' upper atmosphere which give rise to the 4-day rotation period of the cloud tops do not extend down to ground level. Venera 8 also detected the base of the cloud deck 25 kilometers above the planet's surface. Venus' cloud cover therefore begins at a higher altitude than the highest clouds on earth and ends well above the surface. Most of the earth's cloud layer is restricted to within 10 kilometers of the surface.

The first close-up pictures of Venus were obtained by Mariner 10 when it passed within 5800 kilometers of the planet's surface on its way to Mercury in 1974. At visible wavelengths the planet appeared as a featureless yellow disk, but at ultraviolet wavelengths a striking pattern of light and dark areas was evident (Fig. 7.11). Near the equator, the pattern was mottled, suggesting that convection was occurring at the lower latitudes, while further away from the equator the clouds became organized into broad bands that spiraled towards the poles. The contrast between the light and dark areas seen in ultraviolet light is probably due to variable concentrations of carbon monoxide (CO) in the upper atmosphere.

Figure 7.11 *Mariner 10 photograph of Venus in ultraviolet light. (National Aeronautics and Space Administration photograph.)*

In the early 1960s spectroscopic observations of Venus at infrared wavelengths revealed strong absorption bands but the origin of these absorption bands remained a mystery until the 1970s when it was suggested that sulfuric acid (H_2SO_4) was responsible. The presence of sulfuric acid (better known as the battery acid which is used in cars) in Venus' atmosphere would explain many of the most puzzling features of the atmospheric composition. Sulfuric acid has a strong affinity for water, resulting in H_3O^+ and HSO_4^- ions. This explains the previously unidentified absorption lines, since H_3O^+ absorbs light between 3 and 4 microns and HSO_4^- absorbs light at a wavelength of 11.4 microns. It also explains why Venus' atmosphere is so dry, since any water in Venus' clouds would be attacked by the sulfuric acid. The low abundance of free oxygen (O_2) in Venus' atmosphere is also explained since oxygen readily combines with the sulfur.

Measurements of the polarization of light reflected from Venus' atmosphere indicates that spherical droplets about 1 micron in diameter are found at the base of the cloud layer. Mariner 10 detected strong microwave absorption at this level in Venus' atmosphere, indicating that these droplets are composed of a concentrated sulfuric acid solution. Unimaginable as it may seem, the base of Venus' cloud layer consists of a hot rain of sulfuric acid droplets. To make matters worse, trace amounts of hydrogen fluoride detected in Venus' atmosphere probably combine with sulfuric acid, forming fluorosulfuric acid (HSO_3F), one of the most corrosive substances known. Flourosulfuric acid not only dissolves lead, tin, and mercury, but most rocks as well. Fortunately for any future attempts at exploring the surface of Venus, these highly corrosive raindrops boil away at the base of the cloud layer and are recirculated. The surface is therefore spared from the effects of this deadly rain.

Mankind's first look at the surface of Venus came in 1975, when two probes landed on the planet's surface and sent back pictures of the surrounding landscape. Venera 9 recorded a temperature of 760 K and a pressure of 90 atmospheres, while Venera 10, which lasted 65 minutes, found a lower temperature, 740 K, but a slightly higher pressure of 92 atmospheres. Only a slight breeze was detected at the surface. Only 1% of the sunlight striking the planet was found to reach the surface.

The terrain around Venera 9 (Fig. 7.12, top) consisted of a rock-strewn plain which stretched to the horizon about 250 meters away. The rocks were sharp and angular and measured from 30 to 40 centimeters in diameter, giving the appearance of a geologically young terrain. On the other hand, the rocks around Venera 10 (Fig. 7.12, bottom) were worn down to large smooth pancake shapes, indicating a greater age for this region of Venus' surface. The ground around both areas had radioactive element contents indicative of basaltic rocks. The rocks in these pictures probably formed from volcanic eruptions, just like the basalts on the earth and the moon.

Figure 7.12 *The surface of Venus as seen by Venera 9 (top) and Venera 10 (bottom). The rounded rocks around Venera 10 indicate that region to be older than the Venera 9 site where the rocks are flat and angular. (Novosti from SovFoto.)*

The most recent space probes sent to Venus by the United States were launched in 1978 and consisted of an orbiter, called Pioneer Venus 1, and an assemblage of atmospheric probes, called Pioneer Venus 2. The orbiter studied Venus' atmosphere from above and mapped its surface by radar, revealing features on the back side of the planet for the first time. The atmospheric probes plunged through the atmosphere and made careful measurements of its changing properties as they descended. Near the surface the atmospheric probes analyzed the composition of Venus' atmosphere and found 96.4% carbon dioxide, 3.41% molecular nitrogen, and 0.135% water vapor. Molecular oxygen, argon, neon, and sulfur dioxide were detected in trace amounts.

The Soviet Union also launched probes to Venus in 1978, which soft landed on the surface and returned more pictures of the planet's surface. During their descent, Venera 11 reported as many as 25 lightning flashes per second below the cloud cover on Venus, while Venera 12 similarly detected about 1000 flashes between the altitudes of 11 and 5 kilometers. In addition to all its other perils, the atmosphere of Venus may be constantly flickering with lightning near the surface! Pioneer Venus 1 observations reveal a significant airglow in the upper reaches of the Cytherean atmosphere. Neither the lightening nor the airglow were totally unexpected, having been predicted in 1959.

Mars, the Red Planet

Since the dawn of recorded history, Mars, like Venus, has engaged the imagination of mankind. To the ancients, the red color of Mars evoked images of battles and of bloodshed. The name *Mars* is derived from the Latin word for the Roman god of war. In more recent times, Mars has attracted attention because it is generally regarded as the most likely planet in the solar system to be inhabited by extraterrestrial life. Mars is the fourth planet from the sun and the next planet in order beyond the earth. At its average distance from the sun of 1.5 astronomical units, it is not inconceivable that liquid water, an essential requirement for life, could exist on its surface. In addition, its atmosphere is not sufficiently extensive or clouded to hide its surface, and telescopic observations of its surface appearance suggested to many nineteenth century astronomers that life was thriving on the red planet. The persistent idea that Mars supports life remains unresolved to this day, even though two Viking spacecraft were landed on its surface in 1976 with the primary objective of detecting life.

Mars is a small planet. It has only $\frac{1}{10}$ of the mass of the earth. Its diameter of 6800 kilometers is only about $\frac{1}{2}$ the diameter of Venus and the earth and about $1\frac{1}{2}$ times the diameter of Mercury. Although the gravity on Mars is almost as weak as it is on Mercury, Mars had been able to retain an atmosphere because it receives less radiation from the sun. The rather eccentric orbit of Mars causes its distance from the sun to vary by 19%, or 42 million miles.

7.5 Mars through the telescope

Mars is best observed telescopically at oppositions, because it is then closest to the earth and its disk it almost completely illuminated. Mars orbits the sun in 687 days, which makes the Martian year nearly twice as long as the earth's. Since the earth orbits the sun almost twice in the same time, the earth and Mars are lined up at average intervals of 780 days. This is about 50 days longer than 2 earth years, so oppositions with Mars occur about 50 days later in successive alternate years. Because of the eccentricity of its orbit, Mars varies considerably in its distance from the earth at different oppositions, from less than 56 million kilometers at perihelion to more than 100 million kilometers at aphelion.

Favorable oppositions occur when the planet is near its perihelion. At these unusually close approaches, which are always in the late summer, Mars becomes the most brilliant starlike object in the heavens, with the single exception of Venus, and also attracts attention then because of its red color. With a telescope magnifying only 75 times, its disk appears

Table 7.3
Oppositions of Mars, 1980–1999

Year	Opposition	Nearest Earth	Millions of km	Diameter	Magnitude
1980	25 Feb	26 Feb	100.7	13".8	−0.9
1982	31 Mar	5 Apr	95.1	14".7	−1.1
1984	11 May	19 May	79.7	17".6	−1.6
1986	10 Jul	16 Jul	60.5	23".2	−2.4
1988	28 Sep	22 Sep	58.9	23".8	−2.5
1990	27 Nov	20 Nov	77.4	18".1	−1.7
1993	8 Jan	3 Jan	93.7	14".9	−1.2
1995	12 Feb	11 Feb	101.2	13".8	−0.9
1997	17 Mar	20 Mar	98.6	14".2	−1.0
1999	29 Apr	1 May	86.6	16".2	−1.4

Abstracted from material furnished by Dr. R. Duncombe, U.S. Naval Observatory.

as large as does the moon's disk to the unaided eye. Favorable oppositions of Mars occur at 15 to 17 year intervals. The dates of oppositions of Mars from the latest to the next favorable oppositions are given in Table 7.3 and Fig. 7.13.

As viewed through a telescope, Mars possesses a number of permanent surface features, the most conspicuous being its white polar caps. Large dark areas also appear on the globe (see color plate). By observing these dark markings, the rotational period of Mars was shown to be 24 hours and 37 minutes, very close to the 23 hour and 56 minute rotational period of the earth. The rotational axis of Mars is also inclined 24° to its orbital plane, which is nearly identical to the 23°5 inclination of the earth's rotational axis to its orbital plane. These similarities are merely coincidental, but they suggested to early observers that Mars was a planet much like the earth.

The inclination of the rotational axis of Mars to its orbital plane gives rise to seasons on Mars, just as it does on the earth. Because of the high eccentricity of its orbit, the seasons in the northern and southern hemisphere are unequal in duration. Mars reaches perihelion when it is late spring in the southern hemisphere. Since the planet moves through its orbit most rapidly at this time, spring and summer in the southern hemisphere last only 305 days, whereas in the northern hemisphere (when the planet is at aphelion), spring and summer last 382 days. Furthermore, Mars receives 69% more radiation from the sun at perihelion than at aphelion, so that spring and summer are hotter in the southern hemisphere than they are in the northern hemisphere. These two effects combine to make the climate in the southern hemisphere more extreme. The polar ice cap in the southern hemisphere therefore becomes larger in its winter season than the north polar cap does. In the summer, the south polar cap also becomes much smaller.

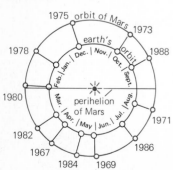

Figure 7.13 *The varying distance of Mars from the earth at oppositions between 1967 and 1988. Mars is at perihelion late in August so the most favorable oppositions always occur then.*

The lighter areas of the Martian surface are a reddish orange to a yellowish brown in hue. Spectral analysis shows that this coloration is due to the presence of hydrated iron oxides, such as limonite and goethite, in the Martian soil. Darker areas on Mars, known as maria by analogy with the moon, appear bluish green through the telescope. These dark surface markings show seasonal variations in size and contrast; as each polar cap alternately retreats with the return of spring to that hemisphere, the dark blue-green areas increase in extent. This wave of darkening on Mars suggested to many hopeful observers that plant life was flourishing on the planet during the summer months. However, recent investigations have put to rest the notion that green plants are responsible for the wave of darkening. It is more likely that light-colored surface material in the form of dust and fine-grain sand is blown by seasonal winds across the planet, alternately covering and uncovering darker underlying rocks on the planet's surface.

Planet-wide dust storms are periodically observed on Mars. When seen with a telescope, they obscure all the surface features on the planet, rendering its disk a featureless yellow globe. In order to support the huge quantities of dust involved, wind velocities during a dust storm must exceed 200 kilometers/hour, which is comparable to conditions found in hurricanes on earth. Global dust storms of this magnitude usually start in the southern hemisphere when Mars is at perihelion because the effect of solar heating is then the greatest.

Telescopic observations also reveal the presence of white clouds in the Martian atmosphere. Smaller clouds occurring over regions of higher elevation on Mars are seen in the spring and summer at lower latitudes. These water–ice clouds appear regularly in the early afternoon and increase in brightness for several hours before disappearing. They form from ground frost that evaporates in the morning sun, producing water vapor which rises and cools by expansion to crystallize as ice in the upper atmosphere. Larger clouds sometimes form at intermediate latitudes which last for weeks before disappearing. In the fall and winter in each hemisphere, a thick layer of haze, known as the **polar hoods,** forms over the polar caps.

During the favorable opposition of 1877, Giovanni Schiaparelli made a map of the Martian surface and named many of its prominent surface features. Schiaparelli also discerned a complex network of fine dark lines crisscrossing the Martian surface, which he said looked like "the finest thread of spider's web drawn across the disk." Schiaparelli believed that these dark linear features were natural waterways bordered by vegetation, so he called them "canali," the Italian word for "channel." However, the word "canali" was translated into English as "canals," carrying with it the implication that intelligent beings had excavated them in order to transport water across the Martian deserts. After Schiaparelli announced his findings, many other observers claimed to have seen the canals as well, while others searched for them in vain. Attempts to photograph them were unsuccessful and it was surmised that they could only be seen during brief instances of exceptionally clear atmospheric visibility. A great controversy arose over their disputed existence. Percival Lowell was a particularly enthusiastic proponent of the canals and in 1894 he established an observatory in Flagstaff, Arizona for the purpose of studying them. Lowell's popularization of the notion of an advanced civilization of Martians that were digging a vast system of irrigation ditches to support plant life on their dry planet led to the widespread belief that Mars was inhabited by little green people.

We now know that the canals were an illusion. Close-up pictures of the planet taken with the Mariner space probes revealed no canals on the planet's surface. In fact, very little correlation between the canali seen from earth and the close-up photographs could be found. Whatever their cause, the canals did much to inspire a popular interest in the red planet.

7.6 The Mariner missions to Mars

The first close-up pictures of Mars were obtained in 1965 by Mariner 4, a space probe which passed within 9790 kilometers of the Martian surface. Mariner 4 was equipped with a television camera which photographed 1% of the Martian surface. The best resolution attained was 2 kilometers, which represented a 50-fold improvement over the 100-kilometer resolution obtainable from ground-based observations. Instead of finding canals on Mars, Mariner 4 revealed only craters. To the disappointment of scientists and laymen alike, Mars looked much like the moon. Although the possibility that Mars might be cratered had been suggested by a few perceptive scientists many years before their existence was confirmed, most people were surprised by this discovery, and optimism about the possibility of life on Mars faded considerably.

The prospects became even dimmer after Mariner 4 examined the Martian atmosphere. Earth-based observations had indicated an atmospheric pressure at the surface of about 85 millibars, which is about one-tenth of the atmospheric pressure at the surface of the earth (1017 millibars). In addition, spectroscopic observations of Mars had only indicated the presence of carbon dioxide in the atmosphere, although in insufficient quantities to account for more than a fraction of the estimated atmospheric pressure. Since other gases had not been identified, it was assumed that molecular nitrogen, which cannot be observed spectroscopically, was the major constituent, just as it is on earth.

After photographing the surface of Mars, Mariner 4 passed behind the planet, sending its radio signal back through the Martian atmosphere as the planet occulted the spacecraft. Analysis of the refraction of the radio signals by the Martian atmosphere indicated that the atmospheric pressure ranged between 5 and 7 millibars at the surface of Mars. This low atmospheric pressure removed the necessity of postulating an unidentified atmospheric component, and therefore carbon dioxide is the primary constituent of the Martian atmosphere. In the extremely thin atmosphere on Mars, liquid water could not exist at the surface. Mars seemed to be a very dead terrestrial planet.

Mariner 4 was equipped with a magnetometer to detect magnetic fields. Since no magnetic field was observed, this placed severe limits on the size of the Martian core. Lacking a magnetic field, radiation belts analogous to the Van Allen belts do not develop around Mars.

In 1969, two more space probes, Mariners 6 and 7, flew past Mars. Mariner 6 examined the equatorial region of Mars, while Mariner 7 took a close look at the south polar cap. Together the two probes increased the photographic coverage of Mars to 10% of its surface area. Mariner 7 photographed an area which Schiaparelli had called Hellas and found that it was almost entirely devoid of any surface features. Presumably, wind-blown dust has obliterated all the craters in this region, indicating that wind was an active erosional agent on at least one part of the planet. In addition, Mariner 6 discovered a region of jumbled hills near the equator which became known as *chaotic terrain*. It appeared to have formed as a result of melting of subsurface ice and the subsequent collapse of overlying surface layers. However, most of the surface was heavily cratered, and the general impression remained that Mars was a geologically lifeless planet like the moon.

Another space probe was launched towards the red planet in 1971. Unlike the previous Mariner spacecraft, which merely surveyed small portions of the planet as they flew by, Mariner 9 went into orbit around Mars and photographed its entire surface. The results from this probe were astounding. Volcanoes, canyons, and dry river beds abound on Mars! Just by chance, the first three Mariner probes had photographed the more uninteresting areas of its surface. Although part of the Martian surface has been geologically dormant for billions of years, the remainder has been shaped by dynamic geological processes right up to the present. Mars turned out to be a geologically active planet with an exciting history after all.

Near the northern hemisphere, Mariner 9 found a vast volcanic plateau which is dotted by four huge volcanic peaks. In fact, when Mariner 9 first went into orbit around Mars, a planet-wide dust storm was in progress which hid the entire surface of the planet from view for two months, except for the south polar cap and these four immense peaks protruding out of the dust-laden atmosphere. The largest of the volcanoes, Olympus Mons, towers 25 kilometers over the surrounding plains (Fig. 7.14).

The base of the volcano is 500 to 600 kilometers in diameter and it covers an area larger than the state of New Mexico. These impressive

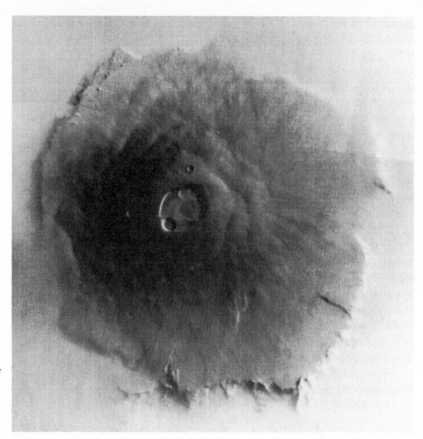

Figure 7.14 *Olympus Mons. This great volcano rises 25 kilometers and measures about 600 kilometers at its base. The lack of impact craters shows that it is relatively young. (National Aeronautics and Space Administration photograph.)*

dimensions make Olympus Mons the largest known volcano in the solar system—5 times larger than the largest volcano on earth! Olympus Mons is topped by a complex crater, known as a caldera, which formed when lava sank back into the vent after each erruption. Lava flows extend radially down the flanks of the volcano. Around its base an immense cliff up to 4 kilometers in height separates the volcanic peak from the surrounding plains. This cliff has long been known from earth-based observations as a circular feature called Nix Olympica ("the Snows of Olympus"), but its significance had not been appreciated.

The three other volcanic peaks first spotted above the Martian dust storm lie about 1000 kilometers to the east of Olympus Mons, within the same volcanic province. This entire area, known as the **Tharsis region,** forms a bulge on the Martian crust which rises 6 kilometers above the average elevation of the planet (comparable to a continent on earth). The volcanoes in the Tharsis region are among the youngest features on the planet, so we can surmise that volcanic activity has occurred in the recent geologic past on Mars. Other volcanoes on Mars are heavily cratered and highly eroded, which implies an extreme age for these features, possibly as old as the volcanic plains on the moon.

East of the volcanic peaks of the Tharsis region, Mariner 9 discovered an immense canyon straddling the Martian equator. Although this feature had been identified as a dark marking called Coprates by earth-based observers, its true nature was not appreciated until close-up pictures were obtained. Now called **Vallis Marineris** in honor of the Mariner missions, this canyon is more than 5000 kilometers in length. At places it is as wide as the Grand Canyon is long. The Grand Canyon would only be the size of one of the small tributaries leading into the Vallis Marineris (Fig. 7.15).

One of the most surprising results of the Mariner mission was the discovery of old dried-up river beds on the Martian surface (Fig. 7.16). This was unexpected on the basis of previous studies because water cannot exist in the liquid state at the low atmospheric temperatures and pressures on Mars; any water released onto the Martian surface would either quickly evaporate (if it was near the equator) or freeze (if it was closer to the poles). Moreover, there is not a sufficient amount of water vapor suspended in the atmosphere to cause extensive stream erosion on Mars, even if it were to condense out all at once. It is therefore a puzzle to explain where the water that carved the channels came from.

One possibility is that water bound up as subsurface ice, known as **permafrost,** is occasionally melted catastrophically, unleashing torrential

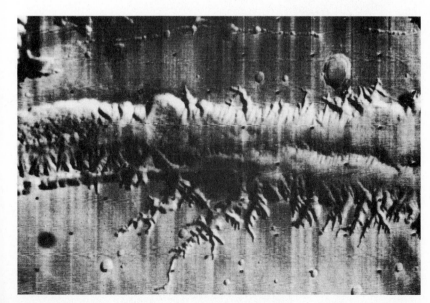

Figure 7.15 *Mariner 9 photograph of Vallis Marineris—a hugh canyon on Mars. This particular photograph shows a 480-kilometer length of the 5000-kilometer long canyon. (National Aeronautics and Space Administration photograph.)*

Figure 7.16   *A dry stream bed on Mars. Braided channels similar to this form from sediment-laden streams on earth. (National Aeronautics and Space Administration photograph.)*

floods which momentarily inundate the surface. The river beds would then be carved quickly before the water could evaporate or refreeze, without a drop of rain ever falling on the surface. This idea is attractive because it could occur on Mars today without any change occurring in the atmosphere at all. If volcanic magma in the interior of Mars were to rise close to the surface, it would melt the overlying permafrost and cause the catastrophic floods.

The discovery of water-carved channels on the Martian surface did much to buoy hopes that life could exist on Mars. If water once flowed on the surface, organisms could have evolved which later adapted to the harsh surface conditions found today. Thus, the bleak picture of Mars painted by the first few Mariner missions was erased by the discoveries of Mariner 9. In the course of a decade, our view of Mars essentially went full circle.

Once the Mariner 9 photographs were all assembled, a map of the planet's surface was drawn by the *United States Geological Survey* (Fig. 7.17). The immediate conclusion that is reached by studying the map is that the northern and southern hemispheres of Mars are very different; while the southern hemisphere is heavily cratered, the northern hemisphere consists of smooth sparsely cratered plains. Most of the riverbeds on Mars run from the cratered terrain in the southern hemisphere to the smooth northern plains. The average elevation of the northern plains is 2 to 3 kilometers below the southern cratered terrain. Thus, if Mars had large bodies of water covering its surface, the northern hemisphere would be one vast ocean, while the southern hemisphere would be one big continent.

Three multiringed basins are found on Mars which resemble the ringed basins found on the moon. Like the lunar basins, they probably formed from the impact of large protoplanetary bodies early in the history of the solar system. The largest basin, known as **Hellas Planitia** from earth-based observations, ranges up to 2000 kilometers in diameter and 6 kilometers in depth. This depression acts like a huge dust trap during dust storms, and the continual deposition of sediments on the basin's floor has produced the smooth featureless plains first photographed by Mariner 6. Argyre Basin is the best preserved of the multiringed basins on Mars (Fig. 7.18). The third large basin, Isidis Basin, is partially destroyed by the advance of the northern plains. The existence of these basins indicate that the southern hemisphere of Mars has been preserved since the last major debris left over from the formation of the planets was swept up nearly 4 billion years ago. Apparently, all the terrestrial planets were bombarded at this time. Only earth and perhaps Venus have been internally active enough to eradicate evidence of this early stage of planetary development.

Wind is the major erosional process occurring on Mars today. Comparison of dark and light markings on the pictures taken previously by Mariner 6 with those taken by Mariner 9 showed slight changes in the

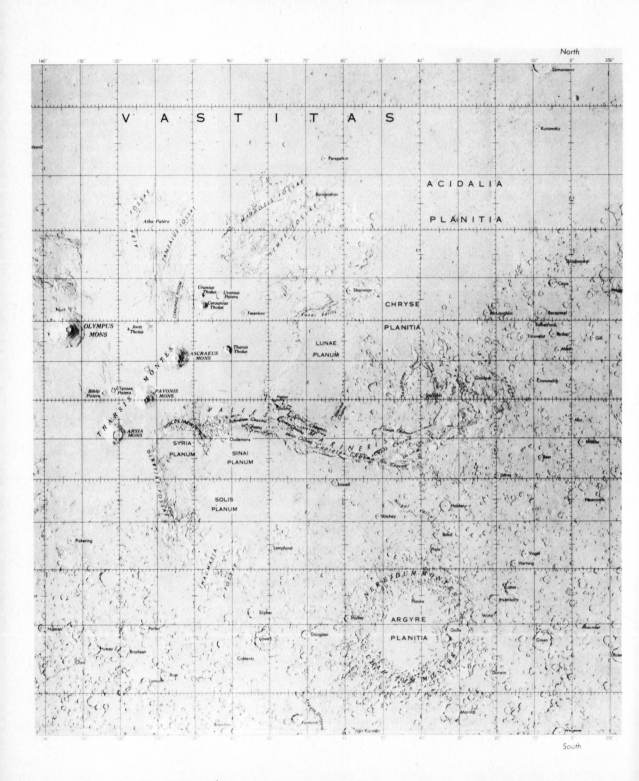

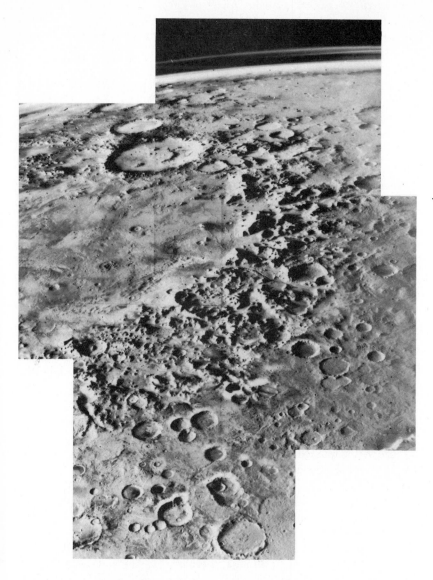

Figure 7.18 *Agyre Basin photographed by the Viking 1 orbiter. The mountains around the basin were formed by the impact of a large object. A thin layer of haze can be seen on the horizon which causes the limb of the planet to look extremely bright. The layers of clouds above the horizon are at altitudes of 25 to 40 kilometers. (National Aeronautics and Space Administration photograph.)*

configuration of the wind-blown material in the course of only four years. In addition, large dune fields were found on Mars which resembled areas containing wind-blown sand in deserts on the earth. Craters on Mars show signs of wind erosion by their filled in interiors and worn-down rims.

Before the Mariner missions, it was debated whether the polar ice caps on Mars consisted of water ice or frozen carbon dioxide. This question was partially answered by the various Mariner spacecraft. Temperatures

Figure 7.17 *Map of Mars (opposite page) based upon Mariner 9 photographs. Note how heavily cratered the southern hemisphere is compared to the northern hemisphere. (Map by the United States Geological Survey.)*

as high as 280 to 290 K (7 to 17°C) were found along the Martian equator during the day, but they fell to 200 K (−73°C) at night. At the south pole of Mars, Mariner 7 measured a temperature of 160 to 170 K. This is the temperature which frozen carbon dioxide is expected to have at the atmospheric pressures found on Mars, indicating that the Martian polar caps are composed primarily of "dry ice." This would explain the rapid retreat of the polar caps, since a thin layer of carbon dioxide would sublimate quickly when the temperature of the air rose above the freezing point of dry ice during the local spring and summer. However, observations of the south polar cap of Mars by Mariner 9 early in 1972 showed that a residual polar cap remained at the south pole right through the local summer. This residual polar cap could only survive the heat of summer in the southern hemisphere if it consisted of water ice. During the summer of 1972, Mariner 9 observed the retreat of the northern polar cap of Mars. Again, the thin carbon dioxide ground cover retreated rapidly at the start of spring in the northern hemisphere, but then slowed until a permanent water–ice region was exposed. The polar ice caps were therefore found to consist of a permanent water–ice cap overlain by a much more extensive carbon dioxide layer during the winter.

7.7 Viking missions to Mars

The Viking missions consisted of two spacecraft that were launched towards Mars in 1976. Known as Viking 1 and Viking 2, they each consisted of a lander and an orbiter. The orbiters and landers journeyed to Mars together. After going into orbit around Mars, they then separated and the landers descended to the planet's surface, while the orbiters continued to observe the planet from above. Both orbiters and landers were equipped with two television cameras and each camera had six color filters. By taking six pictures at different wavelengths and superimposing the monotonal images, true color pictures of the planet were obtained, both from orbit and on the ground (see color plates).

Viking 1 arrived at Mars first and immediately began searching for a safe landing site. Photographic reconnaissance centered on Chryse Planitia ("the Golden Plains") because radar studies had indicated that this area of Mars is relatively smooth on a scale of meters to centimeters and Mariner 9 pictures showed that it was a smooth region free of extensive craters. It was crucial to the success of the mission that the landers touch down in boulder-free terrain, because a sizable rock could easily overturn the lander, rendering it inoperative.

High-resolution pictures of Chryse Planitia provided ample evidence of a series of sudden floodings by large quantities of water on this portion of the planet (Fig. 7.19). The plain lies about 2½ kilometers below the mean level of the Martian surface in an area where a number of channels

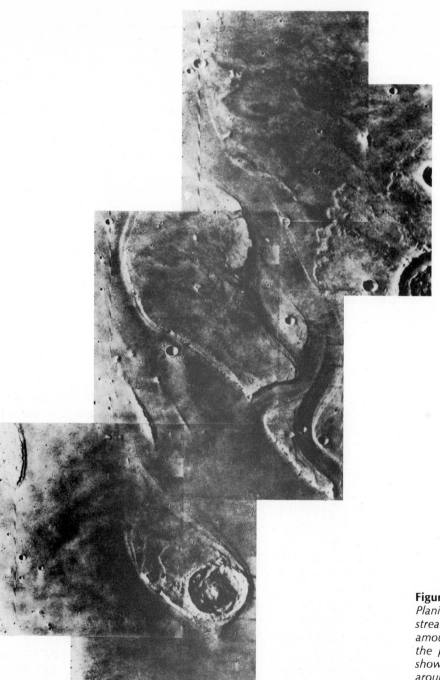

Figure 7.19 *Channels in Chryse Planitia. The broad intertwining stream beds indicate that large amounts of water once flooded the plains. The teardrop shapes show how the water flowed around the islands. (National Aeronautics and Space Administration photograph.)*

originating in Vallis Marineris converge. The plains themselves are covered with stream beds with meandering patterns indicative of rapid water erosion. Where the stream channels encounter crater rims or other obstacles, they wrap around the protrusions, forming teardrop-shaped islands. All the evidence suggested that episodic flash flooding has occurred on Chryse Planitia in the past.

The Viking orbiters also provided excellent photographic evidence for subsurface melting of permafrost in the equatorial regions of Mars. Figure 7.20 shows an example of an area where subsurface ice melted and a "lake" of water formed. When the ponded water reached the brink of the slope on the right of the photograph, the dam restraining it broke, and water surged out the channel to the left. The sediment which once permeated the ice was left as the hilly terrain in the resulting canyon, while the water which gushed from the reservoir was dispersed over wide areas of the planet, sinking back into the soil and refreezing or evaporating into the atmosphere.

On 20 July 1976, the Viking 1 lander separated from the orbiter and descended safely onto the Chryse plain. Shortly afterwards, Viking 2 went into orbit around Mars. After searching for a landing site, the Viking 2 lander descended on 3 September 1976 in Utopic Planitia, a large basin on the opposite side of the planet from Chryse Planitia. Utopia plain had been chosen as a good landing area because of the high water

Figure 7.20 *Collapsed terrain on Mars where subsurface ice melted to form a lake. The melting continued until the water breached the natural dam holding the water in. Once the dam broke, the water flooded the plains for more than 300 kilometers. (National Aeronautics and Space Administration photograph.)*

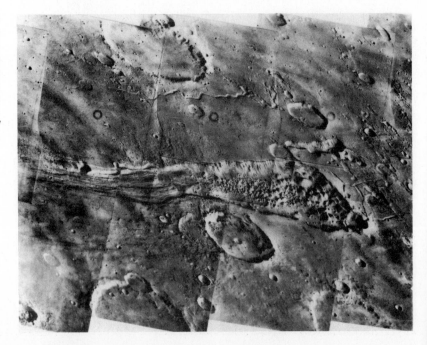

vapor content measured in the air above it. Since Viking's primary task was to search for life on Mars, the presence of water at the landing site was an important requirement.

The views from the landers looked like a rock-strewn desert on earth (Fig. 7.21). The rocks around the Viking 2 lander were more rounded and wind worn than those at the Viking 1 landing site. The rocks around the Viking 2 lander were also highly vesicular, indicating a volcanic origin. Several types of igneous rocks as well as some breccias were seen at the Viking 1 site, indicating that some of the rocks had been thrown there by meteorite impacts. Wind-blown sand and dust were scattered around the rocks at both sites, although large sand dunes were seen only on Chryse Planitia. Chryse Planitia showed considerable irregularities in its surface. By contrast, Utopia Planitia turned out to be almost perfectly flat, with a nearly straight horizon line off in the distance.

Although the first color pictures of Mars showed a ubiquitous red surface and sky, more accurate calibration of the lander cameras revealed that the surface of the red planet is actually yellowish brown in color. This color is attributable to the presence of iron oxides and iron-rich clay minerals in the Martian soil which formed when water reacted with the iron-rich rocks at the surface. The occasional presense of dark grey rocks in the lander photographs, however, shows that not all surfaces have been weathered by water. The Martian sky is yellowish brown in color because of fine dust particles suspended in the atmosphere.

Instruments aboard the Viking landers measured the composition of the atmosphere both during descent and while on the planet's surface. The composition at the surface was 93.32% carbon dioxide, 2.7% nitrogen,

Figure 7.21 *The view from the Viking 1 lander (top) and the Viking 2 lander (bottom). (National Aeronautics and Space Administration photograph.)*

1.6% argon, 0.13% oxygen, and 0.07% carbon monoxide. In addition, variable amounts of water vapor and trace amounts of krypton, xenon, and ozone were detected. With increasing height in the Martian atmosphere, the composition steadily changed towards the lighter constituents, as it does on earth.

The Viking landers came equipped with complete meteorology stations. Atmospheric temperatures were miserably cold at both landing sites. The temperature was lowest just before sunrise (190 K = −83°C) and the highest in early afternoon (240 K = −33°C). A steady breeze blew by the landers at all times. The air was the calmest at midnight when wind speeds were 1 to 3 kilometers/hour. Wind speeds with occasional gusts of 120 kilometers/hour were recorded.

Although the atmospheric pressure on Mars was about 7 millibars, as expected, it was observed to steadily decline for several weeks after the Vikings first landed. This is because it was winter in the southern hemisphere at the time, and carbon dioxide was being removed from the atmosphere as it accumulated on the polar cap. With the return of spring, the cap began to melt and the atmospheric pressure began to rise again. In general, the weather on Mars is amazingly predictable. The winds change direction in a regular pattern, the temperature varies in a continuous diurnal and annual cycle, the atmospheric pressure changes steadily with the seasons, and most predictable of all, it never rains.

The mean density of Mars (3.96 g/cm³) is greater than that of the moon (3.3 g/cm³) but much less than that of Mercury, Venus, and the earth (5.5 g/cm³). This implies that the interior of Mars is less abundant in iron. Although the overall abundance of the iron on Mars is small compared to the other terrestrial planets, it is generally believed that its mantle is relatively rich in iron, giving rise to the iron-rich rocks observed on its surface. It is possible that Mars was not thoroughly differentiated early in its history and much of the iron in its interior remained in the mantle, whereas on Mercury, Venus and the earth, it was concentrated in the core.

One of the major goals of the Viking mission was to find life on Mars. Towards this end, four experiments were carried by each lander which searched for signs of metabolism in the Martian soil. One of these experiments gave a strong response that at first seemed to indicate the presence of biological activity. However, the same response could be obtained from complex inorganic chemical reactions. The other three experiments gave negative results. The Viking landers therefore failed to produce unequivocal evidence for microbal life on the Martian surface, but this is hardly surprising, considering the limited sampling afforded the landers and the difficulty in designing experiments that would be sensitive to unknown life forms.

It remains for future missions to resolve the lingering possibility that life may exist on the red planet.

7.8 The two satellites of Mars

Mars has two moons which were discovered by Asaph Hall during the opposition of 1877. Hall named them after the sons of Mars in Greco–Roman mythology, **Phobos** (fear) and **Deimos** (terror). Their existence had been anticipated 150 years earlier in Johnathan Swift's satire *Gulliver's Travels.* Swift based his speculation on a suggestion by Kepler that Mars should have two moons, since the earth has one and at the time, Jupiter was known to have four. There is of course no compelling reason why the number of satellites around Mars should fit such a neat geometrical progression (1, 2, 4) but Kepler was a great believer in the harmony of the cosmos, and this numerical progression appealed to him.

Phobos and Deimos revolve in nearly circular orbits in the equatorial plane of Mars in a prograde direction (counterclockwise as seen from above the north pole). Like our moon, their rotation and revolution are synchronous so that one face is always pointing toward the planet. Phobos is the larger of the two moons and orbits closest to the surface of the planet. Its shape is irregular, with a maximum diameter of 27 kilometers and a minimum diameter of 19 kilometers. At its average distance from Mars of 9000 kilometers, Phobos only requires 7 hours and 39 minutes to orbit the planet. Deimos is nearly half the size of Phobos, with a maximum diameter of 15 kilometers and minimum diameter of 11 kilometers, it requires 30 hours and 18 minutes to complete its orbit. Although neither moon is impressive as seen from the Martian surface, the smaller size and greater distance of Deimos combine to make it much dimmer than Phobos. To an observer on Mars, Deimos would only appear as bright as Venus does to an observer on earth.

The proximity of Phobos and Deimos to Mars would produce an unusual celestial procession for an observer on the planet's surface. Since Phobos orbits Mars in less time than the planet takes to rotate once on its axis, it would appear to move backwards through the Martian sky, from west to east. Deimos, on the other hand, orbits with nearly the same period as the planet rotates, so that it would require about 60 hours to creep across the sky from east to west. Both satellites orbit so close to the Martian equator that Phobos cannot be seen at latitudes on Mars greater than 69°, and above 82°, both moons would be unobservable.

Pictures of the Martian satellites were obtained by the Mariner 9 and Viking space probes (Figs. 7.22 and 7.23). Both moons are saturated with craters. Phobos' surface is streaked with chains of craters and linear striations (Fig. 7.24). The linear striations on Phobos are difficult to explain.

Figure 7.22 *Viking 1 orbiter photograph of Phobos. The surface is heavily cratered and numerous grooves radiate from the large crater, 10 kilometers in diameter, at the bottom of the photograph. (National Aeronautics and Space Administration photograph.)*

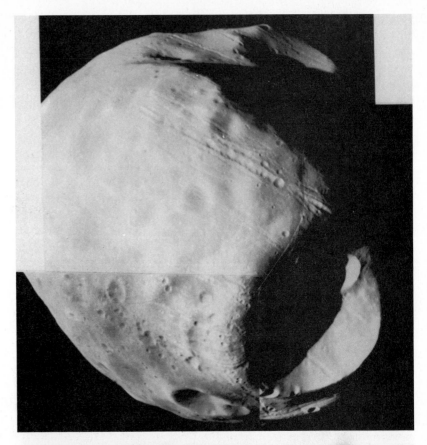

Figure 7.23 *The best photograph of Deimos reveals a heavily cratered surface on this irregularly shaped satellite. (National Aeronautics and Space Administration photograph.)*

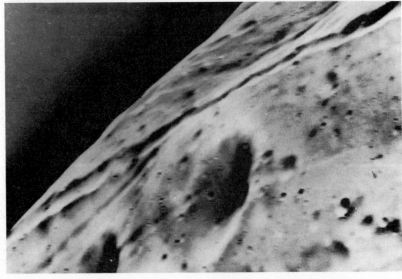

Figure 7.24 *A close-up view of Phobos taken by the Viking 2 orbiter. The largest crater is 5 kilometers in diameter and the chains of craters and striation grooves are clearly seen. (National Aeronautics and Space Administration photograph.)*

They are oriented parallel to each other and perpendicular to the direction between Phobos and Mars. It has been suggested that they formed as a result of the tidal forces exerted on Phobos by Mars, which tend to pull the moon apart. In fact, if Phobos were just a little closer to Mars, tidal forces would rip the moon to pieces, creating a ring around Mars. It is not known whether Deimos has similar surface features.

Like the moon, both Deimos and Phobos are covered with a colorless gray dust formed from meteorite impacts on its surface. The albedo of the two satellites is about 0.05, which is lower than that of the moon. Properties of light reflected from their dark surfaces indicate that the satellites are composed of material resembling carbonaceous chondrites (the most common type of meteorite). Both moons are believed to be captured asteroids.

The Asteroids

The **asteroids,** or **minor planets,** revolve around the sun mainly between the orbits of Mars and Jupiter. Invisible to the naked eye, with the occasional exception of Vesta, they are "starlike" in the sense that they appear as point sources of light instead of disks even through large telescopes. The majority have periods of revolution between 3.5 and 6 years.

7.9 The orbits of asteroids

Toward the close of the eighteenth century, the astronomer Bode invited his colleagues to share in a search for a planet between the orbits of Mars and Jupiter. He explained that a series of numbers, which later came to be known as the Titius–Bode relation (Section 6.9), represented the relative distances of the known planets from the sun with a single exception. No planet had been found corresponding to the number 2.8.

While the search was being organized, the "missing planet" was discovered incidentally by Piazzi in Sicily on 1 January 1801, because of its motion among the stars. The mean distance of the new planet, which Piazzi named **Ceres,** proved to be almost exactly 2.8 times the earth's distance from the sun. There was greater surprise when other minor planets were discovered later at about the same distance as Ceres.

Many thousands of asteroids have been detected by the generally short trails they leave in the photographs by their motions against the background of the stars (Fig. 7.25). More than 2100 have had their orbits determined and have accordingly received permanent running numbers and names in the catalogs. Many of these, however, could readily become hopelessly lost because of considerable and rapid perturbations of their

Figure 7.25 *Trails of three asteroids. (An early photograph by M. Wolf, Königstuhl–Heidelberg.)*

orbits by the attraction of Jupiter except for continuous efforts to keep track of them.

The asteroids are small when compared with the principal planets. Ceres, the largest, is about 1000 kilometers in diameter. Pallas, Juno, and Vesta are also among the larger asteroids. Only about 35 asteroids are larger than 100 kilometers. Some are known to be scarcely a kilometer in diameter. Their albedos run from 0.03, less than that of Mercury, to 0.3 which is nearly that of the earth.

The motions of a small percentage of the asteroids depart considerably from the regularities we have noted (Section 6.10) in the movements of the larger planets. Although they all have direct revolutions, a few have orbits so much inclined to the ecliptic that they venture far outside the zodiac. A few have rather highly eccentric orbits; one asteroid, Hidalgo, has its aphelion as far away as Saturn, and another comes at its perihelion nearer the sun (30×10^6 kilometers) than the orbit of Mercury and is appropriately named Icarus. Another, Chiron, has its orbit lying entirely between Saturn and Uranus and may rival Ceres in size.

The asteroids are not distributed at random through the zone they occupy between the orbits of Mars and Jupiter. They avoid distances from the sun where the periods of revolution would be simple fractions, particularly ⅓, ⅖, and ½, of Jupiter's period. There they would be subject to frequent recurrences of disturbances by Jupiter. These gaps in the asteroid distribution are known as **Kirkwood's gaps.** Where the periods are equal

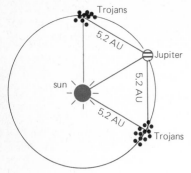

Figure 7.26 *A sketch of Jupiter's orbit showing the locations of the Trojan asteroids. The leading group is called the Greek planets and the trailing group the Trojan planets.*

Figure 7.27 *Orbits of four unusual asteroids. Light lines indicate parts of the orbits below the ecliptic.*

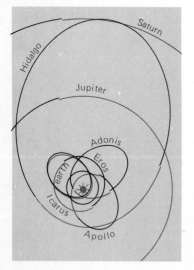

to Jupiter's, however, there are two regions in which asteroids congregate. The **Trojan asteroids** oscillate around two points in Jupiter's orbit, which are equidistant from that planet and the sun (Fig. 7.26). The leading group is called the **Greek planets** and have been named after Greek Homeric heroes of the Trojan wars. One of the Greek planets has a Trojan name. The following group is called the **Trojan planets** and has Trojan names except for one which has a Greek name—a spy in each camp.

Several known asteroids come within the orbit of Mars and make closer approaches to the earth than any of the principal planets. Eros, for example, can come within 22×10^6 kilometers of the earth, at which time this 24-kilometer wide object appears as bright as a star of the seventh magnitude. These favorable oppositions, when Eros is near perihelion, occur rather infrequently; the latest one was in 1975, and the next one is scheduled for 2019. The large parallax on such occasions permits accurate measurements of their distances, which have been valuable for verifying the scale of the solar system.

Examples of asteroids that come even closer are Apollo and Adonis (Fig. 7.27). The perihelion distance of Adonis is only slightly greater than Mercury's mean distance from the sun; this asteroid passes a little more than 1.6×10^6 kilometers from the earth's orbit and about the same distance from the orbits of Venus and Mars. Another neighboring asteroid is Icarus.

Asteroids which have been observed within a few million kilometers of the earth are about a kilometer in diameter. At closest approach they appear as faint stars moving so swiftly across the heavens that they can easily be missed. Most of them have vanished before their orbits could be reliably determined.

7.10 Asteroids as fragments

The erratic orbits of some asteroids suggest that these are fragments that have been propelled in various directions by the collisions of larger bodies. This theory may account not only for the smaller asteroids that pass near us but also for the meteorites that fall on the earth and are occasionally large enough to blast out craters. If this theory is correct we should expect a considerable amount of dust and very small fragments throughout the asteroid belt. While the inner portion of the belt does have somewhat more dust than the average in the ecliptic plane in the vicinity of the earth, the outer portion of the belt is relatively free of dust.

Many asteroids fluctuate periodically in brightness, as would be expected of rotating fragments having irregular shapes. Eros is an example. In its rotation once in 5^h16^m, this asteroid, shaped roughly like a brick, presents its larger sides and smaller ends to us in turn. Thus it becomes brighter and fainter twice in each period. The light variation is greatest when the equator is presented edgewise to us, and it becomes less in other parts of the orbit when a polar region is turned more nearly in our direction.

Extensive studies of representative asteroids have shown that over 90% vary periodically in brightness, having two maxima and minima in each period of rotation. The periods range from 2^h52^m to about 20 hours. The rotation axes have random orientations. In the case of the asteroid Eunomia the direction of rotation is definitely retrograde.

In one case, the change in brightness of an asteroid may be due to the object being two asteroids orbiting each other. This is the first suggestion that asteroids may form binary systems.

Reflection curves as a function of wavelength of many asteroids resemble those of many meteorites that strike the earth. Some resemble meteorites that are called **carbonaceous chondrites** and are characterized by small micron size spherules held together by a fine-grained material. Others resemble basalt rocks and others resemble iron meteorites.

Questions

1 Explain how Mercury's elongations can differ by as much as 10°.

2 Early observations showed that Venus went through crescent phases and that the crescent extended all the way around the disk at new phase. What two facts can we deduce from this information?

3 How was the rotation period of Mercury determined?

4 What is thought to have caused Mercury's weird terrain?

5 Why is the atmosphere of Venus so dense?

6 Why was Mars a reasonable choice for searching for signs of extraterrestrial life?

7 Why is Olympus Mons a young volcano?

8 What is unusual about Phobos?

9 Why do asteroids avoid certain regions of the asteroid belt?

10 Explain how we deduce that certain asteroids must be very old.

Other Readings

Arvidson, R. E., A. B. Binder, and K. L. Jones, "The Surface of Mars," *Scientific American* **238**(3), 76–89 (1978).

Horowitz, N. H., "The Search for Life on Mars," *Scientific American* **237**(5), 52–61 (1977).

Loudon, J., "Pioneer Venus: a First Report," *Sky & Telescope* **57**, 119–123 (1979).

Weaver, K. F., "Mariner Unveils Venus and Mercury," *National Geographic* **147**, 848–864 (1975).

Young, A. and L. Young, "Venus," *Scientific American* **233**(3), 70–81 (1975). "Mars Viewed from Viking I Orbiter," *Sky & Telescope* **52**, 171 (1976). "Mars As Viking Sees It," *National Geographic* **151**, 3–8 (1977).

chapter 8

The Gaseous Planets

The planets beyond the asteroid belt are characterized by their low densities and, except for Pluto, their large sizes. Space probes have visited Jupiter and Saturn yielding new information and raising new questions. One of Jupiter's satellites is active geologically and Saturn's moon Titan may have supported life. Jupiter, Saturn, and Uranus along with their moons resemble miniature solar systems.

Jupiter, the Giant Planet

Jupiter is the largest planet in the solar system. Its equatorial diameter, 142,800 kilometers, is 11 times as great as the earth's diameter. Its mass of 318 times that of the earth exceeds the combined mass of all the other planets, but its mean density is only 1.34, very nearly that of the sun. With the exceptions of Venus and occasionally Mars, this planet is the brightest starlike object in our skies. Even a small telescope shows its four bright satellites and cloud belts clearly. Jupiter has a thin ring and **16 known satellites, the greatest number attending any planet.**

At the distance of almost 778×10^6 kilometers from the sun, Jupiter revolves around the sun once in 11.89 years, so that it advances one sign of the zodiac from year to year. The period of its rotation, about 9^h50^m, is the shortest among the principal planets.

8.1 Jupiter observed from earth

The markings on the disk of the giant planet, which run parallel to its equator, are features of its atmosphere. Bright zones alternate with dark belts. The broad equatorial zone is bordered by the north and south tropical belts. Then come the north and south tropical zones, and beyond them a succession of dark and bright divisions extending to the polar regions. Bright and dark spots appear as well; they often change in form quite rapidly, as atmospheric markings might be expected to do. Yet some of them are of surprisingly long duration. The **Great Red Spot** (Fig. 8.1 and color plates) is an extreme example; this oval spot 48,000 kilometers long has been observed for at least two centuries. The various markings go around the planet at different rotational rates, owing to the unequal horizontal movements of the clouds themselves.

The atmosphere consists mainly of hydrogen and helium. Its refraction effect on the light of a star was recently observed as the planet began to occult the star, and was interpreted to mean that Jupiter's outer atmosphere is very dense. Methane and ammonia contaminate the atmosphere, as is indicated by the presence of their bands in the spectrum. At the temperature of 123 K, methane is still gaseous, whereas ammonia is mainly frozen into crystals. The observed temperature is somewhat higher than that expected from absorption and reradiation of solar energy.

Radiation from Jupiter in radio wavelengths is of three observed types:

1   Thermal radiation consistent with the atmospheric temperature already stated. Here the intensity of the radiation is proportional to the second power of the frequency.

2   Nonthermal emission that issues in short blasts from the planet's ionosphere.

Figure 8.1 *A Voyager 1 photograph of Jupiter showing the Great Red Spot and Io and Europa. (National Aeronautics and Space Administration photograph.)*

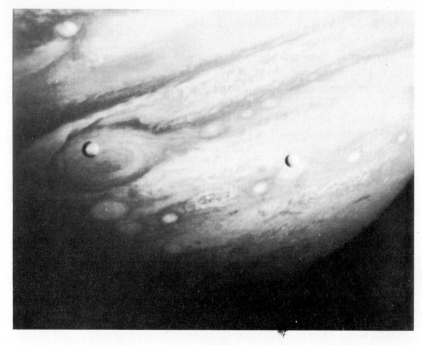

3   Nonthermal emission of a different type, which is weak and is recorded only at microwave frequencies. Here the intensity of the radiation is inversely proportional to some power of the frequency.

The short blasts are analogous to static from lighting discharges and occur when one of Jupiter's moons, Io, is in a certain position in its orbit as seen from earth. The other nonthermal component is synchrotron radiation originating from a concentration of energetic charged particles, trapped by the Jovian magnetic field similar to our terrestrial Van Allen belts. Jupiter's magnetic poles are inclined to its rotational poles similar to the earth's, but unlike the terrestrial situation, the magnetic axis does not pass through the center of Jupiter.

The bulging of Jupiter's equator, which is clearly shown in photographs, provides one clue about conditions in the interior, which is hidden beneath the clouds. With its swift rotation the planet would be even more oblate if its mass were not highly concentrated toward its center. The temperature and the low average density of the whole planet require very light material in the outer parts.

Jupiter, because it dominates the solar system's planets, is of extreme importance to our eventual understanding of the origin and evolution of the solar system. Thus it has received considerable attention in the past ten years.

8.2 Jupiter observed from space

In 1973 the spacecraft Pioneer 10 flew past Jupiter at a distance of 14,000 kilometers and while doing so studied Jupiter's atmosphere, the Great Red Spot, and Io, one of Jupiter's moons. Its encounter with Jupiter was programmed in such a way that it was accelerated by Jupiter and is now leaving the solar system. In 1980 it was between Uranus and Neptune. At this great distance, its transmissions are too weak to be detected by even our largest radio telescopes. It will take Pioneer 10 about 80,000 years to cover a distance equal to that of the nearest star. Since it is not targeted towards any known star it is unlikely that it will be intercepted by other **humanoids** (intelligent life forms). In the event that it is, it carries an appropriate pictogram (Fig. 19.16).

On the way to its encounter with Jupiter, Pioneer 10 observed streams of rather high-energy particles passing it at intervals of 4 hours and 55 minutes, one-half Jupiter's rotation period. Evidently these were particles being accelerated by Jupiter's magnetic field. Later, studies of previous cosmic-ray observations revealed this period in records as far back as 1939. In any case, Pioneer 10 encountered Jupiter's magnetosphere at about 150 Jovian radii, much farther from the planet than expected. In fact, it left the field and entered it again in such a way as to indicate that the field is highly flattened in its outer regions and very floppy like the brim of a very large hat.

Pioneer 11 confirmed Pioneer 10's findings and added a few of its own. Pioneer 11 then proceeded to Saturn, which it passed on 1 September 1979, and was accelerated by Saturn and is leaving the solar system, just like Pioneer 10. It carries a pictogram like Pioneer 10's.

Two more space probes, Voyager 1 and Voyager 2, encountered Jupiter in 1979. Both returned stunning pictures of Jupiter's atmosphere and moons, many of them in brilliant color (see color plates). Because the Voyager spacecraft provides a more stable platform from which its television cameras can focus on their target, the quality of the Voyager photographs were much better than those obtained by the Pioneer spacecraft. The Voyagers could also take time-lapse photographs of Jupiter's rotation and its atmospheric circulation, which were later assembled into motion pictures. These show the Great Red Spot to be a super hurricanelike storm. Gases swirl into the storm and are sucked down into the middle of the vortex (Fig. 8.2).

Voyager 1 showed that Jupiter has a ring around it. Voyager 2 confirmed the ring which is much smaller than the ring around Saturn, but it probably shares a common mode of origin. It consists of boulder-sized particles tens to hundreds of meters in diameter which orbit in the planet's equatorial plane. The outer edge is about 50,000 kilometers from the surface (1½ times its radius).

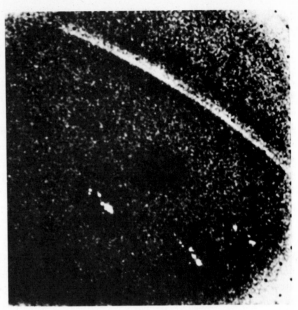

Figure 8.2 *The Great Red Spot showing its spiraling hurricane pattern (photo left) (National Aeronautics and Space Administration photograph.)*

Figure 8.3 *Auroral glow above the north polar limb of Jupiter. The bright spots on the disk are due to lightning (photo right). (National Aeronautics and Space Administration photograph.)*

Photographs of Jupiter's night side revealed a continual play of lightning above the planet's cloud cover. An extensive auroral glow due to the charged particles spiraling onto Jupiter's magnetic poles was also seen on the limb of the dark side of the planet (Fig. 8.3). Fireballs produced by meteors streak through Jupiter's atmosphere on the night side. By the frequency of the meteor falls, scientists were able to determine the abundance of interplanetary debris around Jupiter.

After passing within 172,424 kilometers of Jupiter's surface, Voyager 1 was accelerated towards an encounter with Saturn in November 1980. Voyager 2, which examined Jupiter from twice as far away as Voyager 1, is also headed for Saturn. If all goes well, the Voyager spacecraft may venture on to Uranus and Neptune before leaving the solar system.

8.3 Jupiter's 16 satellites

Jupiter is surrounded by 15 satellites which constitute a sort of mini-solar-system within the solar system. Jupiter itself is so large that it almost qualified as a star in its own right, and four of Jupiter's moons are comparable in size with the terrestrial planets. These four moons were discovered in 1610 by Galileo as soon as he pointed his telescope towards Jupiter. In fact, if it were not for their proximity to Jupiter, these four bright moons, known as the **Galilean satellites,** would be visible with the naked eye. All four Galilean satellites have nearly circular orbits in the equatorial plane of Jupiter. They also orbit synchronously around Jupiter, which means that they keep the same face towards their primary at all times, just as the moon always has the same face pointing towards the earth. The Galilean satellites were named after Jupiter's lovers in Greco-Roman mythology. In order from the planet outward, they are **Io, Europa, Ganymede,** and **Callisto.**

The internal constitution of the Galilean satellites has been inferred from their densities and observed surface compositions. Io has a density of 3.52 g/cm³, which is greater than the density of the moon (3.34 g/cm³), so it must be composed of solid and molten rock throughout. Europa is slightly less dense than Io (3.17 g/cm³) and its infrared-absorption spectrum indicates that most of its surface is covered with water ice, so its rocky core must be surrounded by an ocean of frozen water. Its surface temperature at its great distance from the sun is a frigid 127 K. Ganymede is about twice as dense as water (1.99 g/cm³), indicating that its rocky core is surrounded by a thick mantle of water. The infrared spectrum of Ganymede shows that its crust of water ice is about half covered with a layer of rocky debris. The surface of Ganymede only reflects 44% of the sunlight which strikes it, as compared with 65% for Io and Europa. The greater amount of sunlight absorbed by the rocky debris at its surface raises its temperature to 154 K. Callisto is the least dense of the Galilean satellites. Its density of 1.76 g/cm³ implies a very small rocky core for this satellite, surrounded by a thick mantle of water. The rocky water-ice crust on Callisto only reflects 19% of the sunlight that strikes it, raising its surface temperature to a chilly 167 K, the highest surface temperature of any of the Galilean satellites. Thus, with increasing distance from Jupiter, the density of the Galilean satellites decreases, indicating an increasing content of low-density material, mostly water. The reflectivity of the Galilean satellites also decreases with increasing distance from Jupiter, resulting in an increase in their surface temperatures.

In 1979 when the Voyager space probes flew past Jupiter, they obtained the first high-resolution pictures of the Galilean satellites. Earth-based studies of reddish orange Io had forshadowed some surprises from this unusual moon, but scientists were hardly prepared for what was seen. Black splotches and large white regions abounded on Io's predominately red and yellow surface (Fig. 8.4 and color plate). The white areas may be due to the presence of salts on Io's surface, such as sodium chloride (common table salt). The red color is probably due to sulfur. It has been suggested that Io was once enveloped by water like the other Galilean satellites, but that its oceans evaporated away, leaving behind a layer of evaporite deposits much like the salt flats found in arid regions on earth.

Perhaps the biggest surprise of the Voyager missions was the discovery of intense volcanic or geyser-like activity on Io. Eruptions were first recognized as plumes of dust and gas seen along Io's limb (Fig. 8.5). Further

Figure 8.4 *Io from Voyager 1. Io's mottled surface is red and yellow, but is also highlighted by black and white areas. (National Aeronautics and Space Administration photograph.)*

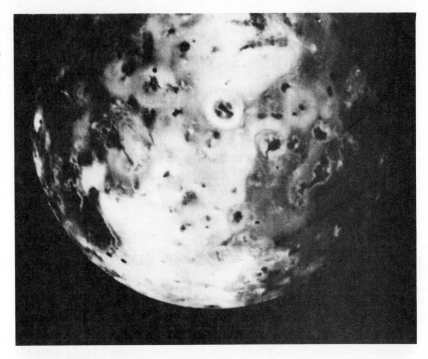

Figure 8.5 *A volcanic plume seen on the limb of Io by Voyager 1. (National Aeronautics and Space Administration photograph.)*

inspection revealed at least seven such events occurring all at once on the surface of Io, making this moon-sized satellite a very active body. The plumes extend to a height of 500 kilometers above the satellite's surface, indicating that the material was ejected with velocities of the order of 1 kilometer/second, which is a considerably higher velocity than is imparted to material ejected from terrestrial volcanoes (a mere 50 meters/second at the most). Another indication of activity on Io is the discovery of three hot spots on Io's otherwise frigid surface. These regions are believed to be ponds of molten lava, sulfur, or sodium overlain by a crust. The temperature of the surface of these hot lakes is almost 300 K, which is comfortable in comparison to the 127 K temperature found elsewhere on Io. The largest of the lakes was estimated to have a greater surface area than the entire state of Hawaii.

The large discharge of material from Io explains many of the puzzling features of this unusual satellite. Io is amazingly smooth for a body not shielded from meteorite impacts by a thick atmosphere. Apparently the continual deposition of material adds about a millimeter of material to Io's surface each year, erasing 1-kilometer craters within a few million years. The extreme activity of Io may also explain another strange feature of this odd world—its thin metallic atmosphere. Earth-based observations of Io have shown that sodium atoms are escaping from Io and going into orbit around Jupiter along Io's orbital path (Fig. 8.6). In addition, Pioneer 10 detected a cloud of hydrogen around Io and showed that there were 10,000 times as many hydrogen atoms in this cloud as sodium atoms. However, the hydrogen cloud disperses more quickly than the sodium

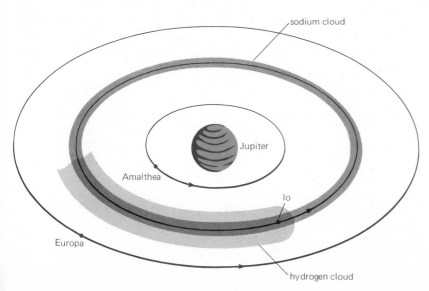

Figure 8.6 *Io's sodium and hydrogen clouds. Atoms which escape from Io cannot escape the gravitational field of Jupiter so they orbit Jupiter along Io's path.*

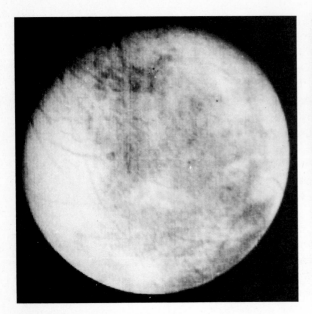

Figure 8.7 *Europa, the second Galilean satellite. Europa is light orange in hue and crossed with streaks, some of which wrap more than halfway around the satellite. (National Aeronautics and Space Administration photograph.)*

Figure 8.8 *Ganymede, the third Galilean satellite is lightly cratered and shows fault lines on its surface. (National Aeronautics and Space Administration photograph.)*

cloud, so that it does not form a complete loop around Jupiter. Potassium and ionized sulfur have also been identified in the vicinity of Io. The glow of these gases is so bright that a continual aurora is seen from Io that would rival the Aurora Borealis as seen from earth.

In order to maintain this cloud, many sodium atoms would have to be ejected from the surface of Io each second. It was formerly thought that these atoms were sputtered off the salt deposits on the satellite's surface by the high-energy electrons and protons in Jupiter's radiation belts. However, Io's plumes are now deemed a more likely source. If the sodium cloud does indeed arise from the plumes on Io, this would indicate that Io has been erupting continuously.

Io's heat source has been explained by calculations showing that the recurrent alignment of Io with the other Galilean satellites caused Io's orbit to deviate considerably from a circular path around Jupiter. Because the size of the tidal bulges raised in Io by Jupiter's immense gravitational pull varies with its distance from the planet, Io's tidal bulges are pumped up and down as the satellite approaches and recedes from Jupiter. Calculations indicate that the internal friction generated by this pumping action has probably melted the majority of Io's interior. The other Galilean satellites are not so severely pumped by Jupiter's gravity because of their greater distance from the planet and the lack of proper orbital resonance with the other Galilean satellites. There is another explanation for Io's heating: It is believed that Io's crossing of flux lines of Jupiter's magnetic field generates electrical currents that heat the satellite much as electricity flowing through a wire heats the wire.

Beyond Io lies Europa, an off-white to light-orange-hued moon without stark contrasts across its surface (Fig. 8.7). Europa is the smallest of the Galilean satellites and the only one that is smaller than the earth's moon. The photographs of Europa taken by the Voyagers reveal long continuous streaks on Europa's surface, some of which wrap more than halfway around the planet. The origin of these streaks is unknown, but they may represent extensive fractures in the icy crust enveloping Europa's rocky core. Like Io, Europa has few craters so its surface is being smoothed over regularly.

The next of the Galilean satellites is Ganymede, a moon larger than the planet Mercury. Voyager photographs of Ganymede showed that its surface is more heavily cratered than the lunar maria, but it is not totally saturated with craters, like the lunar highlands (Fig. 8.8). Ganymede's crater ejecta is much lighter in color than the surrounding areas, and bright rays are frequently seen radiating from the craters (Fig. 8.9). The white material exposed by the meteorite impacts is probably water ice which has been buried beneath a layer of rocky debris. This view is supported by the large number of craters on Ganymede with pits at their centers. These central depressions have also been seen on Mars, whose surface is underlain by permafrost in many places (Fig. 7.20).

Voyager 1 discovered an array of linear features on Ganymede's surface which appear to be transverse faults. Transverse faults such as the San Andreas fault in California produce lateral movements in the crust of the earth which can result in offsets of thousands of kilometers. If the linear features on Ganymede are due to transverse faulting (Fig. 8.10),

Figure 8.9 *Rayed craters on Ganymede. The light material is ice excavated from beneath the rocky surface by a meteorite impact. (National Aeronautics and Space Administration photograph.)*

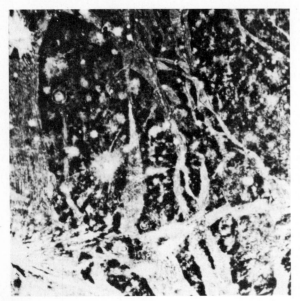

Figure 8.10   *Transverse faults on Ganymede. The 50-kilometer off-set (left) in the white streak indicates transverse faulting. The intertwined fault lines (right) suggest that complex internal processes have occurred on the satellite. (National Aeronautics and Space Administration photograph.)*

Figure 8.11   *Callisto, the outermost Galilean satellite. It is heavily cratered and has ringed basins, one of which is the largest known. (National Aeronautics and Space Administration photograph.)*

this large satellite may be the only other body in the solar system besides the earth which experiences plate tectonics, the driving mechanism behind continental drift.

Callisto is the outermost Galilean satellite (Fig. 8.11). It has 10 times as many craters as Ganymede, which indicates that its crust is considerably older. When Jupiter formed, it released a tremendous amount of energy. In fact, heat left over from this period in Jupiter's history is still being emitted from Jupiter's interior at a rate equal to the amount of radiant energy from the sun striking Jupiter today. This heat from Jupiter would have prevented the water crust of Ganymede from freezing before the crust of Callisto, since Callisto orbits Jupiter at twice the distance of Ganymede. The solid crust of Callisto has therefore had more time to accumulate the scars of meteoritic bombardment. In addition to craters, Callisto has large ringed basins like the moon, giving further testimony to the great age of its surface. The largest basin is more than 1500 kilometers in diameter, and probably formed from the impact of an asteroid or other large body early in the history of the solar system (Fig. 8.12).

Besides the Galilean satellites, Jupiter has twelve small moons which are numbered in the order of their discovery. Three of them orbit inside the Galilean satellites and outside Jupiter's ring. Amalthea is so small that almost 200 years elapsed after the discovery of the Galilean satellites before this fifth satellite of Jupiter was observed. Amalthea is only about 15 kilometers long and is red in color. It was photographed by Voyager 1 and proved to be surprisingly elongated (Fig. 8.13). It is nearly twice

Figure 8.12 *A large ringed basin on Callisto. (National Aeronautics and Space Administration photograph.)*

as long as it is wide, making it the most elongated moon known in the solar system. A few asteroids are as elongated as Amalthea and we speculate that it is probably a captured asteroid.

The nine remaining satellites beyond the Galilean satellites can be divided into two groups. The 6th, 7th, 10th, and 13th satellites have direct revolutions at mean distances of 11 million kilometers from Jupiter. Their orbits are not quite coplanar and are highly elliptical. The 8th, 9th, 11th, 12th, and 14th satellites have retrograde revolutions at about twice the distance of the first group. Their orbits are completely irregular. All nine of Jupiter's outer satellites are probably captured asteroids.

The data on the satellites of Jupiter appear in Table 6.2. The order of distance of the outermost five satellites has little significance because disturbances in their motions by the sun's attraction may change that order in a few years. The diameters of the satellites that do not appear as disks in a telescope because of their small size are estimated from their brightness.

Saturn, the Ringed Planet

Saturn is the most remote of the bright planets and is, therefore, the most leisurely in its movement among the constellations. It revolves once in 29.5 years at the mean distance of 1425×10^6 kilometers from

Figure 8.13 *Amalthea, Jupiter's innermost satellite. It is bright red in color and is highly elongated. (National Aeronautics and Space Administration photograph.)*

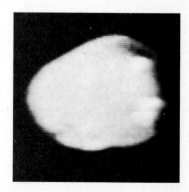

the sun. It appears to be a bright yellow star in our skies, and it ranges in brightness from equality with Altair to twice that of Capella. This planet ranks second to Jupiter in size and mass; its equatorial diameter is 120,000 kilometers. It has the lowest average density, 0.7 times the density of water, and the most prominent bulge at the equator of any of the planets. Its large system of rings makes it one of the most impressive celestial sights with the telescope.

8.4 Saturn and its rings

Saturn resembles Jupiter except for its smaller size, mass, and density. Methane bands are stronger in its spectrum, and there is less evidence of ammonia, presumably because this ingredient is frozen out of its atmosphere, which is at a temperature of 93 K.

From its broad, yellow equatorial zone to its bluish polar caps, the cloud markings show less detail than do those of Jupiter. Although enduring spots are rarely seen on Saturn, those that have been observed seem to show that the planet's rotation period is longer in the higher latitudes. Thus a period of 10^h14^m was derived for a bright spot in the equatorial zone in 1933 and for another in 1960, whereas a period of about 10^h40^m was found for a group of spots around latitude $+60°$ in 1960. Some astronomers conclude that Saturn rotates in these two basic periods instead of having a steady increase of period with increasing latitude, as in the case of the sun and Jupiter. Saturn's magnetic field may be quite strong but it has not been observed by the usual techniques. It seems that the rings of Saturn may be conductive and hence restrict the field to a very limited space. Bursts of non-thermal radio radiation were detected by Pioneer 11.

Saturn's rings are invisible to the naked eye and were, therefore, unknown until after the invention of the telescope. When Galileo began observing Saturn in 1610, he glimpsed what seemed to be two smaller bodies in contact with the planet on opposite sides. Galileo referred to them as Saturn's ears. The supposed appendages disappeared two years later and subsequently reappeared. This changing appearance of the planet remained a mystery until about half a century later, when Huygens, with a larger telescope, concluded that Saturn is encircled by a broad flat ring. In still later times we have found that there are five concentric rings instead of a single one.

The entire ring system (Fig. 8.14) is 275,000 kilometers across, but is scarcely more than 15 kilometers thick. The width of the outer bright ring, or **A ring**, is 16,000 kilometers. The brightest ring, or **B ring**, is 26,000 kilometers wide. It is separated from the outer ring by the 4800 kilometer **Cassini division,** named after its discoverer; this is the most obvious division in the rings. The third or **crape ring,** referred to as the **C ring,** is actually continuous with the bright ring and is about 19,000

Figure 8.14 *Several views of Saturn from a good view of the north side of the rings (top left) through the ring's edge (bottom left and top right) on to the south side of the rings (bottom right). (Lowell Observatory photograph.)*

kilometers wide. The crape ring is quite faint and often not very clearly shown in pictures but nevertheless rather easily visible with telescopes of moderate size, although it was not discovered until 1850. In 1970 the **inner ring,** or **D ring,** was discovered by P. Guerin and it was noted that there is a clear division between it and the crape ring analogous to the Cassini division. The D ring was not confirmed by Pioneer 11. Pioneer

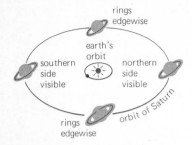

Figure 8.15 *Diagrammatic explanation of the different views of Saturn's rings from above and below. Twice in the course of Saturn's revolution the rings are presented edgewise to the sun and earth.*

11 photographed a less dense ring beyond the A ring. This new ring is referred to as the **F ring.** An earlier ring farther out, the E ring, may not be real. The rings show some remarkable changes in overall reflectivity, occasionally being quite dull for a given opening.

Saturn's rings are inclined 27° to the plane of the planet's orbit, and they keep the same direction during its revolution. They accordingly present their northern and southern faces alternately to the sun, and also to the earth which is never more than 6° from the sun as viewed from Saturn. Twice during the sidereal period of 29.5 years the plane of the rings passes through the sun's position (Fig. 8.15), requiring nearly a year each time to sweep across the earth's orbit. In that interval our own revolution brings the rings edgewise to us from 1 to 3 times, when they disappear if viewed through small telescopes and are only very narrow bright lines through larger telescopes. The next edgewise presentation is in 1980 to 1981.

The widest opening of the southern face occurred in 1974 and the next such opening will be of the northern face in 1989. When the rings are open widest, their apparent breadth is 45% of their greatest diameter and one-sixth greater than the planet's polar diameter. On these occasions Saturn appears brighter than usual, because the rings at this angle reflect 1.7 times as much sunlight as does the planet's disk.

Saturn's rings consist of solid particles that revolve like satellites around the planet in nearly circular orbits in the plane of its equator and in the direction of its rotation. They are mainly icy particles. The light from the separate pieces runs together at a great distance from Saturn, giving the appearance of a continuous surface. We have noted already the changing reflectivity.

If the rings were really continuous, all parts would rotate in the same period, and the outside, having farther to go, would go around faster than the inside. The spectrum shows (Fig. 8.16), however, that the inside has the faster motion, as it should have in accordance with Kepler's harmonic law if the rings are composed of separate pieces.

The outer edge of the outer ring has the longest period of rotation,

Figure 8.16 *Spectrum of the rings (R) and ball (B) of Saturn. The lines of the ball are slanted due to the rotation of Saturn. The lines of the rings have an opposite slant showing that the inner parts revolve more rapidly than the outer parts. Comparison lines (C) of iron are above and below the spectrum. (Lowell Observatory photograph.)*

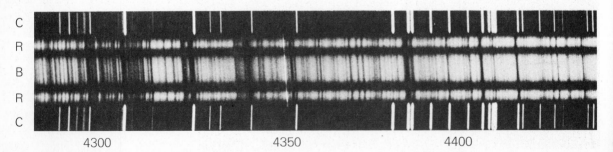

C

R

B

R

C

4300   4350   4400

14^{h}27^m. The inner edge of the bright ring rotates once in 7^{h}46^m, and the material of the crape ring must go around in still shorter time. Meanwhile the planet itself rotates in a period of about 10 hours. Thus the outer parts of the ring system move from east to west across the sky of Saturn, whereas the inner parts seem to go around from west to east, like Phobos in the sky of Mars.

Further evidence that the rings are not continuous is given on the rare occasions when a bright star passes behind the rings. Bright stars can be seen, almost undiminished in brightness, as they apparently move behind the rings. Radar studies suggest that the particles comprising the rings range in size from 10 to 30 centimeters in diameter. The Pioneer 11 measurements confirm this.

8.5   The satellites of Saturn

Titan, the largest and brightest of Saturn's 11 known satellites, is considerably larger than the moon. It is one of the few satellites known to have an atmosphere (Io and Triton are the others). Titan's spectrum shows methane bands. This satellite resembles Mars and Io in its reddish color, probably because of similar action of the atmosphere on the surface rocks. Titan has hydrogen in its atmosphere and at the surface its atmosphere is 10 times more dense than that of Mars. Its temperature, 93 K, is higher than predicted indicating that the greenhouse effect operates. Even against the low surface gravity, hydrogen is present indicating there must be a source to replenish the hydrogen lost to interplanetary space. Perhaps Titan has volcanoes? Titan's mean density of 3.9 g/cm^3 is greater than that of the moon and Io.

Four or five other satellites can be seen with telescopes of moderate size (Fig. 8.17), appearing as faint stars in the vicinity of Saturn. All the satellites have direct revolutions around the planet with the exception of Phoebe, the most distant and the faintest one; Phoebe has retrograde revolution like Jupiter's outer group of satellites. Some of the satellites vary in

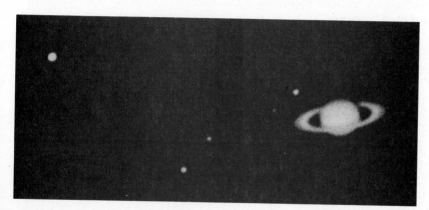

Figure 8.17 *Some moons of Saturn. The moons are, left to right, Titan, Rhea, Dione, Tethys, and Enceladus. (Leander McCormick Observatory photograph.)*

brightness in the periods of their revolutions; they evidently rotate and revolve in the same periods, and are either irregular in form or have surfaces of uneven reflecting power. The very high reflection from the inner satellites suggests that they have icy surfaces, and their low densities may mean that they are composed mainly of ice. Presumably, the inner satellites, Janus and a recently discovered one, will exhibit the same characteristics as the other inner satellites.

Uranus and Neptune

Figure 8.18 *Uranus as seen from the balloon-borne telescope Stratoscope II (top). No markings are visible, but a limb-darkening effect is observed. (Princeton University photograph.) Uranus and its satellites, from left to right: Titania, Umbriel, Ariel, and Oberon. (Lunar and Planetary Laboratory photograph.)*

The discovery of Uranus, in 1781, was accidental and unexpected. William Herschel was examining a region in the constellation Gemini when he noticed a greenish object which seemed to him somewhat larger than a star. The object eventually proved to be a planet more remote than Saturn, and it received the name Uranus. Forty years later, when this planet had gone nearly halfway around the sun, its orbit was calculated from many observed positions, with allowance for the disturbing effects of other known planets. After that, however, the new planet did not follow precisely the course it was expected to pursue. Astronomers finally concluded that its motion in the heavens was being altered by the attraction of another planet, Neptune, which was still more remote and as yet unseen.

Uranus is nearly 51,000 kilometers in diameter; it revolves once in 84 years at 19 times the earth's distance from the sun. It rotates once in about 16 hours, having its equator inclined nearly at right angles to the ecliptic. Its mean density is 1.31 g/cm³. Barely visible to the unaided eye, Uranus shows a small greenish disk through the telescope on which the markings are not clearly discernible (Fig. 8.18). The temperature of the visible disk is 63 K. The spectrum includes a dark band in the infrared showing the presence of molecular hydrogen. This was the first direct evidence of molecular hydrogen in the atmospheres of the major planets. Bands of methane appear in the spectra of both Uranus and Neptune. A haze appears in the upper atmosphere, especially around the poles of Uranus.

One of the remarkable discoveries of 1976 was that Uranus has a set of at least five rings. This discovery was made because of a possible space mission to Uranus in the future. If one of the Voyager spacecraft, after its encounter with Saturn, was to be redirected to Uranus it would be necessary to know the exact size of Uranus. This can be obtained best when Uranus occults a background star; the duration of the occultation being a measure of the diameter. In attempting to observe such an occultation observers from Australia, the South Pacific, and the Indian Ocean astronomers observed that the star winked off and on five times prior to occultation and after occultation. Each ring is not very broad and appears to be very tenuous.

Five satellites revolve around Uranus in nearly circular orbits in the plane of its equator. The orbits were presented edgewise to us in 1966 and will next appear flatwise in 1987. The fifth satellite, discovered in 1948, is the faintest and nearest to the planet.

The existence of Neptune was predicted by Adams in 1841 and Leverrier in 1846. By comparing observed positions of Uranus during the preceding quarter of a century with the predicted ones, they were able to calculate the place at that time of the unseen disturber in the sky. However, both Adams and Leverrier were unable to convince their local colleagues that their calculations should be taken seriously. Leverrier appealed to Galle at the Berlin Observatory where an accurate star map was available. Galle directed the telescope toward the region specified in the constellation Aquarius and soon found Neptune within 1° of the place assigned it by Leverrier. The discovery was acclaimed as a triumph for the law of gravitation, on which the calculation was based.

Figure 8.19   *Neptune and its inner satellite Triton. Neptune is overexposed in order to record the satellite. (Lick Observatory photograph.)*

Neptune, being about 48,500 kilometers in diameter, is almost the same size as Uranus and revolves once in 165 years. It has a direct rotation taking 18 hours and 26 minutes to complete, according to spectroscopic measures. Its temperature is 53 K. Although Neptune is invisible to the naked eye, with the telescope it appears as bright as a star of the eighth magnitude and shows a small greenish disk on which very weak markings have been seen. It seems to closely resemble Uranus. It has an average density of 1.77 g/cm³.

Neptune has two known satellites (Fig. 8.19). The first, Triton, is somewhat larger than the moon and is slightly nearer the planet than the moon's distance from the earth; it is nearly twice as massive as the moon and probably has an atmosphere. Triton has a retrograde revolution, contrary to the direction of the planet's rotation. Triton's mean density is 3.0 g/cm³. The second satellite, discovered in 1949, is much smaller and more distant from the planet. It has a direct revolution in an orbit having an eccentricity of 0.76, the greatest for any known satellite.

Pluto, the most Remote Planet

The discovery of Pluto was announced by Lowell Observatory on 13 March 1930, as the successful result of a long search at that observatory for a planet beyond Neptune. The planet was discovered by C. Tombaugh in his photographs taken in January of that year. The search had been instituted by Percival Lowell, who had calculated the orbit of a trans-Neptunian planet from slight discrepancies between the observed and predicted movements of Uranus, which seemed to remain after the discovery of Neptune.

Pluto is visible with a telescope appearing like a star of the 15th magnitude (Fig. 8.20). Its diameter is about 3000 kilometers and its mass

Figure 8.20 *A long-exposure photograph of Pluto. The protuberance on the upper left elongating the image is Pluto's satellite Charon. (U.S. Naval Observatory photograph.)*

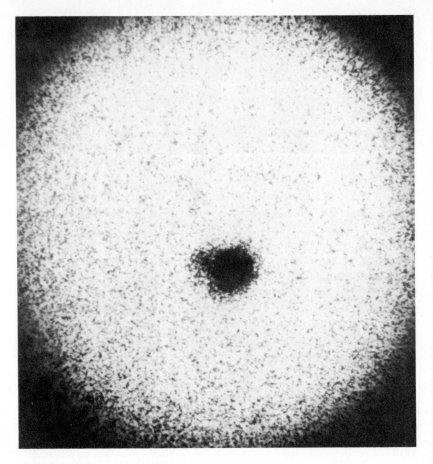

only a few thousandths that of the earth. This yields a density around 0.7 times that of water suggesting that Pluto is composed entirely of frozen volatiles. The planet shows periodic fluxuations in brightness with a period of 6.39 days. Evidently, this is its period of rotation. Pluto is thought to have a gritty snow-covered surface of nitrogen on the morning quarter and an atmosphere that is considerably rarer than ours. At Pluto's distance from the sun, the sun subtends less than 1 minute of arc, hence the surface temperature on the day side does not exceed 40 K.

At its average distance the planet is 39.5 astronomical units, or 5910×10^6 kilometers, from the sun. It revolves once in 248 years, which is half again as long as the period of Neptune's revolution. Its orbit is inclined 17° to the ecliptic, the highest inclination for any principal planet, so that Pluto ventures at times well beyond the borders of the zodiac.

The eccentricity, 0.25, of Pluto's orbit is the greatest for any principal planet. On this account and because of the great size of its orbit, its distance from the sun varies enormously. At aphelion it is 2870×10^6 kilometers beyond Neptune's distance from the sun in its projected orbit, whereas at perihelion it comes 48×10^6 kilometers nearer the sun than the orbit of Neptune; yet in their present orbits the two planets cannot approach each other more closely than 385×10^6 kilometers. At the time of its discovery Pluto was near its ascending node and also near its mean distance from the sun. The distance will diminish until the planet reaches its perihelion in the year 1989. In Fig. 6.8 the part of the orbit south of the ecliptic is indicated by a light line.

Pluto has one moon, Charon, discovered in 1978. The size of the moon is not known, but estimates place its diameter at about 1600 kilometers. This makes Charon larger relative to its planet than any other moon in the solar system. Charon orbits Pluto at a mean distance of 18,000 kilometers.

Questions

1 What is Jupiter's Great Red Spot?
2 What are the two mechanisms that are used to explain the heating of Io?
3 How are surface features on Io erased?
4 Give three observations showing that Saturn's rings are not solid.
5 Calculate the mass of Saturn relative to that of the earth using information in Table 6.2 for the moon and Dione.

6 Saturn's mean density is only 0.7 g/cm³. What two elements account for most of Saturn's mass?
7 How were the rings of Uranus discovered?
8 Explain how studies of Uranus led to the discovery of Neptune.
9 What is unusual about the orbit of Triton?
10 Why do we classify Pluto as a gaseous planet?

Other Reading

Beatty, J. K., "The Far-Out Worlds of Voyager I," *Sky & Telescope* **57**, 423–427 (Part I), 516–520 (Part II) (1979).

Elliot, J. L., E. Dunham, and R. L. Millis, "Discovering the Rings of Uranus," *Sky & Telescope* **53**, 412–416, 430 (1977).

Harrington, R. S. and B. J. Harrington, "The Discovery of Pluto's Moon," *Mercury* **8**, 1–3, 6, 17 (1979).

Hunten, D. M., "The Outer Planets," *Scientific American* **233(3)**, 130–141 (1975).

Pollack, J. B., "The Rings of Saturn," *American Scientist* 30–37 (1978).

"Pioneer 11; Through the Dragon's Mouth," *Sky & Telescope* **49**, 72–78 (1975).

chapter 9

Other Features of the Solar System

Comets and meteoroids revolve around the sun in orbits that are generally more eccentric than those of the planets. Such meteoroids are products of the disintegration of comets; meteor streams are associated with the orbits of comets. Meteorites, which are allied more closely with asteroids, come through the atmosphere to the earth, and very large ones produce meteorite craters. Comets, meteoroids, asteroids, the planets, and the sun originated from a great cloud of gas and dust.

Comets

A conspicuous **comet** visible to the naked eye has a head and a tail. The head consists of a hazy, globular **coma,** sometimes having a brighter **nucleus** near its center. The luminous **tail** is directed generally away from the sun and occasionally extends a considerable distance across the heavens. Many comets, however, are almost featureless telescopic objects. Spectacular comets to the unaided eye, such as Halley's comet, are infrequent.

Figure 9.1 *Comet Ikeya–Seki photographed by H. Giclas in the morning twilight. Note the spiral tail structure. (Lowell Observatory photograph.)*

9.1 Discovery of comets

Comets are likely to be discovered either in photographs of regions of the heavens often taken for other purposes or in visual searches with small telescopes. They are often discovered by interested amateurs, as was the case with the recent comet Ikeya–Seki (Fig. 9.1). The western sky after nightfall or the eastern sky before dawn are most promising for the search. A comet generally appears as a small hazy spot, and its gradual movement among the stars shows decisively that it is not a faint star cluster or nebula. Having found a comet, the observer should report its position, direction of motion, and brightness to the *Central Bureau for Astronomical Telegrams,* Smithsonian Astrophysical Observatory, Cambridge, Massachusetts 02138, which serves as the receiving and distributing station for such astronomical news.

As soon as three positions of the comet have been observed at appropriate intervals, a preliminary orbit is calculated. Then it is usually possible to decide from the records whether it is a new comet or an identified comet returning. A catalog of cometary orbits prepared by B. Marsden of the Smithsonian Astrophysical Observatory in 1979 includes more than 1000 different comets. An average of five or six comets are picked up each year, and generally three or four of them have not been previously recorded.

Comets are provisionally designated by the year of their discovery followed by a small letter in the order in which the discovery is announced. An example is comet Ikeya–Seki 1967n which was the fourteenth comet found in 1967. After the orbit is determined, the permanent designation is the year (not always the year of discovery) followed by a Roman numeral in order of perihelion passage during that year. Thus comet Ikeya–Seki became Comet 1968 I, the first comet to pass its perihelion in 1968. Many comets are known by the name of the discoverer, or discoverers such as comet Kohoutek (1973f) or comet Ikeya–Seki. Occasionally a comet studied carefully by a person receives that person's name. Halley's comet is one such example.

Halley's comet (Fig. 9.2) is named in honor of Edmund Halley, a contemporary of Isaac Newton, who predicted its return. Halley calculated the orbit of a bright comet of 1682 to be a parabola and noted its resemblance to the parabolic orbits of comets in 1531 and 1607, which he had determined from records of their observed places in the sky. Deciding that these were three appearances of the same comet, which must therefore be revolving in an ellipse, he recalculated the orbit and predicted its return to the sun's vicinity "about the year 1758." The comet was sighted on Christmas night of that year and reached perihelion early in 1759. It came around to perihelion again in 1835 and in 1910, its latest appearance.

Figure 9.2   *Halley's comet photographed on 12 May (left) and 15 May (right) 1910. Note the comet's motion as well as the growth of the tail. (Hale Observatories photograph.)*

Twenty-eight returns of this comet are identified from the records as far back as 240 B.C. It was Halley's comet that appeared in the year 1066 at the time of the Norman conquest of England and is depicted on the Bayeux Tapestry. The intervals between returns to perihelion have averaged 77 years, varying a few years because of disturbing effects of the planets.

Halley's comet is not only the first periodic comet to be recognized, but it is also the only conspicuous one of the many periodic comets known today that return to perihelion more often than once in a century. It has a retrograde revolution around the sun in an elongated orbit (Fig. 9.3) at distances from the sun that range from half to more than 35 times the earth's distance. Halley's comet is not visible near its aphelion, which is more remote than Neptune's distance. It will return to the sun's vicinity in 1986. Its perihelion will occur on the side of the sun opposite the earth.

9.2 The orbits of comets

The orbits of comets depart from the regularities we have noted (Section 6.10) in the case of the principal planets. These orbits are generally of large eccentricity and are often much inclined to the ecliptic. With respect to their motions the comets are divided into two groups by a somewhat indefinite dividing line.

1 **Comets having nearly parabolic orbits.** In this more numerous group the orbits are so eccentric that they are not readily distinguished from parabolas in the small portions near the sun where the comets are visible. For example, the eccentricity of the orbit of comet Kohoutek was 0.999. The orbits extend far beyond the region of the planets, and the undetermined periods are all so long that only one appearance of each comet is likely to be found in the records. In this sense the comets are "nonperiodic." About half the revolutions are direct, that

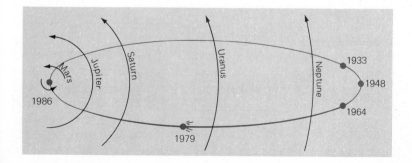

Figure 9.3 *The orbit of Halley's comet. Note how little it moved between 1933 and 1964 and how far it moves between 1964 and 1986.*

is, west to east and the other half are retrograde. Comet Bennett is in this group with an estimated period of 1700 years and an aphelion distance of about 300 AU.

2   **Comets having definitely elliptical orbits.** These "periodic comets" revolve in periods not exceeding a few hundred years. The orbits are allied more closely with the planetary orbits. Although most of their orbits have a large eccentricity, they are frequently more moderately inclined to the ecliptic, and the revolutions are mainly direct. The retrograde revolution of Halley's comet is one of the exceptions, however.

Comet Schwassmann–Wachmann I (1925 II) is unusual in having a nearly circular orbit. This comet revolves around the sun once in about 16 years in an orbit somewhat larger than that of Jupiter. It is also unusual in its occasional surprisingly great and rapid flare-ups, which have been attributed to expansions of a dusty coma; an increase of 100 times in brightness has occurred within less than a day.

Two dozen or more comets revolve around the sun in periods averaging 6 years, or half of Jupiter's period. Their aphelions and one node of each orbit are not far from Jupiter's orbit, so that these comets can come close to the planet itself. They constitute Jupiter's **family of comets** (Fig. 9.4). Their direct revolutions and the low inclinations of their orbits to the ecliptic suggest that Jupiter has assembled the family by capture of comets passing by in originally larger orbits. At successive encounters the planet's attraction has progressively reduced the orbits to their present sizes. The membership is unstable; further approaches of these comets to the planet may occur so that some of them will be removed from the family much as Pioneer 10 has been ejected from the solar system. Three members of Jupiter's family have been especially noteworthy.

Originally discovered in 1786, comet Encke was the first member of Jupiter's family to be recognized in 1819. Its period of revolution, 3.3 years, is the shortest for any known comet. Its aphelion is a whole astronomical unit inside Jupiter's orbit, having gradually been drawn in by this amount. Like other members of the family, comet Encke never becomes more than faintly visible to the unaided eye.

Biela's comet, having a period of 6.5 years, came to an end in a spectacular way. At its return in 1846 it was divided into two separate comets traveling side by side, and at the next return the separation had increased 2.4×10^6 kilometers. The comet was never seen again, but a stream of meteors in its orbit, the Andromedids or Bielids, gave fine showers when they encountered the earth in 1872 and 1885. The Giacobini–Zinner comet, having a similar period, is associated with a meteor stream that in 1933 and 1946 provided one of the most abundant showers of this century.

Figure 9.4 *Orbits of four comets of Jupiter's family. Comet Enke has the smallest orbit of all comets.*

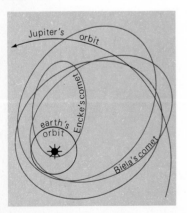

9.3 The nature of a comet

A comet's nucleus consists of a conglomerate of frozen material embedded with dust and small particles. The ices in this "dirty snowball" are mostly methane, ammonia, and water. As the comet approaches the sun, the ices evaporate and the released gases are swept away by the sun's radiations. The dust and debris locked up in the ices are then set free. Because of the various components in the comet's nucleus, three different types of tails are formed:

Type 1 Gaseous tails that glow with blue fluorescence stimulated by the sun's radiations. These are directed almost straight away from the sun by the pressure of the sun's radiations and are generally more conspicuous for comets relatively near the sun.

Type 2 Dusty tails that glow by reflected sunlight. These are curved and are likely to depart more from the direction opposite the sun's position as the comets are farther away.

Type 3 Material fanning out behind the revolving comets (Fig. 9.5) consisting of meteoric particles originally embedded in the ices of the nuclei, which have been released by evaporation of the ices and are being scattered along the comets' orbits as meteor streams.

Most comets have tails of both types 1 and 2, but it is the dust tail that is most prominent because it appears as a long smooth arc. For example, comet Bennett's dust tail was fully 20° long. The gas tail tends to twist and sometimes gives a corkscrew appearance.

The spectra of comets reveal a difference in the chemical constituents of the coma and the tail. The coma consists of unstable neutral molecules of carbon, cyanogen, ammonia, and hydroxyl radicals. By the time these gaseous species reach the tail of the comet, they are converted to the more stable ionized molecules of nitrogen, carbon monoxide, and carbon dioxide. Occasionally a comet will approach very close to the sun and it is then called a **sun grazer.** The coma of sun-grazing comets show the spectral lines of metals such as iron, sodium, nickel, and chromium. These spectral components apparently arise from the solid particles in the comet's nucleus. Ultraviolet spectrographs aboard spacecraft have detected a huge hydrogen cloud around the nucleus of comets which are ten times the size of the visible coma.

The brightness of a comet depends on its distance from the sun and the earth as well as on its mass and composition. A comet's tail attains its greatest length when it is closest to the sun. However, if the earth is far from the comet at that time, it will not appear brilliant to an observer on earth. Other comets will pass close by the earth but are not close enough to the sun to provide an impressive display. Still other comets

Figure 9.5 *Comet Arend–Roland (1956h). The meteoroidal material fanning out below the comet's head gives the appearance of a spike or antitail. (Lowell Observatory photograph.)*

have lost most of their volatile constituents on former visits to the inner solar system and never appear bright regardless of their position relative to the sun and the earth. Although comets visible through a telescope are common objects, a spectacular comet to the unaided eye is rare.

In spite of the impressive size which comets can attain, they are extremely diffuse objects. The space occupied by the tail of a comet is a better vacuum that can be produced in any laboratory on earth. When a comet passes close by a planet or moon, it exerts no detectable influence on that body at all, even though it is itself greatly perturbed. This indicates that the mass of a comet must be very small, less than a billionth the mass of the earth. In addition, when a comet passes in front of the sun, its nucleus is too small to be seen. The size of the nucleus must be smaller than a few kilometers. Comets are therefore very insubstantial objects.

The aphelion of a nonperiodic comet usually lies somewhere between 750 and 2500 times the distance to Pluto. This surprising regularity in the distance from which comets originate led to speculation that a spherical **comet cloud** surrounds the solar system at an average distance of 50,000 AU. At this large distance, comets are influenced by the gravitational attractions of nearby stars. Acting over millions of years, the gravitational perturbations of the stars convert the nearly circular orbits which send the comets into the inner solar system. At its first pass by the sun, a comet is exceptionally brilliant because the frozen gases in its nucleus have never been exposed to the warming rays of the sun. However, with each approach towards the sun, the comet becomes dimmer and evaporates to a smaller size. Once a comet enters the inner solar system, it is also influenced by the gravitational attractions of the planets. Half of the comets entering the inner solar system for the first time are drawn into tighter orbits around the sun. Jupiter is particularly influential because of its large mass, and many comets enter "Jupiter's family" when they happen to pass near this giant planet. Eventually a comet which remains in the solar system boils away and the solid particles in the nucleus become dispersed along its orbital path. A comet is therefore very short lived once it assumes an elliptical orbit within the solar system. In order for comets to be continually supplied to the inner solar system there must be many billions of comets in the comet cloud. Otherwise, all the comets in the solar system would have been expended by now. If every star has as many comets around it as the sun does, comets may well be the most numerous celestial bodies in the universe.

Meteors and Meteor Streams

9.4 Meteors and meteoroids

Interplanetary space contains innumerable small particles, called **meteoroids,** which revolve in elliptical orbits around the sun. When these particles encounter the earth, they produce a bright **trail** across the sky, known as a "shooting star" and are then called **meteors.**

In the vicinity of the earth, the greatest speed which a meteoroid can attain is 42 kilometers/second. Since the earth's orbital speed is 30 kilometers/second, a meteoroid colliding head on with the earth will enter the atmosphere at 72 kilometers/second. On the other hand, if the meteoroid must catch up with the earth it will plunge into the atmosphere at 12 kilometers/second. Meteors can have a velocity ranging anywhere between these two extremes.

The majority of meteoroids are the size of dust particles or grains of sand. Meteoroids that produce a conspicuous meteor trail to the unaided eye may be the size of a pea. The brighter meteors sometimes leave **dust trains** that remain from a few seconds to half an hour. Unusually bright meteors are known as **fireballs.** Fireballs emit so much light that they can cast shadows and may even be visible during the day. A meteoroid which is larger than about 20 centimeters will survive the fall to earth and it is then called a **meteorite.**

If a meteor is observed from two places about 30 kilometers apart, its path through the atmosphere can be determined. Meteors only become visible at about 100 kilometers above the earth's surface. At greater altitudes the air is too thin to heat them to incandescence. The trails of faint meteors usually fade out at a height of 80 kilometers, but the brighter meteors extend down to an altitude of 55 kilometers. Early visual observations of the height of meteors gave the first indication of the extent of the earth's atmosphere. If the two stations are properly equipped (Fig. 9.6), with the aid of photographic techniques the velocity of meteors, both the speed and direction, can be determined. From this information it is then possible to calculate the orbit of the meteoroid which produced the meteor. In every instance it is found that the meteoroid was a member of the solar system.

Radar has contributed substantially to our knowledge of meteors. Radar pulses sent from the ground are bounced off the ionized trails of

Figure 9.6 *A bright meteor passing near the pole star. A shutter chopper interrupts the photograph at known intervals and allows us to calculate the angular motion of the meteor. A similar photograph with another camera would enable us to determine the orbit and velocity. (Harvard College Observatory photograph.)*

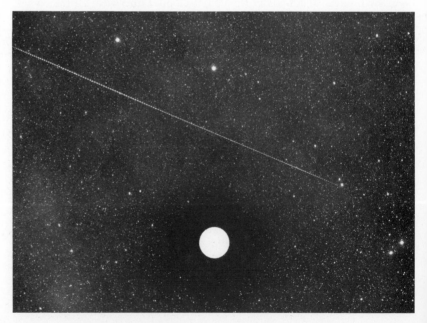

meteors and reflected back to the ground. Since this method does not depend on the visual brightness of a meteor, meteors can be detected during the day as well as at night. Radar is also capable of detecting much smaller meteors than can be observed by visual methods.

The number of meteors that are visible to the unaided eye average about 6 to 8 per hour. However, this is only the number that can be seen from any one location. The total number of meteors that would be visible to the unaided eye over all the earth's surface is of the order of 100 million per day. One must then add to this number the many millions of meteors which are invisible to the eye. The result is an estimated 10,000 metric tons of material that is added to the mass of the earth each day.

Meteors are best observed between midnight and sunrise. This is because we are on the rear side of the earth in the evening and only those meteoroids traveling fast enough to overtake the earth are seen as meteors. However, after midnight we are on the front side of the earth. All the debris in the earth's path is then swept up and meteors are seen in abundance. Because the north pole points in the direction in which the earth orbits in the autumn and winter, these are also the best seasons for viewing meteors in the northern hemisphere.

Occasionally a meteoroid will strike the earth's atmosphere at a very shallow angle and skip like a stone on water back into interplanetary space. This happened in August 1972 when a meteoroid was visible as a meteor in broad daylight (Fig. 9.7) from southwestern Utah to Edmonton, Canada. It was seen and photographed by hundreds of late summer campers.

Figure 9.7 *The grazing passage of a meteoroid in daylight on 10 August 1972.*

9.5 Meteor streams and showers

A **meteor stream,** or more correctly, a **meteoroid stream,** consists of many meteoroids revolving around the sun in about the same orbit. Many meteor streams have been identified as the debris left by the tails of comets. A **meteor shower** occurs where the orbit of the stream crosses the earth's orbit at either one or two points and whenever part of the stream and the earth arrive together at an intersection. The shower occurs around the same date, either annually if the stream is far extended or at longer intervals for a short stream. Only rarely have the showers been spectacular enough to attract the attention of people who are not watching for them.

The trails of the meteors in a shower are directed away from a small area of the sky, the center of which is the **radiant** of the shower. Since the paths are nearly parallel, they spread out over the sky from the radiant (Fig. 9.8), just as the parallel rails of a track seem to diverge from a distant point. Showers of meteors and the streams that produce them are named from the positions of the radiants among the constellations at the heights of the displays. Examples are the Perseids and the delta Aquarids. The more abundant meteor showers are listed in Table 9.1. In each case the table gives the date of maximum display in universal

Figure 9.8   *The radiant of a meteor shower. The drawing shows how parallel moving meteors appear to radiate from the radiant point. The 1966 Leonid shower is shown in the photograph. The presence of a nonshower meteor convincingly demonstrates the shower. Stars in the picture can be identified with the aid of the star maps in Chapter 11. (Photograph by D. McLean.)*

Table 9.1
Meteor Showers and Associated Comets

Shower	Maximum Display (UT date)	RADIANT AT MAXIMUM (EQUINOX OF 1950)		Associated Comet
		R.A.	Decl.	
Quadrantids	3 Jan.	15^{h}20^m	+48°	
Lyrids	21 Apr.	18 0	+33	1861 I
η Aquarids	4 May	22 24	0	Halley (?)
Arietids[a]	8 June	2 56	+23	(=δ Aquarids)
ζ Perseids[a]	9 June	4 8	+23	
β Taurids[a]	30 June	5 44	+19	Encke
δ Aquarids	30 July	22 36	−11	(two streams)
α Capricornids	1 Aug.	20 36	−10	1948n
Perseids	12 Aug.	3 4	+58	1862 III
Draconids	10 Oct.	17 36	+54	Giacobini–Zinner
Orionids	22 Oct.	6 16	+16	Halley (?)
Taurids	1 Nov.	3 28	+17	Encke (two streams)
Andromedids	14 Nov.	1 28	+27	Biela
Leonids	17 Nov.	10 8	+22	Temple
Geminids	14 Dec.	7 32	+32	
Ursids (Ursa Minor)	22 Dec.	13 44	+80	Tuttle

[a] Shower in daytime.

time, the position of the radiant, so that it may be located in the star maps, and the name of the parent comet when known.

In August the Perseids provide the most familiar of the annual showers; the display from this wide stream extends through two or three weeks and the trails are frequently rather bright. The showers of the Orionids and Geminids are also among the most faithful of the annual showers. There are some two hundred minor streams. These are not obvious to anyone less than a skilled observer.

9.6 The zodiacal light and the gegenschine

The triangular glow of the **zodiacal light** can be seen extending up from the west horizon after nightfall in the spring and from the east horizon before dawn in the autumn in our northern latitudes. Broadest and brightest near the horizon it tapers upward, leaning toward the south (Fig. 9.9). The glow is nearly symmetrical with the ecliptic and is most conspicuous when the ecliptic is nearly vertical.

Near the equator, where the ecliptic is nearly perpendicular to the horizon, the zodiacal light can be observed all year round. Here it is said to have been seen extended as a faint, narrow band encircling the

Figure 9.9 *The zodiacal light and comet Ikeya–Seki, 31 October 1965. Note how much brighter the comet's tail is compared to the zodiacal light. Use the star maps in Chapter 11 to identify the region and locate the ecliptic. (Photograph by H. Gordon Solberg, Jr.)*

sky. The light is mainly sunlight scattered by interplanetary dust, which forms a ring around the sun in the plane of the earth's orbit.

The **gegenschein,** or **counterglow** (Fig. 9.10) is a faint, roughly elliptical glow in the sky extending about 20° along the ecliptic and centered nearly opposite the sun's position. Barely visible to the unaided eye in favorable conditions, this glow is observed more clearly when photographed with very wide angle cameras or when recorded with the photoelectric cell. It is believed to be somewhat variable in form and position and can be easily observed from above the earth's atmosphere. Among several proposed interpretations, the gegenschein has been tentatively attributed to sunlight scattered by a dust tail of the earth.

Meteorites and Meteorite Craters

9.7 Falls of meteorites

Near noon one day in November 1492, a number of stones came down in a field near Ensisheim, Alsace. The largest one, weighing 118 kilograms, was placed in a church in that town; a smaller stone is exhibited

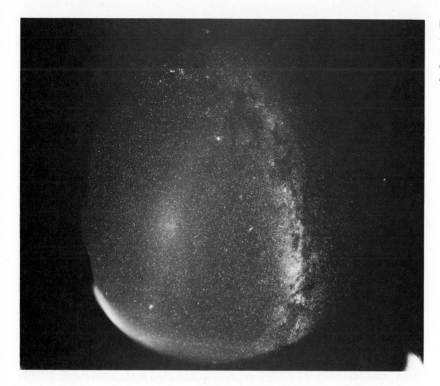

Figure 9.10 *The gegenschein can be seen in the lower left center of this print. The Milky Way arcs across the top. (Photograph by Shohei Suyama.)*

in the Chicago Natural History Museum. This is the oldest observed fall of meteorites on record of which samples are still preserved.

The idea that stones fall from the sky goes back to very early times. There were stones preserved in some of the ancient temples that were doubtless of celestial origin, and these "stones from heaven" were objects of veneration. In later times, however, all reports of stones falling from the sky came to be regarded with suspicion. The stones seemed to choose remote places where there were no reliable observers. It may be, too, that the accounts of terrified spectators of some of the falls were so exaggerated that no one could believe them. Finally, in April 1803, a shower of two or three thousand stones fell at L'Aigle, France, and it was reliably reported. Yet the news spread so slowly that when 136 kilograms of meteorites came down near Weston, Connecticut, in December 1807, the first recorded fall in the United States, many people were reluctant to believe that it was true.

Meteorites are masses of stony or metallic material, or both, which survive their flights through the air and fall to the ground. They arrive either singly or in many pieces. Several thousand individual pieces have come down in one fall, and in such cases they are likely to be distributed over an elliptical area having a major axis in the direction of the flight that is several kilometers long. Although many meteors are products of the disintegration of comets, most meteorites are fragments of shattered asteroids.

Their speeds greatly reduced by air resistance, most meteorites cool before they reach the ground. In their brief flights through the air the heat has not gone far into their cold interiors, and the melted material has been swept away in droplets from their surfaces. They are usually cool enough to be handled comfortably immediately after landing and they do not penetrate far into the ground. Larger meteorites are less impeded by the air; some very massive ones have struck at such high speeds that they have blasted out large craters in the earth's surface (Section 9.9).

Micrometeorites are particles of meteoritic dust, so small that they are not much altered when they fall through the atmosphere. In addition to the ones found on the ground, many micrometeorites have been collected in the opened nose cones of rockets and have been recovered for examination. These samples range from rather compact spheres to irregular dust balls.

9.8 Composition of meteorites

Meteorites are essentially of two kinds, the stones (**aerolites**) and the irons (**siderites**). There are gradations between them (**siderolites**), from stones containing flecks of nickel–iron to sponges of metal with stony fillings. Inside their smooth, varnishlike fusion crusts, stony meteorites are often grayish. Most stony meteorites contain small crystalline spheres, called **chrondrules**, which are about 1 millimeter in diameter. Chondrule-bearing meteorites are known as chondrites, whereas stony meteorites which lack chondrules are called **achondrites**. Although chondrites are unlike any terrestrial rocks, the achondrites are much like the basaltic rocks found on earth.

Iron meteorites are silvery under their blackened exteriors. They are composed mainly of alloys of iron and nickel, which are affected by acid in varying degrees. When they occur in crystalline forms, a characteristic pattern of intersecting bands parallel to the faces of an octahedron may be etched with dilute nitric acid on a polished section. The resulting pattern is referred to as the **Widmanstätten structure** (Fig. 9.11).

The Widmanstätten structure results from the slow cooling of the iron–nickel alloy through a temperature range from 925 to 620 K. In order for iron–nickel crystals of the size observed in some meteorites to have formed, a cooling rate of 1 to 100 K per million years is required. This slow cooling rate indicates that the original size of the parent body must have been between 150 and 400 kilometers in diameter, which is the size of a large asteroid. The range of cooling rates that are observed

Figure 9.11 *Etched section of the Knowles, Oklahoma meteorite. The banded Widmanstätten pattern is characteristic of most iron meteorites. (American Museum of Natural History, New York.)*

implies that the iron meteorites were derived from at least six and possibly as many as 11 different parent bodies. Meteorites are therefore thought to have formed in a number of separate asteroids which subsequently collided and broke up into small fragments, some of which later fell to the earth.

A meteorite that is recovered after being seen or heard while passing through the atmosphere is called a **fall.** By contrast, a meteorite that fell to the earth unobserved and was subsequently discovered by chance is called a **find.** About 84% of all falls are chondrites, 9% are achondrites, 6% are irons, and 1% are stony irons. However, most finds are irons. A stony meteorite can easily be mistaken for an ordinary rock, whereas the high density and metallic luster of iron meteorites make them items of curiosity. Iron meteorites are also very weather resistant, and can remain intact at the earth's surface for tens of thousands of years. Stony meteorites, on the other hand, are quickly weathered away.

Until the lunar samples were returned from the moon, meteorites constituted the only samples of extraterrestrial material that were available for study on earth. They are still the most abundant extraterrestrial samples. Meteorites are particularly valuable to scientists because they provide clues to the nature of the interior of the earth. Thus, iron meteorites are thought to represent the composition of the earth's core and chondrites are used to model the composition of the earth's mantle. The great age of meteorites, dated at 4.6 billion years by radioactive isotopes, also fixes a minimum age for the solar system.

One class of meteorites, the **carbonaceous chondrites,** appear to be the primitive unaltered material from which the solid silicate bodies in the solar system formed. They are very similar in elemental composition to the sun (except for the extremely volatile gases like hydrogen and helium) and contain large amounts of organic compounds. Some carbonaceous condrites also contain up to 16 percent water by weight. Because they are so well preserved, carbonaceous condrites are providing scientists with insights into the conditions which prevailed early in the history of the solar system.

9.9 Meteorite craters

The Barringer meteorite crater, near Canyon Diablo in northeastern Arizona, is a circular depression, 1.3 kilometers across and 174 meters deep (Fig. 9.12). Its rim, which rises 40 meters above the surrounding plain, is composed of debris thrown out of the pit from fine rock dust to blocks of limestone and sandstone weighing up to 6.3×10^3 metric tons apiece.

This crater is a scar left by the fall of a great meteorite probably not less than 50,000 years ago. The meteorite is estimated, at the minimum, to have had a diameter of 61 meters and a weight of 9×10^4 metric tons. It was only slightly retarded by the air because it was so massive, striking the earth with a mighty blow. The intense heat of the collision partly fused the meteorite and the rocks in contact with it; the gases

Figure 9.12 *The Barringer meteorite crater near Winslow, Arizona. The diameter of the crater is 1.3 kilometers. (Photograph by Meteor Crater Society, Winslow, Arizona.)*

expanded explosively, scattering what was left of the meteorite over the surrounding country and blasting out the crater. Within a radius of 10 kilometers around the crater, 27 metric tons of meteoritic iron have been picked up. The largest individual piece, weighing more than 640 kilograms, is exhibited in the museum at the north rim of the crater. Samplings indicate that the total amount of crushed meteoritic material around the crater is 11,000 metric tons.

The Wolf Creek crater in West Australia, having a diameter of 0.85 kilometers at the bottom and a depth of 49 meters, is second in size among the readily recognized meteorite craters. These craters were produced by impacts of meteorites within the last million years. A number of "fossil craters" of meteoritic origin have been detected in aerial photographs of Canada. Examples are the Brent and Holleford craters in Ontario; they are a few kilometers in diameter and their ages are estimated as 500 million years. It is probable that during the earth's history it has been hit by innumerable large meteorites. If it were not for the weathering and erosive action of wind and water, the land areas would look like the moon and Mars. As we learn more about the moon's impact craters we will learn more about our own.

The Origin of the Solar System

Theories of the origin of the sun's planetary system are related to the problem of the origin of the sun itself and of stars in general. According to current opinion, planets evolve as a natural consequence of the formation of stars (Chapter 15) from contracting masses of cosmic gas and dust. Thus, planetary systems are thought to exist around many stars.

Numerous theories have been advanced during the past two centuries concerning the origin of the solar system. As our knowledge increased many of these ideas were seen to conflict with the evidence and were rejected. The theories which survived have undergone repeated modifications to accomodate additional observations. We are now left with a relatively complex picture of the origin of our planetary system.

Any theory of the origin of the solar system must be able to account for the regular motions which the system displays. We have already noted some of these features: the orbits of the planets are nearly circular and lie in the plane of the sun's rotation, the rotation of the sun and planets (with the exception of Uranus and Venus) are in the same direction, and the satellites lie in the plane of rotation of their planets and generally revolve in the same sense as their planets.

Other important features of the solar system are the Titius–Bode relation, the terrestrial- and gaseous-planet division, and the random distribution of the nonperiodic comets. We also know of various tidal locks between the sun, the planets, and their satellites. Any adequate theory must make these observations intelligible.

9.10 The nebular hypothesis

The first scientific account of the formation of the solar system was offered by René Descartes in 1644. He suggested that the planets coalesced from a gaseous nebula enveloping the sun. The subsequent observation of nebula through telescopes led Immanuel Kant to speculate that the solar system had condensed from one of these disk-shaped clouds. The most concise of the early versions of the **nebular hypothesis,** however, was presented by P. L. Laplace in 1796.

Laplace proposed that the solar system began as a slowly rotating cloud of gas and dust. As the cloud contracted under the influence of gravity it rotated faster and flattened out into its equatorial plane. Eventually a critical stage was reached around the periphery of the disk-shaped nebula in which the outwardly directed centrifugal effect was as great as the gravitational attraction toward the center. The forces acting on the dust and gas around the margin of the nebula were therefore balanced, and further contraction in the interior of the cloud left an equatorial ring of material orbiting outside of the nebula. This halo of gas and dust then coalesced into a planet. Each time this process was repeated another planet was formed from the abandoned ring. The remaining gas concentrated at the center of the solar system formed the sun.

This hypothesis explained many of the observed features of the solar system. Because the system inherited the initial rotation of the flattened nebula, the planets all revolve in the same direction in the equatorial plane. A replication of the process around the planets also accounted for the regular motions of their satellites. Finally, the rotation of the sun in the same sense as the revolution of the planets was viewed as a consequence of their simultaneous formation.

The simplicity of the account and the authority of Laplace concerning the mechanics of the solar system gave the nebula hypothesis a leading place among the scientific theories of the nineteenth century. What caused its demise was the theory's inability to account for the distribution of angular momentum between the sun and the planets. The theory predicted that the majority of the angular momentum of the original nebula would be concentrated in the sun. In fact, only 2% is actually found there.

9.11 The protoplanet hypothesis

The **protoplanet hypothesis** was proposed by G. P. Kuiper around 1950. It is essentially a refinement of the nebular hypothesis in which the solar system is regarded as a degenerate double star whose smaller mass was spread out to form the planets rather than condensing into a second star. Instead of rings of stability forming in the equatorial plane, as in Laplace's theory, Kuiper envisioned spheres of stability forming in orbits in the equatorial plane at discrete distances from the sun. These stable regions formed because the cloud became dense enough in these areas for the influence of self-gravitation to isolate them from the rest of the cloud. The protoplanets then formed out of these gravitationally bound spheres by gravitational self-attraction and condensation of the constituent gas and dust.

Once the protoplanets formed, the heavier elements in them gradually settled to form solid cores while the lighter elements remained in the gaseous state as extensive atmospheres. In the meantime the sun, which was originally large and cold at its surface, had become smaller and very hot. The intense radiation which it generated blew the lighter gases that the protoplanets could not retain out of the solar system. The large size of Jupiter and Saturn enabled them to retain most of their original mass while Uranus and Neptune lost a considerable portion of their volatile gases. The inner planets were stripped to little more than their cores.

The problem of insufficient angular momentum in the sun can be resolved within the framework of this theory by the large mass loss which occurred from the protoplanets. If a magnetic field generated in the sun swept the excess gases out of the solar system after the protoplanets formed, angular momentum would have been transferred from the sun to the material which left the solar system. The rotational period of the sun would have thereby slowed to its present rate.

In this picture of the early solar system the sun raised immense tides in the large protoplanetary atmospheres. Since solar tidal bulges only form on the sides of a planet which face toward and away from the sun, all the planets were forced to rotate in the same direction with periods the same as their revolutionary periods around the sun. Later, as they contracted further, the planets slipped away from their tidal lock and began to rotate in shorter periods but generally in the same direction as before. The orbital characteristics of the satellites were influenced in the same fashion by their planets and their regular patterns may be similarly explained. A number of *regular* satellites have never succeeded in slipping away from the tidal lock because they are relatively near their parent planet. Thus, they rotate in the same period in which they revolve around their primary, keeping one face toward their planet, as the moon does around the earth.

A dozen satellites are *irregular*. Their orbits are more inclined to the ecliptic or more eccentric than the others. Half of them have direct and half have retrograde revolutions. Kuiper explains that these and other satellites withdrew from the control of their parent bodies and began to move independently around the sun in orbits like those of their primaries.

The 12 irregular satellites were later recaptured with the aid of the nebula remaining around the planets. Other satellites were diverted into completely different orbits. Hidalgo and the Trojan asteroids may have originally been satellites of Jupiter, and Pluto may have originated in Neptune's family.

At the outer margin of the nebula the density of the gas was insufficient to coalesce into a single protoplanetary body. Instead, small flakes of ice and dust aggregated into billions of small comets revolving in nearly circular orbits around the sun. The gravitational effect of nearby stars eventually caused them to enter the interior regions on highly eccentric orbits in totally random directions.

The protoplanet hypothesis goes a long way towards explaining many of the observed features of the solar system. However, certain aspects of the theory make it less attractive now than it was 25 years ago. Our ideas about the origin and evolution of stars have changed and we now know much more about the interstellar medium.

9.12 The massive solar nebula hypothesis

A new, more comprehensive theory has been developed by A. G. W. Cameron and M. R. Pine that retains some of the general features of the protoplanet hypothesis but removes many of the objections. The theory immediately takes the separation of the terrestrial and gaseous planets as significant and explains this difference through the occurrence of a **protosun** which encompassed all the inner planets.

According to this theory, the solar system began as part of a large, cold interstellar cloud which eventually formed hundreds of stars. The principal components of the cloud were molecular hydrogen, gaseous hydrogen, and helium. In addition, the cloud contained smaller amounts of dust, silicates, and ices of organic and inorganic compounds. We see clouds of this composition in regions where star formation is taking place.

Under the force of its own gravitation the cloud began to contract and fragment into smaller, separate clouds. One of these fragments eventually collapsed into a rotating disk of ice and dust enveloped in a roughly spherical gas cloud. Because the random turbulent energy of the interstellar cloud was partially converted to rotational motion in the nebula, the center of the disk possessed too much angular momentum for a central star to form immediately at the spin axis. Instead, the disk consisted of a central region, or protosun, which extended as far as the asteroid belt, and a protoplanet region extending beyond the present orbit of Pluto.

During the collapse, part of the gravitational potential energy of the cloud was converted into thermal energy which vaporized all the ice and dust in the central region. In time the temperature started to drop inside the protosun and the vaporized material began to condense, the ensuing condensation process progressing from the asteroid belt inward. The specific sequence in which the gases solidified accounts for the compositional differences between the inner planets.

The **accretion** of the planets by collision and aggregation of clumps of solidified material followed different courses with increasing distance from the center of the nebula. In the outer portion of the protoplanet region the nebula never became hot enough to melt the dust and ice so that cold clumps of rock and ice rapidly accreted to form the cores of Uranus and Neptune. Once the core reached a sufficient size the gases in the nebula around them collapsed to form their atmospheres. Closer to the protosun where Jupiter and Saturn formed the temperature was high enough to melt the ices so only rocky cores formed. However, the greater densities found further inside the nebula enabled these giant planets to acquire an atmosphere much faster than Uranus and Neptune, and as a result their final masses were much larger. Inside the protosun, where the terrestrial planets formed, convection stirred up the rocky debris in the disk and dispersed them to large distances out of the midplane of the nebula. Since it was much more difficult for the dispersed material to collide and stick together, the rate of accumulation of the planetary cores was much slower and the gaseous nebula dispersed before appreciable atmospheres could form around these planets.

After the planets solidified, the sun began to shine due to the initiation of nuclear reactions in its center. The start of nuclear burning sent a blast of intense radiation through the cloud, removing all the remaining gases and smaller particles by driving them out of the solar system. Only relatively large objects remained. The planets rapidly collected this debris. The cratered surfaces of Mercury, the moon, and Mars attest to the final accretion stage.

This theoretical picture can be tested in several ways. It predicts that planets closer to the sun should be composed of heat-resistant materials and be relatively free of ice components. That is, planets near the sun should have greater mean densities than those farther away from the sun. Indeed, worked out in detail, the theory predicts that the earth should have a slightly higher mean density than Venus and even predicted in advance that Venus would have a high sulfur content.

The theory predicts a mean density of 2.2 g/cm³ for Jupiter and 2.0 g/cm³ for Saturn. At first this seems to refute the theory but we must remember that the giant planets were able to collect and retain a large amount of gas from the original cloud. Their densities should, therefore, be abnormally low. Indeed, the mean density of Jupiter's Galilean satellites is very near 2.2 g/cm³ so perhaps they are more representative of the early solar material at that distance. The mean density for Saturn's satellites is about 2.0 g/cm³.

Thus we see that the massive solar nebula hypothesis accounts not only for the size but also for the composition of the planets. There can be no doubt that this theory is on the right track. Details may change in the sense of refinements, but the overall picture will probably remain the same.

The recent observations by the Pioneer and Voyager spacecrafts have extended our knowledge of planet formation further. Jupiter, Saturn, and Uranus seem to have replicated the solar system in miniature. Jupiter, as it contracted, produced enough heat to melt Io and Europa and to a certain extent Ganymede. For a few hundred million years, Europa may have been warm enough to be an earth-like object with oceans and a very active weather pattern. If our ideas are correct, planet forming on a large to small scale is quite common.

Questions

1 What are the two groups of comets?
2 What is the composition of the coma of a comet?
3 Why is the application of radar to meteor observations so effective?
4 Distinguish between meteoroids, meteors, and meteorites.
5 What is the gegenschein?
6 What are carbonaceous chondrites and why are they important?

7 List some regularities that a theory for the origin of the solar system must explain.
8 Why do many of the early theories for the origin of the solar system fail?
9 How does Kuiper's theory explain the "irregular" satellites?
10 What are two significant predictions made by the Cameron–Pine theory?

Other Readings

Grossman, L., The Most Primitive Objects in the Solar System," *Scientific American, 232(2),* 30–38 (1975).

Hartmann, W. K., "Cratering in the Solar System," *Scientific American, 236(1),* 84–99 (1977).

Head, J. W., C. A. Wood, and T. A. Mutch, "Geologic Evolution of the Terrestrial Planets," *American Scientist, 65,* 21–29 (1977).

Millman, P. M., "Quadrantrid Meteors from 41,000 Feet," *Sky & Telescope, 51,* 225–228 (1976).

Schramm, D. and R. N. Clayton, "Did a Supernova Trigger the Formation of the Solar System?," *Scientific American, 239(4),* 124–139 (1978).

Wetherill, G. W., "Apollo Objects," *Scientific American, 240(3),* 54–65 (1979).

chapter 10

The Sun with Its Spots

The sun is the dominant member of the solar system. It is a fairly average star. It is also the only star near enough to us for its features to be examined in detail. Our account of the sun is accordingly associated both with the descriptions of the solar system in the preceding chapters and with those of the stars in the following chapters.

Observing the Sun

As the only intrinsic source of light in the solar system and the well-spring of life on earth, the sun is the most important object in the heavens. It sustains the temperature of our planet's surface and supplies the radiant energy needed for photosynthesis in plants. Any minor changes in the sun would therefore profoundly effect our climate and livelihood. With the dawn of new technologies, the activity of the sun also affects us by jamming our radio transmissions. For these reasons the sun deserves our closest scrutiny.

10.1 The sun through the telescope

To the unaided eye the sun appears as the featureless luminary of the daytime sky. In its path along the ecliptic it seems to present the same undistinguished face from day to day. However, when viewed through a telescope, it is seen to be marked by blemishes and protrusions that are perpetually changing. Its turbulent surface possesses an infinite complexity of detail. The sun is the liveliest spectacle in the sky.

Since the sun only traverses a limited portion of the sky in declination, a telescope may be specially designed for viewing the sun. The tube of the telescope can therefore be constructed in a fixed position. This enables objectives with long focal lengths to be used and large images can be formed. For example, the McMath solar telescope (Fig. 10.1) has a focal

Figure 10.1 *The R. R. McMath solar telescope at the Kitt Peak National Observatory. A heliostat at the top directs the sunlight down the sloping tube to the right. The tube is fixed in position parallel to the earth's axis of rotation. (Kitt Peak National Observatory photograph.)*

length of 91.4 meters (the length of a football field). A 1.52-meter concave mirror near the bottom of the tube forms an image of the sun averaging 0.85 meters in diameter in the observing room at ground level. At the top of the telescope is a rotating-plane mirror 2 meters in diameter, called a **heliostat,** which follows the sun in its course across the sky and directs its light down the tube of the telescope.

In order to observe the sun with a regular telescope, special precautions must be taken to protect the sensitive retina of your eye from the blinding intensity of the sun's rays. One must never look directly at the sun through the eyepiece of a telescope. Instead, a plain piece of paper should be held behind the eyepiece to serve as a projection screen on which to view the sun. This has the added advantage that many people can observe at the same time.

Since the sun must be observed during the day, the air is always more turbulent than it would be at night and the resolution that can be obtained is often considerably reduced. Under the very best conditions a resolution of only ½ second of arc is possible. This corresponds to a distance of 300 kilometers on the sun's surface.

Individual wavelengths must be examined in order to discern the various processes occurring in the sun. An instrument which takes pictures of the sun at just one wavelength is a **spectroheliograph.** The highlighting of detail that can be obtained by this device is seen in Fig. 10.11. Filters are also used to selectively view the sun at individual wavelengths.

The distance to the sun can be determined by measuring the solar parallax. The **solar parallax** is defined as the angle subtended by the earth's radius as seen from the sun (Fig. 10.2). By knowing the earth's radius and using the measured solar parallax, simple trigonometry yields a distance to the sun of 146 million kilometers. This distance is known as one astronomical unit. It is the astronomer's yardstick for the solar system.

The sun spans 32 minutes of arc in the sky. This is almost identical to the angular diameter of the moon, which is a fortunate coincidence, for this produces an almost perfect eclipse, without which our knowledge of the sun's outer atmosphere would have been limited. Ordinarily the sun's atmosphere is invisible because the sunlight scattered by the earth's atmosphere outshines the sun's outer envelope. However, during an eclipse the light from the bright surface of the sun does not reach the observer and the atmosphere of the sun can be seen in all its brilliance.

Solar eclipses are rare events and until recently we had a rather spotty knowledge of the sun's atmosphere. The ability to produce artificial eclipses with an occulting disk in a telescope improved this situation, but this technique is fundamentally limited by the residual brightness of the sky and by imperfections in the telescope itself. Now, however, we can view the sun from orbiting observatories, without the obscuring effects of the atmosphere. As a result, the sun's atmosphere can be monitored continu-

Figure 10.2 *Solar parallax is the angle subtended by the earth's radius as seen from the sun.*

8″8

ally. What is more, without the earth's atmosphere to block our view, the sun is observable at all wavelengths of the spectrum. X-ray and ultraviolet studies of the sun from satellites, rockets, and balloons, as well as ground-based observations with radio telescopes, have been indispensible in increasing our understanding of processes occurring on the sun.

10.2 The interior of the sun

Our sight cannot penetrate beneath the blazing surface of the sun. In order to understand what the interior of the sun is like, we must rely on the laws of physics. The sun is a huge gaseous globe held together by gravity and supported by gas pressure from within. These two forces are in equilibrium so that the tendency for the sun to collapse is counterbalanced by the tendency for the gases to disperse into space. The result is a hot gaseous sphere in which the temperature, density, and pressure all steadily increase towards the center.

The sun is 1.39 million kilometers in diameter. This is 109 times the diameter of the earth, so its volume is more than a million times that of the earth. Knowing the earth's mass and applying Newton's law, we can determine the sun's mass, which is found to be 330,000 times as massive as the earth. Then, by dividing the mass of the sun by its volume we arrive at a density for the sun of only 1.4 g/cm^3 or a little less than 1½ times the density of water. This is an extremely low density for such a large body and we must conclude that the sun is composed primarily of hydrogen, the lightest element.

The density at the surface is less than 10^{-6} g/cm^3 (as compared with the atmosphere of the earth, which is 10^{-3} g/cm^3), yet at this point the sun is opaque. Working inward layer by layer we find that the density, temperature, and pressure must increase; slowly at first, but then rapidly as we approach the center. Halfway into the sun the density is about 1 g/cm^3 and the temperature is somewhat more than 3×10^6 K. In the deep interior of the sun the density is 160 g/cm^3 and the temperature is 15×10^6 K. The center of the sun has a density 15 times greater than that of lead. The pressure there is ten-billion times greater than that of our atmosphere at the surface of the earth.

Under the extreme conditions in the center of the sun liquids and solids cannot exist and matter enters a gaseous state in which all molecules are broken apart, atoms are stripped of their electrons, and atomic nuclei collide directly to be transmuted into other elements. It is through these transformations that the energy which sustains the sun is generated. The nuclear reactions only occur in the core of the sun because only here are temperatures and pressures high enough. We will return to discuss the energy sources of the sun and stars in more detail in Chapter 15.

The Photosphere

The energy which is generated in the sun's interior does not reach us directly. Instead, the energy is slowly degraded to longer wavelengths and diffused to the sun's surface. Only then is it released as the radiation which illuminates the solar system. The interface between the opaque interior of the sun and the surrounding translucent gases of the sun's atmosphere is called the photosphere. The **photosphere** is the visible surface of the sun. It is the layer from which most of the light from the sun is emitted.

10.3 The structure of the photosphere

When viewed at high resolution the surface of the sun is seen to possess a mottled texture of bright areas, called **granules,** surrounded by darker rims (Fig. 10.3). Granules are perpetually expanding, breaking apart, and being pushed aside by other granules. They average about 1000 kilometers in diameter (approximately the size of Texas) and last only a few minutes before being consumed by neighboring granules. They are indicative of an active process of energy transfer occurring on the sun.

The temperature of the sun drops with increasing distance from the

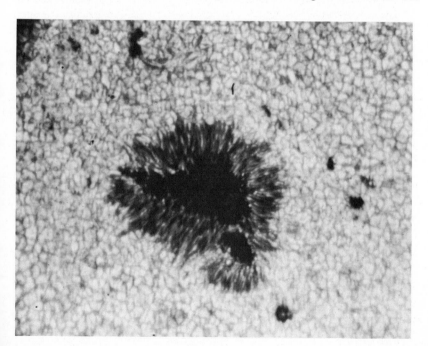

Figure 10.3 *Solar granulation around a sunspot. Each bright area surrounded by a dark rim is a granule. The dark central umbra and lighter penumbra of the sunspot is clearly visible. This picture was taken by Stratoscope I from more than 24 kilometers above the earth. (Princeton University Observatory photograph.)*

center until just below the photosphere the decrease is so rapid that gases which lie at greater depths in this region are less dense than the overlying layers. Consequently, the hot gases from beneath well up in immense swells. Where the expanding gas reaches the surface its temperature is higher than the surroundings and the upwelling gas appears brighter. Once the gas reaches the surface it is able to radiate away its heat so it cools and is pushed aside by more upwelling gas from beneath. The cooled gas then sinks back down into the sun at the dark borders of the granules. This process is perpetuated by the continual flow of heat from the sun's interior. This region of circulation is called the **convective zone** (Fig. 10.4).

When we look at the sun through a telescope the edges of the sun appear to be darker and redder than the center (Fig. 10.5). This is referred to as **limb darkening.** It is caused by the change in depth from which the light emerges as we look closer to the limb. When viewing the center of the sun, the light observed is emitted at the photosphere. However, when we look at the limb we see a higher level of the atmosphere. The limb of the sun appears darker because the gas is cooler at these higher levels. By measuring the degree of limb darkening which occurs, a temperature profile of the sun's atmosphere can be obtained. It is found that the temperature drops from 5900 K at its base to 4400 K and then begins to rise again.

Figure 10.4 *Schematic drawing of the sun's interior and exterior. The various surface and coronal features are indicated. (Adapted from J. M. Pasachoff "The Solar Corona." Copyright © 1973 by Scientific American, Inc. All rights reserved.)*

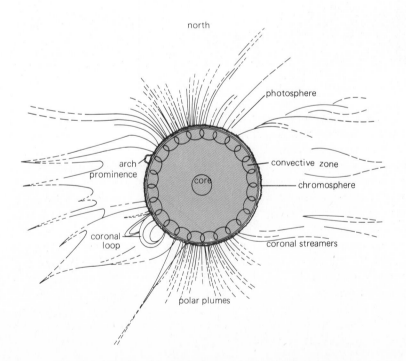

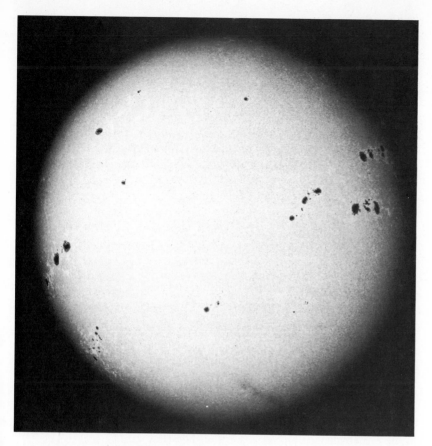

Figure 10.5 *A picture of the sun taken on 7 April 1947. Sunspot groups are seen in two bands north and south of the equator. General limb darkening of the disk and bright faculae near the limb are also noticeable. (Hale Observatories photograph.)*

10.4 The solar spectrum

The spectrum of sunlight consists of the rainbow of colors from violet to red interrupted by thousands of dark lines (Fig. 10.6 and color plate). These lines are not seen in the spectrum formed by a prism alone. They require that a narrow slit be placed before the dispersing element in order to be seen. J. von Fraunhofer first observed the dark lines in 1814. He mapped several hundred of these **Fraunhofer lines** and labeled the stronger ones with letters beginning at the red end of the spectrum. The lines are still known by the letters he assigned them.

The **continuum** of the absorption spectrum is the uniform background of light on which the Fraunhofer lines are imposed. The Fraunhofer lines form at various depths at the photosphere and in the overlying atmosphere by the absorption of photons by various chemical elements. The fainter lines originate at the top of the atmosphere and successively stronger lines arise at lower levels until the strongest lines, the Fraunhofer H and K lines of calcium, are formed at the base of the sun's atmosphere. The

Figure 10.6 *The visible solar spectrum from the ultraviolet to the near infrared. The colors, wavelengths, and Fraunhofer lines (B through K) are labeled above the individual strips which have been cut from the original spectrogram. (Hale Observatories photograph.)*

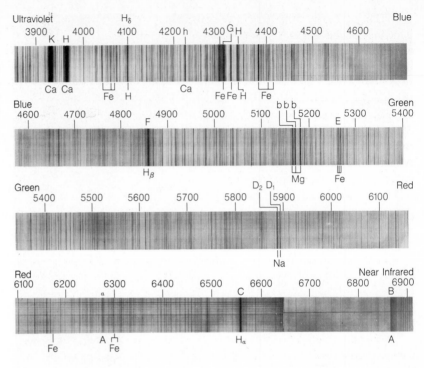

fact that different absorption lines are formed at different levels in the sun enables us to see to different depths in the sun. For example, we can observe the chromosphere, or lower atmosphere, by viewing the sun in the Hα (hydrogen alpha) absorption line (Fig. 10.11).

By comparing the absorption lines in the solar spectrum with laboratory spectra, the chemical elements in the sun can be identified. More than 60 elements have been detected and it is assumed that all 92 natural elements found on earth are present in the sun. Hydrogen is the most abundant element in the sun and helium is second. These two elements predominate in stars and the universe in general. Hydrogen contributes about 71% of the mass of the solar material, helium 26.5%, and the heavier elements the remainder. Exceptions to these proportions occur in the earth and other small bodies because most of the lighter gases have escaped from them.

Some absorption lines in the solar spectrum are not formed in the sun at all. These **telluric bands** arise in the earth's atmosphere. The Fraunhofer *B* band, for example, is formed by terrestrial oxygen molecules. A list of the Fraunhofer lines, their origin and wavelengths is given in Table 10.1. The letters *I* and *J* are not used because they can be mistaken for each other. Fraunhofer himself only observed as far as the *G* line. The remaining letters continue his scheme and were found much later.

Table 10.1
Fraunhofer Lines

Fraun-hofer Letter	Wave-length (Å)	Identi-fication
A	7594	oxygen (telluric)
B	6867	oxygen (telluric)
C	6563	hydrogen
D	5893	sodium (double)
E	5270	iron
F	4861	hydrogen
G	4310	composite blend
H	3968	calcium
K	3934	calcium
L	3820	iron
M	3735	iron
N	3581	iron

The intensity of the sun's radiation is called the **solar constant.** It is defined as the amount of solar energy which falls on 1 cm² per second at the average distance of the earth from the sun. The best measurement of this quantity to date is 1.4×10^6 ergs/cm²/sec. From the geological record we know that the earth has been receiving this much energy from the sun for the last 3 billion years because rocks of that age were formed in water and water can only exist over a very limited temperature range. Since the sun is a fairly average star, it should continue to supply the earth with this quantity of radiation for another 5 billion years to come. Nonetheless, the sun does show some long-term variations. This conclusion comes from the study of sunspots.

Sunspots

Sunspots are dark patches on the surface of the sun. They can occasionally be seen with the unaided eye at sunset or through mist and clouds. It is by this means that sunspots were known to the ancient Greeks and Chinese. However, it was not until the invention of the telescope that their patterns could be extensively documented.

Sunspots usually consist of two distinct parts: the **umbra**—or inner darker part—which is often divided, and the lighter **penumbra** around it (Fig. 10.3). Sunspots also occur in groups; where a single sunspot is seen, it is likely to be the survivor of a group or the precursor of a new group. A normal group consists of two principal spots with smaller spots in between (Fig. 10.5). The preceding spot lies in the direction of the sun's rotation and frequently becomes the larger of the two. The following spot is the largest of the spots in the rear. It usually subdivides and vanishes along with the smaller spots, until only the preceding spot is left to shrink and disappear.

Sunspots appear dark because their temperature is much lower than the surrounding photosphere. In contrast to the average photospheric temperature of 6000 K, sunspots have a temperature of 4000 K. They are still radiating light but it can not be seen against such a bright background. If a sunspot could be placed out in space it would appear hundreds of times brighter than the full moon.

10.5 The sun's rotation

Sunspots last from a few days to a few weeks. The longer-lived sunspots can be followed in their course across the face of the sun and occasionally are seen to reemerge after traveling around the far side of the sun. Since the sun's equator is inclined 7° to the plane of the earth's orbit, the paths of the sunspots across the disk are generally curved. The curvature is

Figure 10.7 *Sunspots show the sun's rotation. This large group in 1947 lasted more than three months. (Hale Observatories photograph.)*

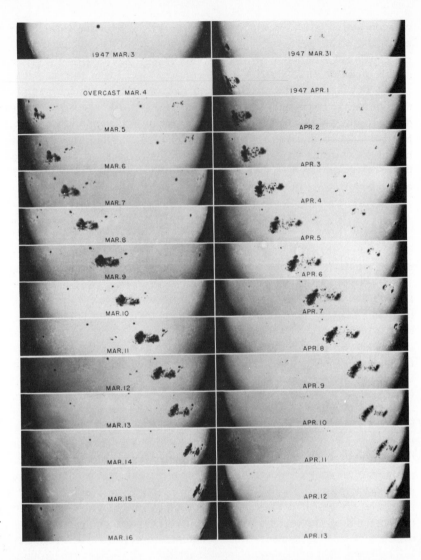

Figure 10.8 *The sunspot cycle from 1749 to 1978. The 11-year cycle is obvious. A longer-term cycle of 88 years is also evident.*

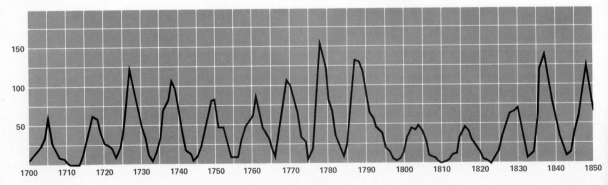

greatest in early March, when the sun's south pole is in the earth's direction and again in early September, when the sun's north pole is towards the earth. By observing the motion of sunspots, the sun's rotational period can be determined (Fig. 10.7).

In the period between the appearance of a sunspot on one limb of the sun and its disappearance on the other limb the earth moves a certain distance around the sun. Since the earth orbits in the same direction that the sun rotates, the apparent rotational period of the sun, or the synodic period, is longer than the true period, or the sidereal period. Consequently, in order to determine the true period of the sun's rotation, the movement of the earth must be taken into account.

The sun is found to rotate faster at the equator than at higher latitudes. At the equator the rotational period is 25 days, whereas at a latitude of 35° the period is about 27 days. Spots seldom occur farther from the equator than this and the rotation rate must be obtained from the Doppler effect at the limb of the sun. At a latitude of 75° the period is 33 days and near the pole the period is 40 days. Both methods for determining the rotation period yield essentially identical results and it must be concluded that the sun is undergoing **differential rotation** just like Jupiter.

In addition to the differential rotation occurring on the sun at all times, the general period of rotation of the sun may change erratically from time to time. Fragmentary records of sunspots prior to 1640 suggest that the rotation period of the sun was about the same as now. However, in about 1640 the sun's period became significantly shorter for a few years.

10.6 The sunspot cycle

Observations of sunspots over the years indicate that sunspots vary in number in a regular cycle called the **sunspot cycle.** In some years the sun's disk is seldom free from spots whereas in other years it may remain unspotted for several days in succession. The intervals between the times of maximum spottedness have averaged 11.1 years since accurate records were first kept, although the average has been closer to 10 years for the

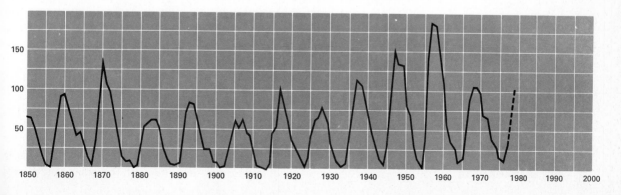

Figure 10.9 *Maunder's butterfly diagram showing the shift in latitude of sunspot groups toward the equator during each sunspot cycle. Each sunspot group is represented by a short dash in the year it was observed.*

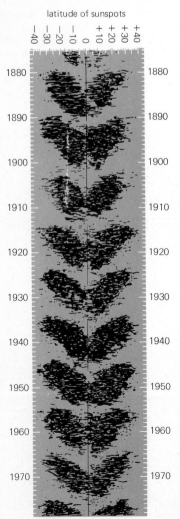

latitude of sunspots

past half century. The number of spots at the different maxima are not the same. Records maintained since 1700 indicate that the number of spots at the maximum also varies in a cyclical pattern. This **long-term cycle** has a period of approximately 88 years (Fig. 10.8).

The sunspot records are spotty from 1610 to 1648, but through that period they confirm the cyclic nature of sunspot occurrences. However, during the interval from 1650 to 1700, few and often no sunspots were observed on the sun despite continuous observation by many reliable observers. This interval is now referred to as the **Maunder minimum.** It was preceded by a period of rather rapid rotation of the sun and a noticeable cooling of the earth followed, presumably due to a change in the solar constant. This unusual behavior indicates that the solar cycle is not as predictable as recent observations would imply.

Sunspots occur mainly between latitudes 5° and 30° north and south of the equator. Very few sunspots have been reported beyond 45°. At any one time the sunspots are likely to appear in two rather narrow zones equidistant from the equator (Fig. 10.5). These zones progressively shift toward the equator in the course of the sunspot cycle. About a year before sunspot minimum, a few surviving members of the fading cycle are seen around latitude 5° while some spots of the next cycle are already visible at the higher latitudes. This pattern of sunspot migration is vividly displayed in Maunder's "butterfly" diagram (Fig. 10.9).

10.7 Magnetic fields

The sun's general magnetic field is beautifully revealed by the polar patterns in the corona during a total eclipse (Fig. 10.19), especially when the eclipse occurs during the sunspot minimum. In the picture the alignment of the electrons and other charged particles along the magnetic field lines is quite clear.

Magnetic field strengths are usually given in units called a **gauss.** The general field strength at the sun's surface is about 2 gauss. By comparison, the earth's magnetic field strength near its surface is about 1 gauss. However, in sunspot regions the local field strength may attain values of 1000 gauss or more. The high magnetic field strengths in sunspots are responsible for their lower surface temperature. In normal areas of the photosphere the temperature of the surface is maintained by the upwelling of hot gas in the granules. However, in sunspots the convection zone is suppressed by the magnetic field. This is because the charged particles in a gas are deflected when they move through magnetic field lines. The gas in the sun is highly ionized so that it consists almost entirely of charged particles. The convection currents in the sunspots are therefore impeded and the temperature of the surface is depressed.

When the image of a sunspot is focused on the slit of a spectroscope and passed through an analyzer, the dark lines of the solar spectrum appear split into two or more parts (Fig. 10.10). This is due to the **Zeeman effect,** named after the physicist who discovered it in the laboratory. He observed that spectral lines formed in a gas under the influence of a magnetic field are split by an amount proportional to the strength of the magnetic field. The effect on the sunspot spectrum enables the magnetic field strength and also its polarity—whether the positive or negative pole of the magnet is towards us—to be determined.

Most sunspot groups are **bipolar,** that is, their two principal spots have opposite polarities. Thus, most sunspot groups have a magnetic field pattern reminiscent of the magnetic field around a bar magnet. This is clearly revealed in Hα photographs of the chromosphere immediately above a sunspot group (Fig. 10.11). The bipolar orientation of sunspots is different between the northern and southern hemispheres of the sun. If the leading spots in the sun's northern hemisphere have their positive poles towards us, and the following spots their negative poles, then in the southern hemisphere the preceding spots present their negative magnetic poles and the following spots their positive poles (Fig. 10.12). A remarkable feature of sunspot magnetism is the complete reversal of this pattern with the appearance of the groups of a new cycle. This polarity reversal means that the true period of the sunspot cycle is 22 years, since it takes two 11-year cycles to reestablish the initial sunspot polarity relationships.

Figure 10.10   *The Zeeman effect shows the increased strength of the sun's magnetic field in sunspots. The line through the sunspot on the left shows the position of the spectrograph slit. The resulting spectral lines on the right are split over the center of the sunspot. The vertical bands are due to variations in the brightnesss of the sun's surface across the sunspot. (Big Bear Solar Observatory photograph.)*

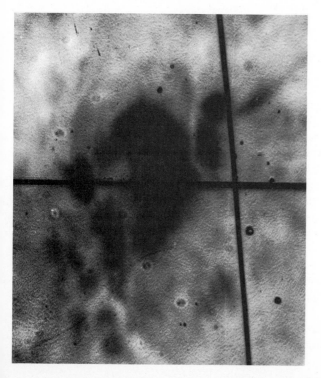
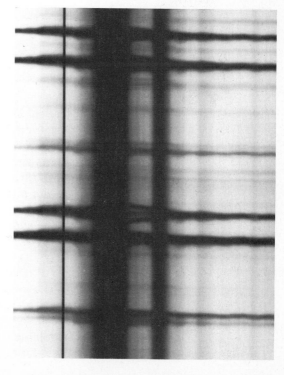

Figure 10.11 *A section of the chromosphere of the sun photographed in red hydrogen light (Hα) on 3, 5, 7, and 9 August 1915. The bright and dark stripes, called flocculi, trace out the pattern of the magnetic field around a sunspot group. The bright area over the sunspot group (not visible) is called a plage and the dark ribbons in the plage are filaments. (Hale Observatories photograph.)*

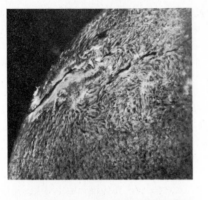
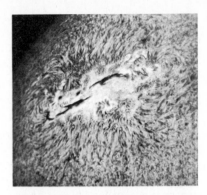

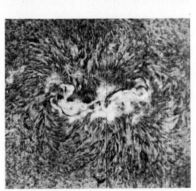

Figure 10.12 *Reversal of polarities of sunspots with the beginning of a new solar cycle. The circles represent the preceding and following sunspots of groups in the northern and southern hemispheres of the sun.*

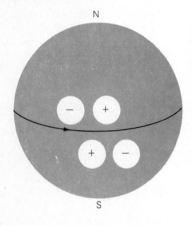

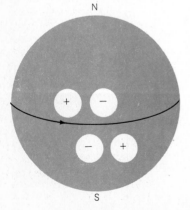

The Solar Atmosphere

The solar atmosphere extends from the photosphere for an indefinite distance out into space, becoming more and more tenuous with increasing distance from the sun. The atmosphere of the sun is not static. Some of its storms are so violent that they have consequences for man 146 million kilometers away. The solar "weather" is therefore watched quite closely. Although the solar atmosphere is ordinarily invisible to the unaided eye, it provides a spectacular display at the time of solar eclipse. The sun's atmosphere is divided into two parts on the basis of their distinctive appearances: the chromosphere and the corona.

10.8 The chromosphere

The **chromosphere** appears as a red fringe around the dark disk of the moon when the moon completely covers the photosphere at the time of a total solar eclipse. On these occasions the bright-line spectrum of the chromosphere can be observed. It is known as the **flash spectrum** because it flashes into view in the spectroscope for a few seconds near the beginning and end of totality. The red color of the chromosphere is due to strong emission by the Fraunhofer C line of hydrogen (Hα).

The flash spectrum is an emission spectrum, in contrast to the absorption spectrum of the uneclipsed sun; the instant that the photosphere is occulted the dark lines in the spectrum become bright lines at these same wavelengths. This is because the gases in the chromosphere selectively absorb light at the wavelengths of the absorption spectrum and then reradiate this light in all directions. When viewing the sun directly the light reemitted in the direction of the observer is only a fraction of the light heading in that direction initially, and this redirection diminishes the intensity of the reradiated light in comparison to the bright background of the photosphere (Fig. 10.13). However, at the time of an eclipse the only background light we see is the sunlight emitted to us obliquely by the chromosphere. The light reemitted by the chromosphere therefore appears bright. The atomic processes responsible for the absorption and emission lines are the same. The difference between the spectra is due to the different intensity of the background against which they are viewed.

As it is photographed with a slitless spectrograph, the flash spectrum consists of a series of crescents of different lengths (Fig. 10.14). Each crescent is a picture of the chromosphere at a particular wavelength. The longer a crescent is, the higher above the sun's surface the atoms or ions responsible for the light being emitted extends. Some emission lines in the flash spectrum originate so far above the photosphere that they completely encircle the occulting disk of the moon and form circles on the flash spectrum. By studying the extent of the crescents in the flash spectrum, the variation in the ionization state of the chromosphere with height can be ascertained.

Although most of the lines in the emission spectrum are the same as those in the solar absorption spectrum, some new lines do appear, particularly at greater heights. For example, neutral and ionized helium and some highly ionized metals are seen in the emission spectrum of the photosphere. This indicates that temperatures in the upper chromosphere are higher than at the photosphere. Earlier we said that the atmospheric temperature drops with increasing height, so this trend must be reversed

Figure 10.13 *The origin of absorption and emission lines in the solar spectrum. The intensity of radiation at a particular wavelength in the photosphere is represented in (a). If the light is absorbed and reemitted by gas in the overlying layers it is reradiated in all directions (b). When viewing the sun directly this results in a reduction of the intensity of the light at that wavelength (c). However, if the sun is observed during an eclipse, the reemitted light is seen against a dark background and the spectrum has a bright line at that wavelength (b).*

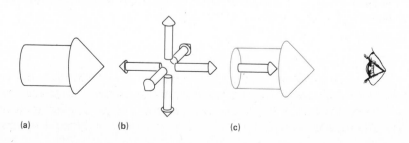

(a)   (b)   (c)

Figure 10.14 *The solar "flash" spectrum: top, just before the chromosphere is covered (second contact); center, with Bailey's beads showing; bottom, just after the chromosphere is uncovered (third contact). The pair of very bright lines on the left are the H and K lines of calcium and the bright line on the right is the F line of hydrogen (Hβ). (Hale Observatories photograph.)*

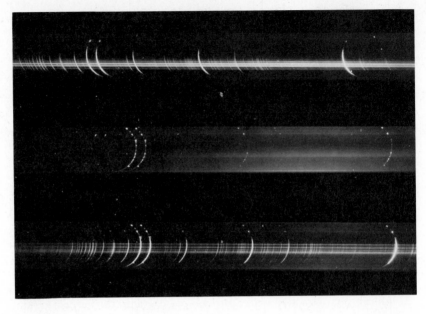

in the chromosphere. From this temperature minimum the chromosphere then becomes progressively hotter until the temperature reaches millions of degrees in the outer atmosphere. We do not yet know what causes the increase in temperature above the photosphere, but it is thought that shock waves generated in the sun travel up into the sun's atmosphere and heat the gases there. Shock waves can carry energy, just as a sonic boom from a jet carries energy. If the energy in the shock waves is dissipated high in the sun's atmosphere, this would account for the high temperatures that are observed.

When viewed under high resolution at the limb of the sun the chromosphere looks like blades of grass or a burning prairie. Each individual blade is a **spicule,** a cylindrical protuberance approximately 6000 kilometers in height. Spicules are short-lived features with a lifetime of only a few minutes, but they reappear at the same rate that they disappear so that their total number remains relatively constant.

When viewing in monochromatic light just off the center of the Hα line, the spicules on the disk of the sun appear to trace out polygonal shapes about 30,000 kilometers in diameter (Fig. 10.15) These patterns are referred to as the **chromospheric network.** They form at the boundaries of **supergranules,** large convection cells in the lower atmosphere of the sun. The supergranules are much larger than the granules of the photosphere, encompassing many hundred of them within their boundaries. Nonetheless, it appears that the same convection process is operating here as in the granules.

The Caracol (foreground), a Mayan Observatory at Chichen Itza, Mexico.

Venus as it appears in ultraviolet light (National Aeronautics and Space Administration photograph).

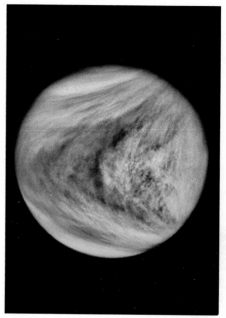

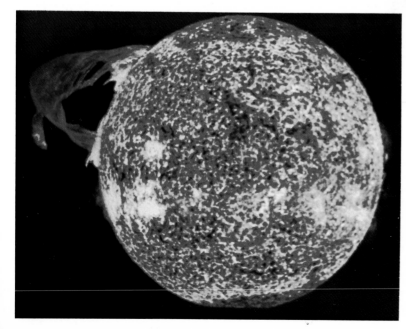

A great twisting prominence on the sun photographed from Skylab. (National Aeronautics and Space Administration photograph).

Exploring the moon (National Aeronautics and Space Administration photograph).

Mars, its polar cap, clouds, and the volcano Olympus Mons (National Aeronautics and Space Administration photograph).

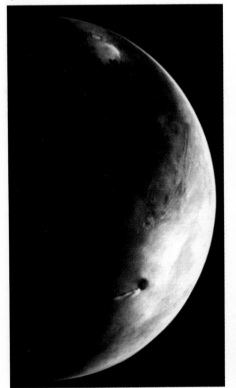

The surface of Mars after a light frost has formed (National Aeronautics and Space Administration photograph).

*Jupiter's great redspot and intri-
cate cloud circulation patterns
(National Aeronautics and Space
Administration photograph).*

*The Galilean satellites set to the same scale. Io (top
left), Europa (top right), Ganymede (bottom left), and
Callisto (bottom right). (National Aeronautics and Space
Administration photographs).*

*Saturn as seen by Pioneer 11. (National Aeronautics
and Space Administration photograph).*

The Trifid Nebula, a site of star formation (Hale Observatories photograph).

The Pleiades Cluster. The nebula in this relatively young cluster shines by reflected light. (Hale Observatories photograph).

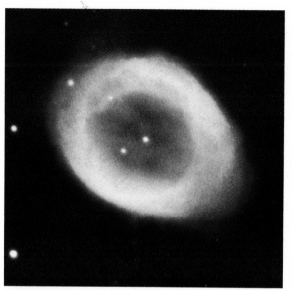

The Ring Nebula in Lyra. Stars like the sun shed mass in this way (Hale Observatories photograph).

The Rosette Nebula. A cluster of young stars has formed and radiation pressure along with the stellar winds are pushing the nebula away (Hale Observatories photograph).

The Crab Nebula in Taurus. This is the remains of the supernova event in AD 1054 (Hale Observatories photograph).

The Veil Nebula in Cygnus. This is a portion of the large wreath-like remains of an ancient super-nova event (Hale Observatories photograph).

A North American Indian pectrograph in Navaho Canyon depicting the supernova of AD 1054 (Museum of Northern Arizona photograph, courtesy of W. C. Miller).

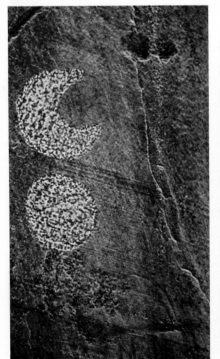

The Large Magellanic Cloud. The great H-II region called S Doradus is seen in the middle left (Hale Observatories photograph).

The Great Andromeda Galaxy. This galaxy is somewhat larger than the Milky Way Galaxy. Two of its elliptical companion galaxies are present. (Hale Observatories photograph).

Centarus A. The galaxy shows signs of a violent internal event and is a very strong radio source. (Cerro Tololo Interamerican Observatory photograph).

The three element radio interferometer at Greenbank, West Virginia. (National Radio Astronomy Observatory photograph).

The Kuiper Observatory. The observatory is a C-141 jet aircraft carrying a telescope designed for use in the infrared (National Aeronautics and Space Administration photograph).

The solar neutrino detector located in a gold mine in Montana. (Photograph by Raymond Davis, Jr., Brookhaven National Laboratory).

Solar spectrum in the visible region. (Photograph courtesy of Goto Instruments, Inc.).

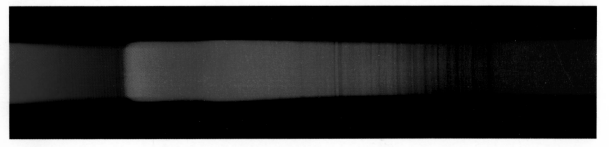

Other prominent features of the chromosphere seen in the monochromatic light of absorbtion lines are flocculi and plages. **Flocculi** are the light and dark markings seen in monochromatic Hα and calcium *K* absorption lines. Their patterns trace out the configurations of magnetic field lines in the chromosphere around sunspots (Fig. 10.11). **Plages** are bright areas above sunspots in the monochromatic light of calcium *K* and hydrogen alpha absorption lines. They are caused by the heating of chromospheric gases in areas of intensified magnetic fields. Because the magnetic field needed to produce a plage is not as strong as the magnetic fields needed to produce a sunspot, plages frequently develop before sunspots appear and linger after the sunspot has disappeared. They therefore serve to predict future sunspots and provide a sensitive index of solar activity. Plages are also prominent in extreme-ultraviolet (Fig. 10.16) and x-ray (Fig. 10.17) photographs of the sun. When seen in white light on the limb of the sun plages are called **faculae** (Fig. 10.5).

The chromosphere extends some 7000 kilometers above the photosphere where it becomes the outer atmosphere or solar corona.

10.9 The corona

The **corona** is the outer envelope of the sun and the chief contributor to the splendor of the total solar eclipse. Its brightness is one-half that of the full moon. The visible corona is characterized by delicate streamers that vary with the sunspot cycle. Near sunspot maximum the form is

Figure 10.15 *The chromospheric network as seen in monochromatic light 1 Angstrom above the center of the H absorption line. The network is revealed by the dark spicules which trace out polygonal patterns on the sun's disk. (Big Bear Solar Observatory photograph.)*

Figure 10.16   *The sun as seen in the extreme-ultraviolet radiation of ionized helium (304 Å) from Skylab. The bright areas on the disk of the sun are active regions above sunspots which have been heated to incandescence by strong magnetic fields. The loop on the upper left is an eruptive prominence. (National Aeronautics and Space Administration photograph.)*

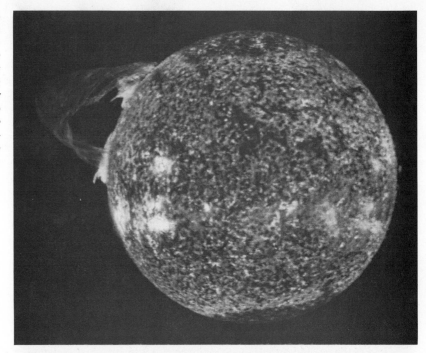

Figure 10.17   *An x-ray picture of the sun. The bright areas are highly excited regions above sunspots. The disk of the sun is dark because temperatures are not high enough in the photosphere to emit x rays. (American Science and Engineering, Inc. photograph.)*

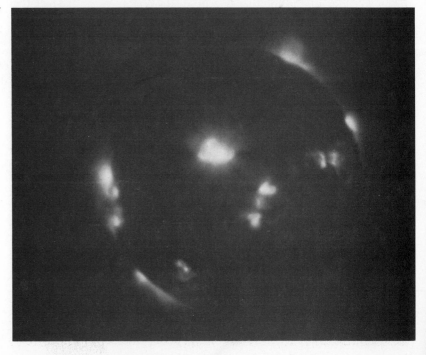

Figure 10.18 *A picture of the total solar eclipse on 20 May 1947. The round appearance of the corona is typical of sunspot maximum.*

roughly circular (Fig. 10.18). Near sunspot minimum the corona is flattened in the polar regions (Fig. 10.19) where short, curved streamers are reminiscent of the lines of force around a bar magnet. Long streamers may reach out more that 1 million kilometers from the equatorial regions.

The spectrum of the corona has three components referred to as the *L, K,* and *F* corona. The **L corona** consists of a bright-line spectrum

Figure 10.19 *A picture of the total solar eclipse on 30 June 1973. The long equatorial plumes and streamers are typical of sunspot minimum. Note the suggestion of the sun's magnetic field in the coronal lines at the sun's north and south poles. (High Altitude Observatory photograph.)*

formed from highly ionized atoms of iron, nickel, calcium, and argon which have been stripped of 9 to 15 electrons. Such high-ionization states could only occur if the corona had a temperature of 1 million degrees Kelvin or more. The *L* corona is stronger near the sun's equator than near the poles and its intensity drops off quickly away from the sun.

The *K* **corona** is photospheric light reflected by electrons in the corona. Since these high-temperature electrons are moving at high speeds, the light they reflect is substantially Doppler shifted and the Fraunhofer lines of the solar spectrum are smeared out to produce a continuum (*K* coming from the German word Kontinuierlich meaning continuous). The *K* corona is the dominant spectral component closest to the sun.

The *F* **corona** is not formed in the corona at all. It is sunlight reflected off the dust orbiting the sun. Since the dust is moving at relatively slow speeds, the sunlight it reflects is not appreciably Doppler shifted and the Fraunhofer lines of the solar spectrum are preserved (*F* standing for Fraunhofer). The *F* corona has therefore an absorbtion spectrum. It slowly decreases in intensity away from the sun and dominates the spectrum further out. At its outer extension the *F* corona becomes the zodiacal light (Section 9.6).

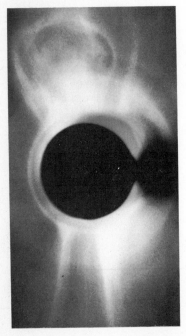

Figure 10.20 *An immense coronal loop and hole observed from Skylab. Several such loops were observed during the one-month mission. (High Altitude Observatory photograph.)*

Features of the inner corona around the uneclipsed sun were first photographed in 1930 by Lyot with an instrument of his own design, the coronagraph. This type of telescope blocks out the light coming directly from the photosphere and minimizes the light scattered in the telescope so that the inner corona can be observed continuously without need of a solar eclipse. However, ground-based coronagraphs are limited in their utility by the obscuring effects of light scattered by the earth's atmosphere. In order to observe the outer corona, a telescope must be put into orbit above the earth's atmosphere.

The corona was studied for 1½ years beginning in 1973 from the manned observatory aboard Skylab. In that period several large coronal loops were formed and two coronal holes appeared. **Coronal holes** are areas of reduced temperature and increased density which appear darker than the rest of the corona (Fig. 10.16). **Coronal loops** are bright areas in the outer corona (Fig. 10.20). The presence of so many loops and holes in the corona came as a surprise because they were only seen during three solar eclipses since the 18th century. By coincidence, all the other eclipses occurred when coronal holes and loops were absent from the sun. The erroneous conclusion was therefore reached that coronal loops and holes are rare phenomena.

The temperature of the corona varies with the number of sunspots but on the average is about 2 million degrees Kelvin. Since the temperature is so high the gases in the sun's outer atmosphere are slowly boiling away. The resulting steady stream of particles emanating from the sun is called the **solar wind.** It consists of electrons and ions of all the chemical elements in the sun but most of the particles are the electrons and protons. The solar wind sweeps past the earth at a rate of 400 kilometers/second and causes the auroral displays in our atmosphere. It is also responsible for directing the tails of comets away from the sun and slowly wearing down the surface features of Mercury and the moon. Since the solar wind is solar material blown off into space, we can determine the composition of the sun by directly sampling the solar wind with space probes. Pioneer 10 detected the solar wind well beyond the orbit of Saturn. At the time of maximum sunspot activity the solar wind may extend beyond the orbit of Pluto. The solar wind therefore pervades all of interplanetary space and is a major component of the **interplanary medium.**

The reception of radio noise from the sun was first detected in 1942 by radar defense stations in Great Britain, where the source of the disturbance was traced to a large spot group near the central meridian of the sun. This accidental discovery provided a new means of studying the sun, which is now being utilized by radio telescopes. The radio emissions are produced by the interaction of fast-moving charged particles from the photosphere with chromospheric and coronal gases. It is strongest at wavelengths of 1 to 10 meters, which originate at different levels in the corona, and it is weakest at centimeter wavelengths, which originate in the chromosphere. The radio emission has its least strength and is fairly constant from the **quiet sun.** From the **active sun,** when sunspots are most numerous, bursts of irregular and much greater strength are superimposed. The radio bursts are associated with solar prominences and flares.

10.10 Filaments, prominences, and flares

Many features in the solar atmosphere are frequently associated with sunspots. **Filaments** are dark ribbons in the disk of the sun around sunspots which can be seen in Hα light (Fig. 10.11). They are concentrations of chromospheric material suspended in the corona by magnetic fields. Filaments gradually grow longer in the course of their development, sometimes attaining lengths of 100,000 kilometers. When a filament rotates to the edge of the disk it looks bright red in contrast to the faint background of white coronal light. It is then called a **prominence.**

Prominences are generally of two types. **Quiescent prominences** last for weeks or months and tend to lie between the positive and negative poles of a sunspot group. Sequences of pictures show that the concentrated gas in the prominences condenses out of the corona and descends into the chromosphere. Quiescent prominences look like curtains or trees (Fig. 10.21). **Active prominences** last only a few hours. They rain down upon the chromosphere in streamers which often assume beautiful shapes. In **loop prominences,** for example, material descends along both sides of a series of loops which follow the magnetic field lines of the underlying

Figure 10.21   *A quiescent prominence. Material condenses out of the corona and descends into the chromosphere along tree-shaped streamers and curtains. (Sacramento Peak Observatory Association of Universities for Research in Astronomy, Inc.)*

Figure 10.22   *A loop prominence. The hot gas flows down from the corona along magnetic lines of force emanating from a sunspot group. (Big Bear Solar Observatory photograph.)*

Figure   10.23   *The   eruptive prominence of 4 June 1946. (High Altitude   Observatory   photograph.)*

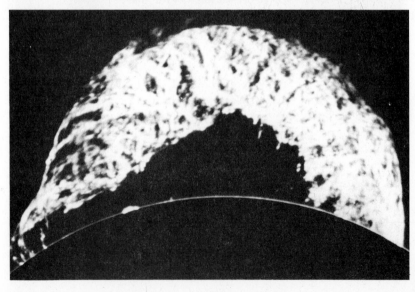

sunspot group (Fig. 10.22). **Eruptive prominences** rise from active material above the chromosphere and attain high speeds and great altitudes before they vanish (Fig. 10.16). A prominence on 4 June 1946 rose to a distance of more than 1½ million kilometers above the sun's surface (Fig. 10.23).

Solar flares are violent storms on the sun which emit bursts of particles and radiation of tremendous energy. They are caused by eruptions of glowing gas brought up from below the photosphere in great upheavals. Solar flares are most visible in Hα light (Fig. 10.24). Only rarely is a flare large enough to be seen in visible light. Large amounts of x rays and radio waves are emitted by flares. Flares also emit cosmic rays—protons, electrons, and atomic nuclei which are extremely energetic owing to their great speed. Most of the electrons in solar cosmic rays emitted

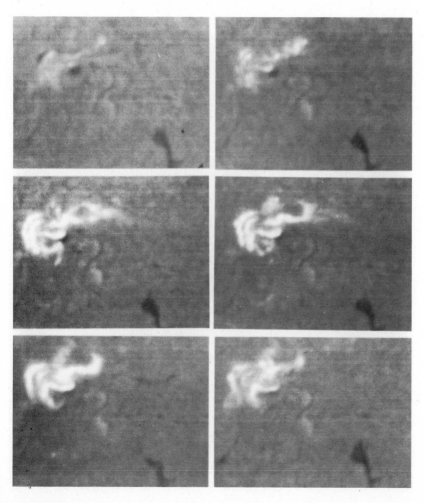

Figure 10.24   *A sequence of pictures of a solar flare taken in Hα light. The flare is moving up from the center of the sun's disk directly toward the earth. The whole event took less than 2 hours. (Hale Observatories photograph.)*

by flares do not reach the earth because they lose their energy in stimulating radio emissions in the corona. The cosmic rays which do reach the earth are mostly protons (hydrogen nuclei) and alpha particles (helium nuclei). They travel to the earth in a blast wave which penetrates the solar wind at a velocity of 1500 kilometers/second. Flares vary in duration from a few minutes to a few hours. A large flare may require up to 15 minutes to attain its greatest brightness. Although really large flares are unusual, small flares occur over one hundred times a day on the sun.

Flares erupt in active sunspot groups with irregular polarities. Only rarely does a flare arise in a sunspot group with distinct positive and negative poles. As a rule flares develop in complex regions where the magnetic field is tangled and confused and changing rapidly. For example, areas where two sunspot groups emerge near each other and then become intertwined are likely candidates for flare activity. By closely watching the sun, flare warnings can be given.

Perhaps the most significant solar study conducted from Skylab was the study of solar **bright points.** The bright points had been seen in pictures from rockets and spacecraft, but seemed to be random, energetic events. They are relatively brief transients where a very small region suddenly emits copious amounts of x ray and ultraviolet radiation. They do not endure as long as a sunspot nor are they nearly as large. However, they follow the sunspot cycle in the sense that when the sunspots are at a minimum there are many bright points. At sunspot maximum there are few bright points. Unlike sunspots, bright points appear anywhere on the sun, even at its poles.

10.11 Solar–terrestrial relations

The earth and the sun interact in several ways. These interactions are both **short term** and **long term.** There is a constant interaction between the solar wind and the earth's magnetosphere and atmosphere. Also, solar ultraviolet radiation maintains the ozone layer and the ionosphere. Superposed on these continuous effects are cyclic interactions keyed to the solar sunspot cycle. When the sunspot cycle is at a maximum, the solar surface is most active and short-term solar–terrestrial interactions occur.

The appearance of an intense solar flare is likely to be followed by a deterioration of our radio communications at higher frequencies. Powerful ultraviolet radiation from the flare arrives with visible evidence of the flare itself, which is about 8 minutes after the onset of the flare. Swift particles that cause radio flashes in the corona reach the earth less than an hour later. These disrupt the layers of the ionosphere which normally reflect our radio signals back to the ground (Section 1.9).

Less swift ionized particles (Section 12.11) arrive here a day or so after the flare is observed. These particles produce an outburst of radio emission in the corona as they pass through it. They then spiral out along the sun's magnetic field lines and eventually enter the earth's magnetic field where they excite the gases of our upper atmosphere and set them glowing in an auroral display. The appearance of an aurora is generally accompanied by unusual gyrations of a magnetic compass.

Long-term solar–terrestrial relations are determined by the sunspot cycle as well. These relations are poorly understood and are only now coming under serious study. The almost total disappearance of sunspots during the Maunder minimum was preceded by a speeding up of the rotation of the sun and the earth subsequently cooled into a "little ice age" in the early eighteenth century. There is evidence of similar events going back some 5000 years.

Questions

1 Explain why solar spectral lines appear as dark absorption lines in the normal solar spectrum, but during eclipse they appear as emission lines.
2 What is the temperature at the very center of the sun?
3 What are granules?
4 Give two methods for determining the rotation period of the sun.

5 What is the visible surface of the sun called? The lower atmosphere? The extended atmosphere?
6 Explain how we think sunspots form.
7 What are coronal loops and coronal holes?
8 Distinguish between prominences and flares.
9 What is the solar wind?
10 List three solar–terrestrial relationships.

Other Readings

Bahcall J., and R. Davis, "Solar Neutrinos: A Scientific Puzzle," *Science* **191**, 264–267 (1976).

Demarque, P., "Models of the Sun," *Physics Today* **28**, 71 (1975).

diCicco, D., "Eclipse at Sea," *Sky & Telescope* **54**, 470–474 (1977).

Eddy, J. A., "The Case of the Missing Sunspots," *Scientific American* **236**(5), 80–88,92 (1977).

Eddy, J. A., P. A. Gilman, and D. E. Trotter, "Anomalous Solar Rotation in the Early 17th Century," *Science* **198**, 824–829 (1977).

Gosling J. T., and A. J. Hundhausen, "Waves in the Solar Wind," *Scientific American* **236**(3), 36–43 (1977).

Gough, D., "The Shivering Sun Opens its Heart," *New Scientist* **70**, 590–592 (1976).

Pallavicini R., and G. Poletto, "Is there Anything New on the Sun?," *Mercury* **VII**, 23–26,33 (1978).

chapter 11

The Stars in Their Seasons

In the original sense the constellations are configurations of stars. The brighter stars form patterns of dippers, crosses, and the like. Some of the star figures we recognize today were well known to the people of Mesopotamia 5000 years ago, who had named them after animals and representatives of occupations, such as the herdsman and the hunter. The plan was later adopted by the Greeks who renamed some of the characters after the animals and heroes of their mythology.

The Constellations

Forty-eight constellations were known to the early Greeks and nearly all of these are described in the *Phenomena,* which the poet Aratus wrote about 270 B.C. The popularity of Aratus' poem did much to perpetuate the imagined starry creatures which he vividly described. Interest in the celestial menagerie was also promoted by the publication of Ptolemy's *Almagest,* about A.D. 150, where the places of the stars are designated by their positions in the mythological figures. Then, too, famous artists of later times vied with one another to produce the liveliest pictures of the imagined creatures (Fig. 11.1).

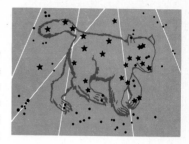

Figure 11.1   *The Great Bear. The Big Dipper makes up part of this famous constellation.*

The original constellations did not cover the skies of the Greeks completely; they did not include the duller areas where there were no striking configurations of stars to claim attention. At that time, only those stars within the imagined creatures belonged to constellations. In addition, the part of the heavens around the south celestial pole that did not rise above the horizon of the Greeks remained uncharted.

Figure 11.2   *Old (top) and new (bottom) boundaries of Orion.*

Celestial mapmakers of later times filled the vacant spaces with new constellations that they named after scientific instruments, birds, and other things having no connection with the creatures of the legends. Not all of them survived. At present we recognize 88 constellations, Table 11.1, which completely cover the sphere of the stars from pole to pole. Seventy of them are visible, either wholly or in part, from the latitude of New York.

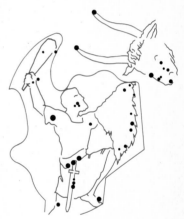

For the purposes of astronomy, the constellations are now definite divisions of the heavens marked off by boundary lines, just as states and nations are bounded. The boundaries first appeared in the star maps at the beginning of the nineteenth century. They were frequently irregular, making detours to avoid cutting across outstretched arms and paws of the legendary creatures. These devious dividing lines were straightened for most of the southern constellations by B. Gould in 1877, and finally for all the constellations by decision of the International Astronomical Union in 1928. The boundaries now run only from east to west and from north to south, although they zigzag considerably so as not to expatriate bright stars and variable stars from constellations with which they have long been associated (Fig. 11.2).

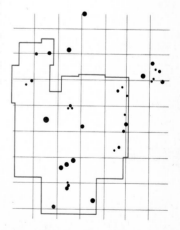

11.1 Names of the stars

The brightest stars and other especially famous ones are known to us by personal names that have been handed down through the ages. Some of these names are of Greek and Latin origin. Some are derived from the Arabic; names such as Algol, Altair (*al* is the Arabic definite

Table 11.1
Names of the Constellations

Latin name	Possessive	English equivalent	Map
Androm'eda[a]	Androm'edae	Andromeda	4,5
Ant'lia	Ant'liae	air pump	
A'pus	A'podis	bird of paradise	
Aqua'rius[a]	Aqua'rii	water carrier	4
Aq'uila[a]	Aq'uilae	eagle	3,4
A'ra[a]	A'rae	altar	6
A'ries[a]	Ari'etis	ram	4,5
Auri'ga[a]	Auri'gae	charioteer	5
Boö'tes[a]	Boö'tis	herdsman	2,3
Cae'lum	Cae'li	graving tool	2,3
Camelopar'dalis	Camelopar'dalis	giraffe	
Can'cer[a]	Can'cri	crab	2,5
Ca'nes Vena'tici	Ca'num Venatico'rum	hunting dogs	2
Ca'nis Ma'jor[a]	Ca'nis Majo'ris	larger dog	5
Ca'nis Mi'nor[a]	Ca'nis Mino'ris	smaller dog	5
Capricor'nus[a]	Capricor'ni	sea-goat	4
Cari'na[b]	Cari'nae	keel	6
Cassiope'ia[a]	Cassiope'iae	Cassiopeia	1,4
Centau'rus[a]	Centau'ri	centaur	2,6
Ce'pheus[a]	Ce'phei	cepheus	1,4
Ce'tus[a]	Ce'ti	whale	4,5
Chamae'leon	Chamaeleon'tis	chameleon	
Cir'cinus	Cir'cini	compasses	
Colum'ba	Colum'bae	dove	5
Co'ma Bereni'ces	Co'mae Bereni'ces	Berenice's hair	2
Coro'na Austra'lis[a]	Coro'nae Austra'lis	southern crown	
Coro'na Borea'lis[a]	Coro'nae Borea'lis	northern crown	3
Cor'vus[a]	Cor'vi	crow	2
Cra'ter[a]	Crater'is	cup	2
Crux	Cru'cis	cross	6
Cyg'nus[a]	Cyg'ni	swan	3,4
Delphi'nus[a]	Delphi'ni	dolphin	4
Dora'do	Dora'dus	dorado	
Dra'co[a]	Draco'nis	dragon	1,3
Equu'leus[a]	Equu'lei	little horse	
Erid'anus[a]	Erid'ani	river	5,6
For'nax	Forna'cis	furnace	
Gem'ini[a]	Gemino'rum	twins	5
Grus	Gru'is	crane	4
Her'cules[a]	Her'culis	Hercules	3
Horolo'gium	Horolo'gii	clock	
Hy'dra[a]	Hy'drae	sea serpent	2
Hy'drus	Hy'dri	water snake	6
In'dus	In'di	Indian	
Lacer'ta	Lacer'tae	lizard	
Le'o[a]	Leo'nis	lion	2

Table 11.1
(cont'd.)

Latin name	Possessive	English equivalent	Map
Le′o Mi′nor	Leo′nis Mino′ris	smaller lion	
Le′pus[a]	Le′poris	hare	5
Li′bra[a]	Li′brae	scales	3
Lu′pus[a]	Lu′pi	wolf	3
Lynx	Lyn′cis	lynx	
Ly′ra[a]	Ly′rae	lyre	3,4
Men′sa	Men′sae	table mountain	
Microsco′pium	Microsco′pii	microscope	
Monoc′eros	Monocero′tis	unicorn	
Mus′ca	Mus′cae	fly	6
Nor′ma	Nor′mae	level	
Oc′tans	Octan′tis	octant	
Ophiu′chus[a]	Ophiu′chi	serpent holder	3
Ori′on[a]	Orio′nis	Orion	5
Pa′vo	Pavo′nis	peacock	6
Peg′asus[a]	Peg′asi	Pegasus	4
Per′seus[a]	Per′sei	Perseus	4,5
Phoe′nix	Phoeni′cis	phoenix	4
Pic′tor	Picto′ris	easel	
Pis′ces[a]	Pis′cium	fishes	4
Pis′cis Austri′nus[a]	Pis′cis Austri′ni	southern fish	4
Pup′pis[b]	Pup′pis	stern	5
Pyx′is[b]	Pyx′idis	mariner's compass	
Retic′ulum	Retic′uli	net	
Sagit′ta[a]	Sagit′tae	arrow	3,4
Sagitta′rius[a]	Sagitta′rii	archer	3
Scor′pius[a]	Scor′pii	scorpion	3
Sculp′tor	Sculpto′ris	sculptor's apparatus	4
Scu′tum	Scu′ti	shield	
Ser′pens[a]	Serpen′tis	serpent	3
Sex′tans	Sextan′tis	sextant	
Tau′rus[a]	Tau′ri	bull	5
Telesco′pium	Telesco′pii	telescope	
Trian′gulum[a]	Trian′guli	triangle	4,5
Trian′gulum Austra′le	Trian′guli Austra′lis	southern triangle	6
Tuca′na	Tuca′nae	toucan	6
Ur′sa Ma′jor[a]	Ur′sae Majo′ris	larger bear	1,2
Ur′sa Mi′nor[a]	Ur′sae Mino′ris	smaller bear	1,3
Ve′la[b]	Velo′rum	sails	2,6
Vir′go[a]	Vir′ginis	virgin	2
Vo′lans	Volan′tis	flying fish	
Vulpec′ula	Vulpec′ulae	fox	

[a] One of the 48 constellations recognized by Ptolemy.

[b] Carina, Puppis, Pyxis, and Vela once formed the single Ptolemic constellation Argo Navis.

article), and many others are survivors from earlier times when astronomy was a favorite study of Mohammendan scholars.

Procyon, meaning "before the dog," precedes Sirius the Dog Star in its rising. Aldebaran means "the follower"; it rises after the Pleiades. Antares is the "rival of Mars," because it is red. These are examples of how some of the stars were designated. Other star names have come to us in a different way.

In the earliest catalogs, such as Ptolemy's, the stars were distinguished by their positions in the imagined figures of heroes and animals. One star was the "mouth of the Fish"; another was the "tail of the Bird." Transcribed later into the Arabic, some of these expressions finally degenerated into single words. Betelgeuse, the name of the bright red star in Orion, was originally three words perhaps meaning the "armpit of the Central One."

11.2 Letter designations of the stars

The plan of designating the brighter stars by letters was introduced by Bayer, a Bavarian attorney, in 1603. In a general way, the stars of each constellation are denoted by small letters of the greek alphabet (see the Appendix) in order of their brightness, and the roman alphabet is used for further letters. Where there are several stars of nearly the same brightness in the constellation, they are likely to be lettered in order from head to foot of the legendary creature. The full name of a star in the Bayer system is the letter followed by the possessive of the Latin name of the constellation. Thus Capella, the brightest star in Auriga, is alpha (α) Aurigae. The letters for some of the brighter stars are shown in the maps that follow.

Some of the fainter stars are known by their numbers according to a plan of numbering the stars in each constellation in order of right ascension developed by John Flamsteed. The star 61 Cygni is an example. Most faint stars, however, are designated only by numbers in one of the many catalogs to which the astronomer can turn for information about their positions, brightness, and other features. Three of these catalogs are most frequently used. One is the *Bonner Durchmusterung*. The second is the *Henry Draper Catalog*. The third is the *Smithsonian Astrophysical Observatory Catalogue*.

As astronomy progressed a series of catalogs developed to serve special purposes. For example, double stars (Section 13.1) are found in numerous catalogs as are variable stars (Section 13.8). X-ray sources are cataloged by constellation in order of their discovery. We will refer to various special catalogs as we go along.

Maps of the Sky

The six maps in this chapter show all parts of the celestial sphere. They are designed particularly for an observer in latitude 40°N, but are useful anywhere in middle northern latitudes. In order to avoid confusion, the many stars only faintly visible to the naked eye are not included, the names of inconspicuous constellations are generally omitted, and the boundaries between the constellations do not appear in the maps. When using Maps 2,3,4, and 5 face south with the north point of the map up; east will be to the left. Locate a few of the bright stars and procede from those stars to other stars or constellations of interest.

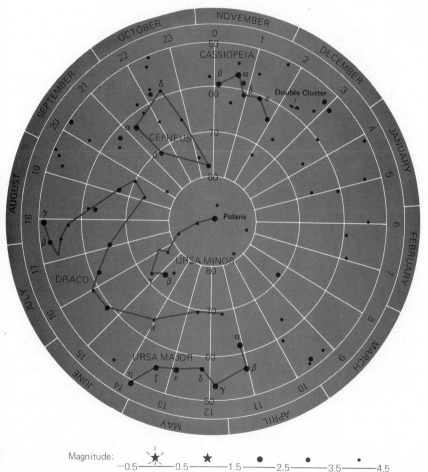

Map 1 North polar stars and constellations.

Magnitude: −0.5 —— 0.5 —— 1.5 —— 2.5 —— 3.5 —— 4.5

Map 2 Spring stars and constellations.

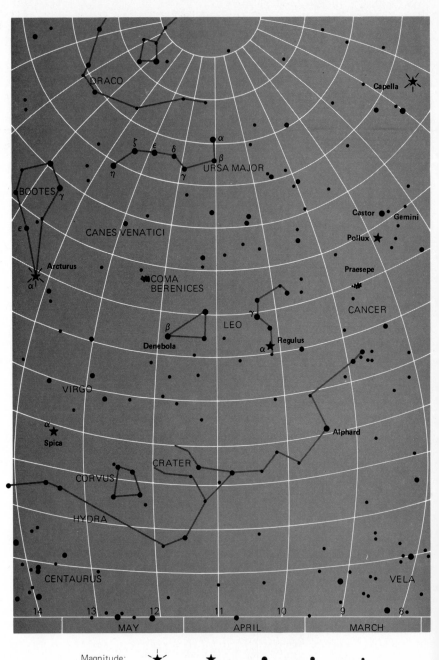

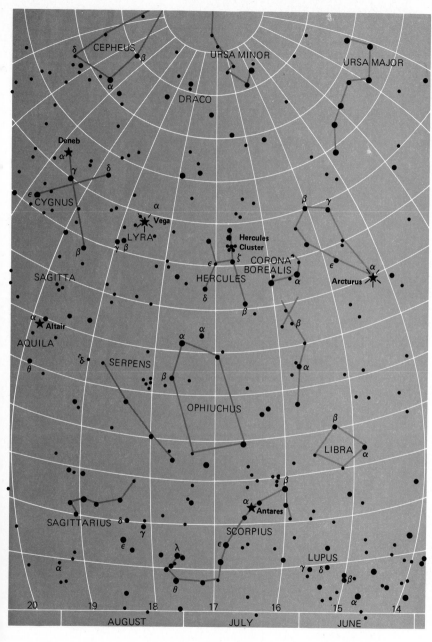

Map 3 Summer stars and constellations.

Map 4 Fall stars and constellations.

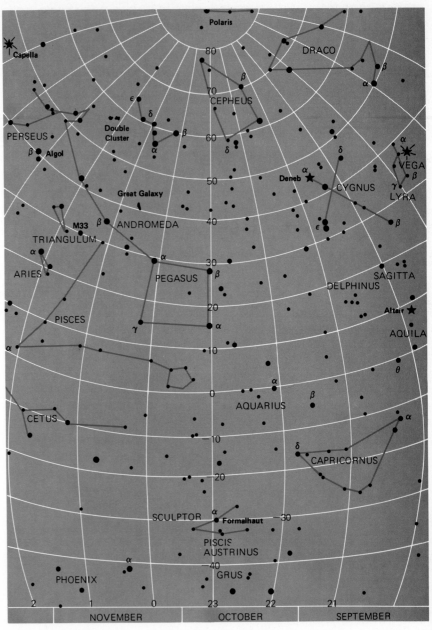

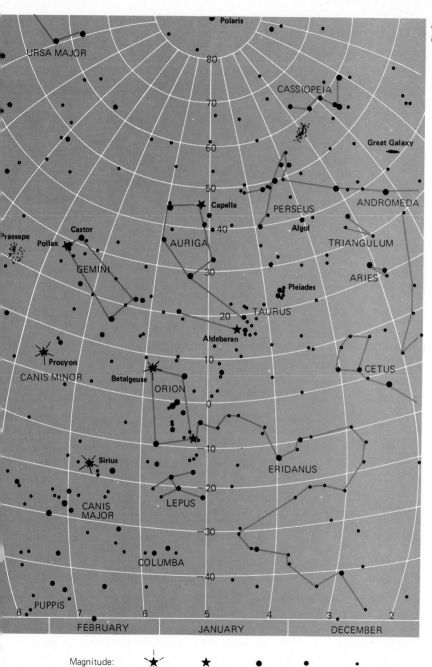

Map 5 Winter stars and constellations.

Map 6 South polar stars and con-stellations.

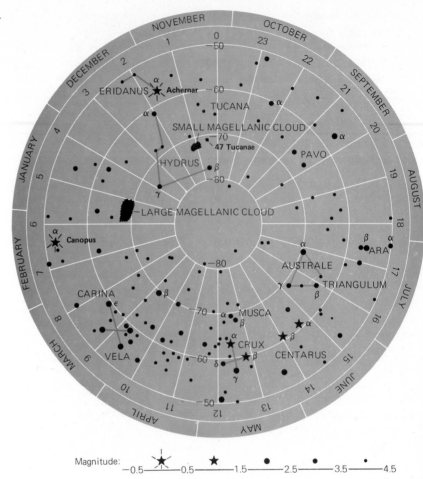

11.3 Directions in the sky

North in the sky is in the direction of the north celestial pole; south is toward the opposite pole. The stars seem to circle from east to west daily. If these rules are remembered, there can be no confusion about directions in the sky.

The position of a star is clearly described when its directions in relation to other stars are given in this way. Its position is also perfectly definite if we determine how it is related to two or more other stars. For example, the star may complete an equilateral triangle with two other stars already identified or it may be in line with them. We have noticed that a line through the pointer stars of the Big Dipper leads to the pole star. Such directions remain unaltered through the day and year.

Directions relative to the horizon, however, change as the stars go around the pole. From a star above the pole, north is down and west is to the left; from a star below the pole, north is up and west is to the right. Note how the Big Dipper changes its position relative to the horizon during the night or from night to night as the seasons progress. (Fig. 11.3). In the winter this figure of seven stars stands on its handle in the northeast at 9 o'clock, in spring it appears inverted above the pole, in summer the Dipper is seen descending bowl-down in the northwest, and in autumn it is right side up under the pole, where it may be lost from view for a time.

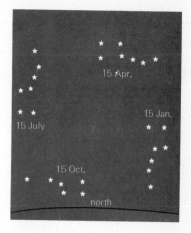

Figure 11.3 *The Big Dipper at 9 o'clock (standard time) in the evening at different seasons.*

Maps 2 through 5 show the constellations that cross the meridian at 9 o'clock in the evening during each of the four seasons, all the way from the north celestial pole to the south horizon of latitude 40°N. The pole is near the top of each map. Hour circles radiating from the pole are marked in hours of right ascension near the bottom of the map, and circles of equal declination go around the pole.

To observe the constellations, select the map for the present season and hold it toward the south. The hour circle above the date of observation coincides with the celestial meridian at 9 o'clock in the evening on that date. Accordingly, the central vertical line in each of the four maps represents successively the positions of the meridian at 9 P.M. on 21 April, 21 July, 21 October, and 21 January. Remembering that the daily motions of the stars are from left to right for these maps and that a star comes to the meridian 2 hours earlier from month to month, we can locate the meridian in the maps for other times and dates. Thus the maps are arranged to show what stars are crossing the celestial meridian at a particular time of the day and year.

The zenith at the place of observation is on the circle having the same declination as the latitude of the place. As we are facing south, it would be necessary to lean backward to view the constellations in the upper parts of the maps. The northern constellations, however, are shown more conveniently in Map 1. They are repeated in the seasonal maps to display more clearly their relation to the constellations farther south.

11.4 The map of the northern sky

Map 1 shows the region of the heavens within 40° from the north celestial pole. The pole is at the center, closely marked by Polaris at the end of the Little Dipper's handle. Hour circles appear in this projection as straight lines radiating from the center; they are marked around the circumference of the map in hours of right ascension. The concentric circles are circles of equal declination at intervals of 10°; their declinations are shown on the vertical line.

This map is to be held toward the north. When it is turned so that the present month is at the top, the map represents the positions of the northern constellations at 9 P.M. standard time or, more exactly, local mean time if the correct part of the month is at the top. The hour circle that is then vertically under the date coincides with the celestial meridian at this time of night on this date. For a different time of night the map is to be turned from the 9 o'clock position through the proper number of hours—clockwise for an earlier time and counterclockwise for a later time.

If we wish, for example, to identify the northern constellations at 9 o'clock in the evening on 7 August, the map should be turned so that this date is at the top (at about the 18-hour circle). The Big Dipper is now bowl-down in the northwest, Cassiopeia is opposite it in the northeast, and so on. These northern constellations appear again in the maps for the different seasons. On all the maps the brightness of the stars is indicated by symbols representing their approximate magnitudes (Section 12.16).

11.5 Stars of spring

The star maps are useful for identifying the prominent constellations in the sky and also for reference during the reading of this book. Our brief inspection continues with the stars of spring (Map 2), which are near the celestial meridian in the early evenings of this season.

In the early evenings of spring the Big Dipper appears inverted above the pole. This figure of seven bright stars is part of the large constellation Ursa Major, the Larger Bear, which extends for some distance to the west and south of the bowl of the Dipper; pairs of stars of nearly equal separations (Figs. 11.1 and 11.4) mark three paws of the ancient creature. Mizar, at the bend in the handle, has a fainter companion visible to the naked eye. With even a small telescope Mizar itself is revealed as a pair of stars.

Figure 11.4   *The constellation of Ursa Major (Great Bear) giving the names of the stars in the Big Dipper. The inset shows Mizar and Alcor as seen in a normal inverting telescope.*

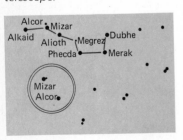

Following the line of the Pointers northward we come to Polaris, the north star or pole star, less than 1° from the celestial pole. This useful star marks the end of the handle of the Little Dipper, the characteristic figure of Ursa Minor.

Leo is peculiarly associated with the spring season. Its familiar sickle appears in the east in the early evening as spring approaches, and it becomes the dominant figure in the south as this season advances. The bright star Regulus marks the end of the handle of the sickle figure, which with the right triangle to the east is the distinguishing feature of this constellation of the zodiac.

To the west of Leo is the dim zodiacal constellation Cancer with its Praesepe star cluster. To the east is the larger zodiacal constellation Virgo containing the bright star Spica. The course of the ecliptic here is very nearly the line from the Praesepe cluster through Regulus and Spica.

Three-fifths of the way from Regulus to Spica it crosses the celestial equator at the autumnal equinox (Fig. 11.5). The four-sided figure of Corvus in this region is likely to attract attention.

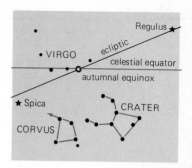

Figure 11.5 *The position of the autumnal equinox roughly halfway between Spica and Regulus.*

11.6 Stars of summer

A procession of familiar constellations, from Boötes to Cygnus, marches through the zenith in the evenings of summer in middle northern latitudes (Map 3). A fine region of the Milky Way, which we examine in a later chapter, extends from Cygnus past Aquila down to Scorpius and Sagittarius in the south (Fig. 11.6).

Boötes is overhead when the stars come out at the beginning of summer. Its stars outline the figure of a large kite with the brilliant Arcturus at the point where the tail is attached. This somewhat reddish star is pointed out by following the curve of the Big Dipper's handle around past its end. Arcturus' only peers in the whole northern celestial hemisphere are blue Vega and yellow Capella (see Chapter 12 for a discussion of star color and brightness).

Eastward from the kite figure and beyond the semicircle of Corona Borealis, the Northern Crown, we find Hercules, which is passing overhead at nightfall in midsummer. Some of its brighter stars may seem to suggest the figure of a rather large butterfly that is flying toward the west. The great cluster in Hercules, scarcely visible to the naked eye, is situated two-thirds of the way from the imagined butterfly's head along the west edge of the northern wing. Farther east on the line from the Crown through Hercules we come to Lyra. The figure here is a small parallelogram with a triangle attached at its northernmost point (Fig. 11.6). Vega marks a vertex of the triangle. The star at the north point of the triangle is the "double-double" epsilon Lyrae. This star is visible as two stars to the unaided eye but is more easily seen with binoculars; each star is again divided into two if observed with a telescope. The Ring nebula in Lyra (Fig. 11.7), visible only with the telescope, is about midway between the two stars in the southern side of the parallelogram.

The formidable figure of Scorpius, the scorpion of the zodiac, dominates the southern sky at nightfall in early summer. Its bright Antares is a red supergiant star enormously larger than the sun.

Directly east of Antares, six stars of the zodiacal constellation Sagittarius outline the Milk Dipper, so named because it is in the Milky Way. In this direction we look toward the center of our Galaxy. The great star cloud of Sagittarius is an outlying portion of the galactic central region that is mainly obscured by intervening heavy clouds or cosmic dust. The Trifid nebula and the globular star cluster M 22 are some of the other features of this spectacular tract of the heavens. The place of the winter solstice is also near the handle of the Milk Dipper (Fig. 11.8).

Figure 11.6 *Cygnus and Lyra. This chart helps to locate the famous double-double star epsilon Lyrae and the Ring nebula.*

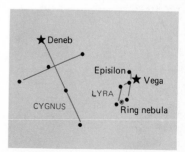

Figure 11.7 *The Ring nebula in Lyra. This beautiful planetary nebula (Chapter 13) is visible in a good small telescope. (Lick Observatory photograph.)*

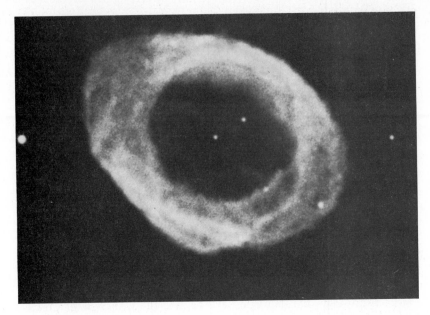

11.7 Stars of autumn

Cassiopeia rises high in the north in our autumn sky (Map 4). The Northern Cross is overhead early in the season and, as it moves along, the Square of Pegasus becomes the dominant figure. Dim watery constellations spread across in the south, having only the star Fomalhaut to attract attention unless bright planets appear there as well.

The Northern Cross is overhead at nightfall in the early autumn in middle northern latitudes. This attractive figure, set in a fine region of the Milky Way (Fig. 11.9), is the characteristic feature of Cygnus, the Swan. Its brightest star, Deneb, marks the northern end of the longer axis of the Cross. Albireo (beta Cygni) decorates the southern end of this axis; it appears double with even a small telescope, a reddish star with a fainter blue companion. The Northern Cross is the direction toward which the sun with its planetary system is moving in the rotation of our Galaxy.

Figure 11.8 *The Milk Dipper in Sagittarius. The winter solstice is located just to the right of the handle near the Trifed Nebula.*

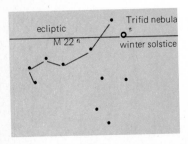

The Square of Pegasus is peculiarly associated with our autumn skies. As this season approaches, we see it in the east in the early evening balanced on one corner. It crosses the meridian at 9 o'clock about 1 November. Four rather bright stars mark the corners of the large square which is the characteristic figure of Pegasus; they are alpha, beta, and gamma Pegasi, and in the northeast corner the star alpha Andromeda. If we imagine that the Square of Pegasus is the bowl of a very large dipper figure, we find its handle extending toward the northeast (Fig. 11.10); its first three

Figure 11.9 *The Northern Cross forms the constellation of Cygnus. Beta Cygni is out of the picture to the lower right. The North American nebula is at the upper left and a wreath nebula is in the lower left. (Yerkes Observatory photograph.)*

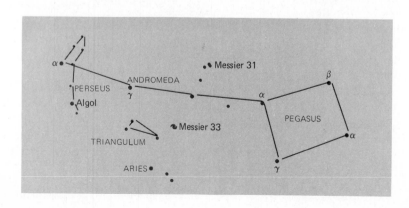

Figure 11.10 *The region of Pegasus. The spiral galaxies M 31 and M 33 are nearby.*

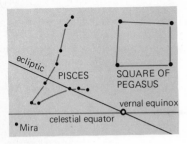

Figure 11.11   *The location of the vernal equinox below Pegasus.*

stars are the brightest members of Andromeda. A feature of this constellation is M 31, which is only faintly visible to the naked eye but is actually a spiral galaxy larger than our own. It is shown in Map 4. Another galaxy, M 33, is marked in the figure in the neighboring constellation Triangulum.

Southeast of the Square two streams of faint stars represent the ribbons with which the Fishes are tied. This dim constellation Pisces of the zodiac contains the zodiacal sign Aries, which has moved westward from its own constellation. Here we find the "first of Aries," or vernal equinox (Fig. 11.11). The line of the eastern side of the Square extended as far again to the south leads to this important point of the celestial sphere.

11.8 Stars of winter

The bright winter scene is grouped around Orion, its most conspicuous figure. The zodiac is farthest north here.

Figure 11.12   *The Pleiades cluster showing the names of the stars. This cluster is easily visible to the unaided eye. (Yerkes Observatory photograph.)*

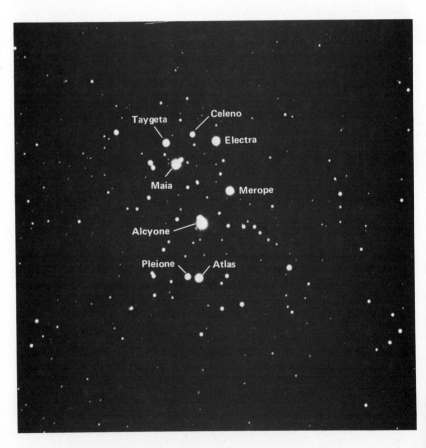

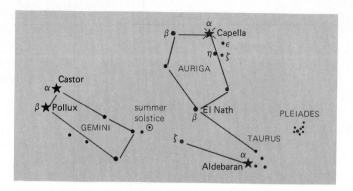

Figure 11.13 *The summer solstice which is located roughly halfway between Castor and Aldebaran.*

Taurus contains two bright star clusters. The Pleiades cluster looks something like a short-handled dipper (Fig. 11.12). Six of its stars are clearly visible to the naked eye, and two or three others occasionally twinkle into view. The conspicuous feature of the larger, Hyades cluster is in the form of the letter V which, including the bright star Aldebaran, represents the head of Taurus. The Crab nebula, the remnant of the A.D. 1054 supernova, can be seen in Taurus with the aid of binoculars or a small telescope. At the tips of the horns of Taurus are two fairly bright stars, the northern one of which is needed to complete the muffin figure of Auriga; its brilliant star Capella is near the zenith at 9 o'clock toward the end of January.

Gemini has an oblong figure. The heads of these Twins of the zodiac are marked by the bright stars Castor and Pollux. The position of the summer solstice (Fig. 11.13) is between their feet and the horns of Taurus. The Milky Way sweeps down from Auriga and Taurus past the feet of the Twins and on to the south horizon.

Orion, the brightest constellation, is peculiarly associated with the winter. Its oblong figure rises in the early evening as winter approaches, appears in the south at 9 P.M. about 1 February, and is setting in the twilight as spring advances.

This bright region of the heavens inspired a lively scene in the old celestial picture book. Orion, a mighty hunter accompanied by his dogs, stands with uplifted club awaiting the charging Taurus. Red Betelgeuse glows below his shoulder. Blue Rigel diagonally across the figure is somewhat the brighter of the two. Three stars near the center of the rectangle mark Orion's belt, and three fainter ones in line to the south (Fig. 11.14) represent his sword. The middle star of the three appears through the telescope as a trapezium of stars surrounded by the foggy glow of the great nebula in Orion.

The stars of Orion's belt are useful pointers. The line joining them directs the eye northwestward to the Hyades and southeastward to Sirius,

Figure 11.14 *The constellation Orion. A blue-sensitive film was used to take this photograph in order to highlight the nebulosity in the area. This explains the brightness of Bellatrix and Rigel (blue stars) as compared with Betelgeuse (a red star) which is brighter than Bellatrix and Rigel to the eye.*

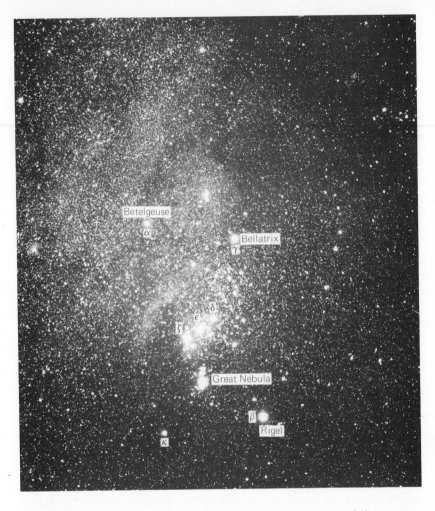

the "Dog Star," the brightest star in the heavens and one of the nearest. Sirius, Betelgeuse, and Procyon across the Milky Way form a nearly equal-sided triangle.

11.9 The southern constellations

Map 6 shows the part of the heavens that does not come up into view in latitude 40°N. It contains the Southern Cross, the two Magellanic Clouds, a number of brilliant stars, and a fine region of the Milky Way.

Crux, the Southern Cross, is a small figure of four stars, which resembles a kite as much as a cross. It becomes entirely visible south of latitude 28°N. Its brightest star, alpha Crucis, is about as bright as Aldebaran. The Magellanic Clouds, companions of our spiral galaxy, do not rise anywhere in the United States. From more southern latitudes they are clearly visible to the unaided eye. Canopus, almost directly south of Sirius, is second to it in brightness. Alpha Centauri is third in order of brightness among all the stars, and is nearest of all to the sun.

The south celestial pole is situated in the dim constellation Octans. There is not a star as bright as Polaris within 20° from this point. Sigma Octantis, a star barely visible to the naked eye, is 50' from the pole.

Planetarium

The word **planetarium** may refer either to the complex instrument (Fig. 11.15) that projects a changing picture of the heavens on the interior of a hemisphere, or else to the building or chamber that houses the apparatus. The planetarium is playing an ever increasing role in public education in astronomy. Many planetariums are open to the public on a regular schedule and their demonstrations change four to six times per year. In

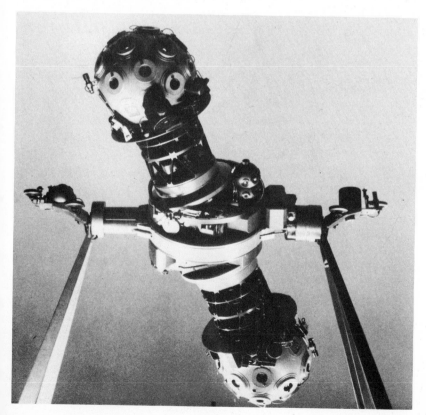

Figure 11.15 *Planetarium projector of the American Museum–Hayden Planetarium, New York City. (Photograph courtesy of Carl Zeiss, Inc., New York, N.Y.)*

addition all of them feature special science exhibits of equipment and demonstrations of basic principles.

Planetariums in increasing numbers are offering instructive replicas of the heavens and of the motion of the celestial bodies. They are often used in connection with introductory astronomy courses having laboratory exercises. Laboratory classes are at the mercy of the weather and are difficult at best in large cities, so the planetarium offers a chance to make certain observations under known conditions. The students can plot the motions of the planets, receive training in meteor counting and plotting, etc. The planetarium is also proving to be a very useful tool in the teaching of mathematics and celestial navigation. One of the novel uses was the training of Apollo astronauts for celestial observations.

Questions

1 How did the brightest stars get their names?
2 What are the three major catalogs used by astronomers?
3 Give the approximate coordinates for α Ursa Majoris, Regulus, Antares, Formalhaut, and the Large Magellanic Cloud.
4 What is the simplest way to locate the pole star?
5 Does Arcturus pass the meridian north or south of your zenith?

6 Why is the star Mizar interesting?
7 What are the colors of Vega and Capella?
8 How far south must you be to be able to just see the Southern Cross?
9 Mark the approximate pole of the ecliptic on Map 6. What object is nearby?
10 Note the large wreath in the lower left-hand corner of Fig. 11.9. Mark its location on the proper star map.

Other Readings

Emlen, S. T., "The Stellar-Orientation System of a Migratory Bird," *Scientific American,* **233(3),** 102–111 (1975).

Milliken, E., "New Hampshire's Student-Run Planetarium," *Sky & Telescope,* **55,** 222–224 (1978).

Norton, O. R., "A Major Planetarium for Tucson, Arizona," *Sky & Telescope,* **49,** 143–146 (1975).

chapter 12

The Stars Around Us

The distances of the nearer stars are shown by the amounts of their slightly altered directions as the earth revolves around the sun. The motions of the stars are revealed by their progress against the background of more remote stars and by displacements of the lines in their spectra. Stars in the sun's neighborhood are often bright enough to permit the study and classification of their spectra. Using such observable data and their relative brightness, stars may be compared with one another and with the sun.

Distances of the Stars

The distance of a celestial object may be found by observing its parallax, which has been defined as the difference between the directions of an object as viewed from two different places a known distance apart. The moon's parallax is large enough to be measured from widely separated stations on the earth. The difference between the directions of even the nearest star, however, observed from opposite points on the earth is not greater than the width of a period on this page if the period were viewed from a distance of about 110 meters. To detect the parallaxes of stars we require the far greater separation of points of observation afforded by the earth's revolution around the sun.

12.1 Parallaxes of the stars

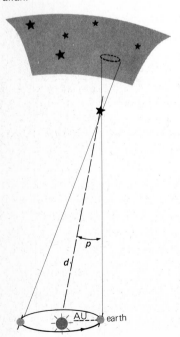

Figure 12.1 *The parallax of a star. As the earth revolves around the sun the nearer stars appear to oscillate relative to the more remote stars. Half of the total angle is the parallactic angle or parallax.*

The earth's revolution causes apparent slight oscillations of the nearer stars (Fig. 12.1). The extent of the displacement for a particular star is found by comparing its positions in a series of photographs taken 6 months apart; the positions are determined with reference to other stars that appear nearby in the photograph but are likely to be too far from us to be considerably displaced. Because part of the observed displacement may be caused by the movement of the star itself, it is necessary to repeat the photographs at half-year intervals until the annual parallax oscillation can be disentangled from the straight-line motion of the star. Since the displacements are extremely small the photographs must be recorded on glass plates.

The parallax of a star, in the usual meaning of the term, is its **heliocentric relative parallax.** It is the greatest difference between the star's directions from the earth and sun during the year, with slight correction to keep the earth at its average distance from the sun, and is about half the greatest parallax displacement of the star. When the parallax has been measured, the star's distance can be calculated (Fig. 12.2).

In this book we shall use the symbol p for **parallax.** When the symbol stands alone it is understood to be the relative parallax of the star, that is, the star's parallax relative to the reference stars used. When a correction is applied to allow for the finite distance of the reference stars, a very small correction obtained from statistical studies, we say that we have obtained the **absolute parallax** (p_{abs}) of the star.

The distances to the stars are so great in ordinary units that new terms must be devised that make the numbers more manageable. A convenient unit is the distance d at which a star will have a parallax of 1 second of arc and is called the **parsec** (pc), an acronym formed from the words parallactic second. By definition, 1 parsec equals 206,265 astronomical units. Since 1 astronomical unit is about 8.3 light minutes, the time it takes light to travel 1 astronomical unit, we can compute that 1 parsec equals 3.26 **light years** (ly).

The parsec as a unit of distance is particularly convenient for two reasons. First it is the reciprocal of a star's parallax, and second, it is roughly the distance between stars. The moon is 1.3 light seconds away. The sun is 8.3 light minutes from the earth. Pluto is almost 5.5 light hours from the sun. The nearest stars, as we will soon see, are the alpha Centauri system with a parallax of 0.760 second of arc. According to the above its distance in parsecs is 1/0.760 or 1.32 parsecs, or 4.28 light years away. Space is essentially devoid of stars and the solar system is miniscule indeed.

An easy analogy for recalling the extreme distances involved is that the solar system has a radius of approximately one school day. The comet cloud has a distance from the sun that light travels in one school year. Finally, the distance to the nearest star is the distance that light travels during the four years of undergraduate school.

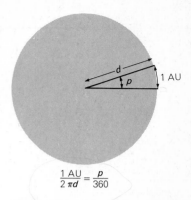

Figure 12.2   *The distance* d *of a star is simply computed from the length of the astronomical unit and the angle this unit subtends as seen from the star.*

12.2   The nearest stars

When Copernicus proposed his heliocentric theory it was realized that if it were true, the stars should reflect the earth's annual motion around the sun. Parallaxes were searched for but could not be detected, therefore the theory was thought to be wrong. People simply did not realize how far away even the nearest stars were. The first measured parallax was not announced until 1838 by F. W. Bessel, long after the heliocentric theory had been proved by other unanticipated methods (Section 3.2). The technique used by Bessel and other early measurers of parallaxes was to measure very accurately the transit times of several faint stars around the suspected parallax star. Over the course of 18 months he noted the minute annual change in position (a total of 0.04 second of time) of 61 Cygni. Current methods replace this technique with a long-focus telescope and the photographic plate, but the principle remains the same.

The sun's nearest neighbor among the stars is the bright double star alpha Centauri which has a distant third companion. The faint companion is situated about 2° from them; it is known as "Proxima," because it seems to be slightly nearer to us than they are. The parallax of these stars is 0.″760, so that their distance is 4.1×10^{13} kilometers. At least 26 stars are known to be within 3.7 parsecs of the sun; a 27th is just beyond this distance. These stars are listed in Table 12.1. The columns list the proper motion, parallax, distance, and apparent magnitude and spectrum of these stars. The terms not yet defined are discussed later in this chapter and the reader should refer back to this table when each term is defined. Note that there are three columns under "Visual Apparent

Table 12.1
Stars Nearer than 3.7 Parsecs

No.	Name	Proper Motion	Parallax	Distance in Parsecs	Visual Apparent Magnitude and Spectrum		
					A	B	C
1	Sun	–	–	–	−26.8 G2	–	–
2	α Centauri	3″.68	0″.760	1.32	0.1 G2	1.5 K5	11 M5e
3	Barnard's star	10.30	.552	1.81	9.5 M5	*	–
4	Wolf 359	4.84	.431	2.32	13.5 M6e	–	–
5	Lalande 21185	4.78	.402	2.49	7.5 M2	*	–
6	Sirius	1.32	.377	2.65	−1.5 A1	7.2 wd	–
7	Luyten 726–8	3.35	.365	2.74	12.5 M6e	13.0 M6e	–
8	Ross 154	0.74	.345	2.90	10.6 M5e	–	–
9	Ross 248	1.82	.317	3.15	12.2 M6e	–	–
10	ε Eridani	0.97	.305	3.28	3.7 K2	*	–
11	Luyten 789–6	3.27	.302	3.31	12.2 M6	–	–
12	Ross 128	1.40	.301	3.32	11.1 M5	–	–
13	61 Cygni	5.22	.292	3.42	5.2 K5	6.0 K7	*
14	ε Indi	4.67	.291	3.44	4.7 K5	–	–
15	Procyon	1.25	.287	3.48	0.3 F5	10.8 wd	–
16	Σ 2398	2.29	.284	3.52	8.9 M3.5	9.7 M4	–
17	Groombridge 34	2.91	.282	3.55	8.1 M1	11.0 M6	–
18	Lacaille 9352	6.87	.279	3.58	7.4 M2	–	–
19	τ Ceti	1.92	.273	3.66	3.5 G8	–	–
20	BD +5° 1668	3.73	.266	3.76	9.8 M4	*	–

* Denotes unseen components.

Magnitude and Spectrum." This is because the stars are arranged in multiple-star systems.

Of the 28 nearest stars including the sun, three form a triple system, and twelve form six double systems. In addition, one of the double systems gives evidence of a third star—an astrometric companion—and four of the seemingly single stars also have astrometric companions. **Astrometric companions** are companion stars, usually too faint to be visible, but which reveal their presence by causing the proper motion of the visible star to move in a wavy line. Thus there are only 9 single stars in this volume of space and at least one of these, the sun, has a group of planets.

The term spectrum refers to the color of the stars and is explained in Section 12.14. Very briefly, most stars seem to arrange themselves quite nicely on a graph from bright, hot blue stars to faint, cool red stars, and this smooth arrangement is called the **main sequence.** Although the bright stars Alpha Centauri, Sirius, and Procyon are included in this list, the majority of the nearest stars are too faint to be seen without a telescope. We conclude that the stars differ greatly in intrinsic brightness, so that a bright star in our skies may not be nearer than a faint one.

The direct method of determining the distances of stars by observing their parallaxes is limited to the nearer ones. Approximately 2200 stars can have their distances determined accurately by the direct parallax method. At the distance of 100 parsecs (326 light years), which is only a small step into space, the parallaxes become too small to be measured reliably by this technique. Ways of finding the distances of more remote stars will be described later.

The Stars in Motion

The motions of the stars in different directions relative to the sun are shown by the very gradual changes in their places in the heavens and by Doppler displacements of their spectral lines. The sun is also moving among the stars around it. Like the sun, the stars are rotating on their axes.

12.3 Proper motions of the stars

In 1718 Edmund Halley, after whom Halley's comet is named, was the first to explain that the stars are not stationary. He observed that Sirius and some other bright stars had drifted as much as the apparent width of the moon from the places assigned them in Ptolemy's early catalog. Since then, the proper motions of all lucid stars and of many fainter ones have become known by comparison of the records of their positions at different times.

The present technique makes use of an astrograph (Section 4.4); photographic plates for this purpose are taken at widely separated times (10 years or more). The plates, often covering fields up to 10×10 degrees, contain the images of many thousands of stars, and when the two plates are alternately superposed optically, the stars that have moved appear to jump back and forth and attract the astronomer's attention.

The **proper motion** of a star is the angular rate of its change of place in the plane of the sky per year. This change becomes so slow at great distances from us that the remote stars can serve as landmarks to show the progress of the nearer stars. Barnard's star has the largest proper motion of any star. In Fig. 12.3 three plates, September 1948, March 1949, and September 1949, are superposed so that the two background stars are fixed. This reveals the east–west parallactic motion of Barnard's star and its large south–north proper motion. Named after the astronomer who first observed its swift flight, this faint star in Ophiuchus moves among its neighbors in the sky at the rate of 10.″3 a year, or as far as the apparent diameter of the moon in 175 years. If all the stars were moving as fast as this and at random, the forms of the constellations would be altered appreciably during a lifetime. The known motions of only about 330 stars exceed 1″ a year and the average proper motion for all stars visible with the naked eye is not greater than 0.″1 a year. In the course of a century the progress of very few stars is enough to be detected without a telescope.

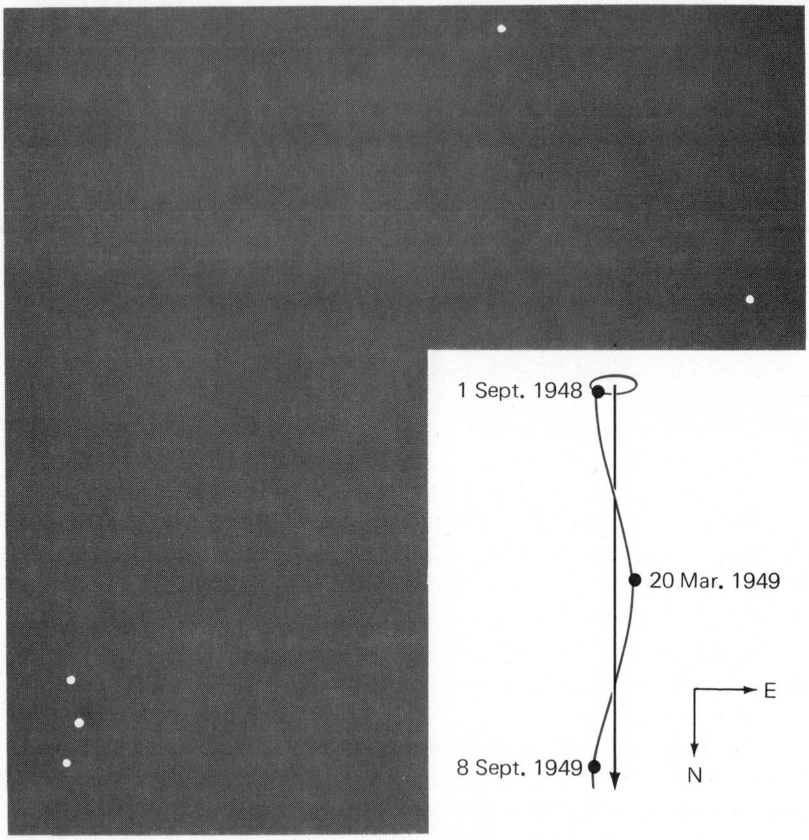

Figure 12.3 *The proper motion and parallax of Barnard's star. This picture is a composite of three different photographs taken at 6-month intervals. The reference stars have been carefully superposed and the picture demonstrates proper motion and parallax. (Sproul Observatory photograph.)*

1 Sept. 1948

20 Mar. 1949

E

8 Sept. 1949   N

Figure 12.4 *Proper motion (μ) tells us the motion across the line of sight. In order to learn the space motion, or true motion, we must obtain the radial motion or motion along the line of sight.*

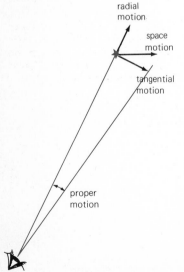

radial
motion

space
motion

tangential
motion

proper
motion

12.4 Radial velocities of the stars

The proper motion of a star tells us nothing of its movement toward or away from us (Fig. 12.4). The **radial velocity,** or velocity in the line of sight, is revealed by the Doppler effect (Section 4.14) in its spectrum. If the lines are displaced toward the violet end of the spectrum, the star is approaching us; if they are displaced toward the red end, the star is receding from us. The amount of the displacement is proportional to the speed of approach or recession.

$$v_r = c\,\Delta\lambda/\lambda$$

Recall that a star's radial velocity (v_r) is equal to the velocity of light (c) times the spectral line shift ($\Delta\lambda$) in wavelengths divided by the rest wavelength (λ) of the line.

The spectra of two stars are shown in Fig. 12.5. This photograph is a negative print, the dark comparison lines and areas are actually bright and the light lines in the stars' spectra are actually dark absorption lines.

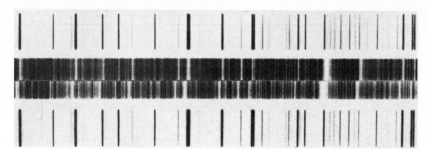

Figure 12.5 *Line of sight or Doppler displacements in the spectra of the stars HD 161096 (top) and HD 66141 (bottom). The comparison spectra have been carefully aligned. (David Dunlop Observatory.)*

This is because on a negative the material struck by light appears dark when the negative is developed. The comparison lines at the top and bottom are from an iron discharge tube and have been carefully aligned so that the two stellar spectra can be directly compared. The star's spectrum at the top shows almost no Doppler shift and the iron lines in its spectrum are easily identified. The star's spectrum at the bottom show its lines Doppler shifted to the left or blue; it is approaching us at about 67 km/sec. These spectra are taken on the earth and hence contain a component of the earth's orbital motion. When the earth's motion has been removed the velocities are different and are said to be heliocentric, that is the velocity as it would be seen from the sun. Of course, to get the true radial velocity we must also then correct for the sun's motion towards or away from the star being observed.

The Doppler effect in a spectrum is a very powerful tool; as long as we can measure it, it gives us the line-of-sight velocity of the object. If an object is one million parsecs away and is bright enough to give us a spectrum equal to that of a star 100 parsecs distant, we can get the radial velocities of both with equal accuracy. This is not true for proper motions and hence the space velocities of stars and other objects. This is because for a given velocity the angle measuring proper motion gets smaller as the distance away from us increases.

12.5 The sun's motion

If we assume that the sun is at rest with respect to the stars in its neighborhood and we then chart the proper motions of these stars with respect to much more distant stars we note a curious effect. The proper motions, on the average, indicate that the stars are moving past the sun, apparently coming from the direction of the constellation Hercules and moving toward the direction of the constellation of Columba.

Rather than assume that the sun is at rest we can assume that the average motion of the stars in our neighborhood is zero and that it is the sun that is moving away from the constellation Columba, called the **antapex** and towards the constellation Hercules, called the **apex.** We illustrate this schematically in Fig. 12.6 where the various stars are assumed

Figure 12.6 *Apparent drift of stars away from the apex. Note that stars near the direction of the apex or antapex and very distant stars have the least drift.*

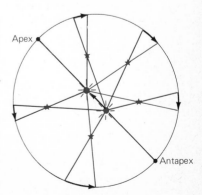

Apex

Antapex

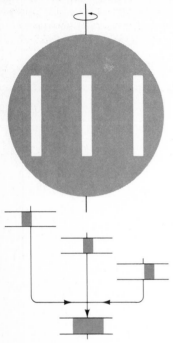

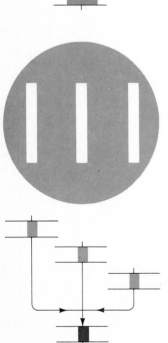

to be at rest while the sun moves from position 1 to position 2. The apparent proper motions projected on the distant circle representing the celestial sphere appear to spread from the apex and converge toward the antapex.

A similar effect is observed in the radial velocities of these neighborhood stars. The radial velocities have their greatest values of approach in the direction of the apex and their greatest values of recession in the direction of the antapex. Evidently, the maximum velocity is the velocity of the sun with respect to the nearby stars which form the **local standard of rest.** Relative to the stars around it the sun with its family of planets is moving at a rate of 19.4 kilometers/second toward a point in the constellation Hercules, about 10° southwest of the bright star Vega.

When the drift of fainter and more distant stars is included, the apex shifts towards Cygnus in which direction the sun's motion is directed in the rotation of the Galaxy (Section 17.5) at about 250 kilometers/second. When the drift of special types of stars and objects is considered, the apex of the sun's way shifts accordingly. This tells us that these types of stars and objects form significant groups differing from the average motion of all stars together. For example, the motion of the sun with respect to the short-period variable stars (Section 13.8) is about 130 kilometers/second whereas its motion with respect to the globular clusters (Section 16.8) is 175 kilometers/second.

12.6 Rotations of the stars

The rotations of stars have been studied extensively since 1930 by means of the widening of the lines of their spectra. Starlight emanating from a spinning source comes from a source that is partly approaching and partly receding from the earth except when the axis of rotation is directed toward the earth. Thus the lines are widened by the Doppler effect by an amount that depends on the speed of the rotation and the direction of the axis. We can picture (Fig. 12.7) the line widening as if our spectroscope could see first the approaching limb of the star, then the center, and finally the receding limb as shown in Fig. 12.7. Actually, since the stars are seen as point sources our spectrograph records the line from every point of the star. The integrated effect is a broadened, shallow line as shown. If the star is not rotating or if its axis of rotation is pointed toward us the integrated effect is a narrow deep line as shown in the lower part of the diagram.

Blue stars of the main sequence (Section 12.19) are likely to have high speeds of rotation, some exceeding 300 kilometers/second at their equators. As an example, note the wide lines in the spectrum of Altair (Fig. 12.8); the period of rotation of this star is 6 hours, as compared

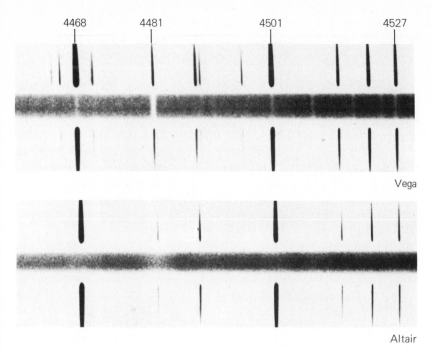

4468   4481   4501   4527

Vega

Altair

Figure 12.8 *Negative prints of the spectra of Vega and Altair. The broadened lines of Altair reveal the rapid rotation of this star. (Yerkes Observatory photograph.)*

with almost one month in the case of the sun. The narrower lines in the spectrum of Vega, also a blue star, would suggest that its axis is directed more nearly toward us if its rotational period is 6 hours like that of Altair.

Yellow and red stars of the main sequence rotate more slowly (Fig. 12.9) except when they belong to close binary systems; perhaps much of the original spins of the single stars now appear in the revolutions of their planets. Giant stars generally have slower rotations than corresponding main-sequence stars. Conservation of their angular momentum requires them to rotate more slowly as they expand in their evolution from the main sequence to the giant stage.

Stellar Spectra

The spectra of stars are characterized by patterns of dark lines, and in some cases of bright lines as well, on otherwise continuous bright backgrounds. The patterns are different for stars having different surface temperatures. As noted earlier, different molecules and atoms produce characteristic lines on any particular spectrum. In order to understand this process more thoroughly, we will first examine the chemical makeup of these substances and then look at some spectra in detail.

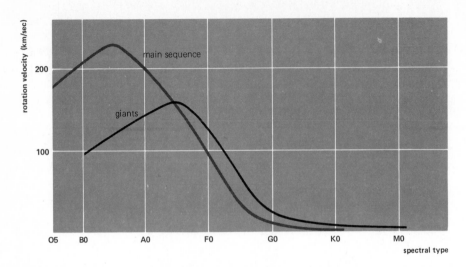

Figure 12.9 *Early type stars rotate more rapidly than late type stars.*

12.7 Constituents of atoms

The smallest particle of any of the 92 chemical elements (103 including man-made elements) that retains the properties of that element is known as an **atom.** Atoms are composed of electrons, protons, and neutrons. The **electron** is the lightest of these constituents and it carries unit negative charge of electricity. The **proton** is 1836 times heavier than the electron and carries unit positive charge. The **neutron** has about the same mass as the proton and is electrically neutral.

The atom consists of a nucleus surrounded by electrons. The nucleus ranges from a single proton in the ordinary hydrogen atom to an increasingly complex group of protons and neutrons in the heavier chemical elements. Each added proton contributes one unit to the positive charge on the nucleus. In the normal atom the nucleus is surrounded by negatively charged electrons equal in number to the protons, so that the atom as a whole is electrically neutral (Fig. 12.12).

Atoms can combine to form compounds that are also electrically neutral. Compounds will have different properties than those of the atoms that comprise them. The smallest part of a compound that still retains the properties of the compound is known as a **molecule.** An example is carbon dioxide, which is composed of an atom of carbon and two atoms of oxygen.

12.8 The chemical elements

Table 12.2 lists the names, symbols, atomic numbers, and atomic weights of the 26 lighter elements, from hydrogen to iron. All elements heavier than helium are referred to as metals by astronomers.

Element	Symbol	Atomic Number	Atomic Weight	Table 12.2 The Lighter Chemical Elements
Hydrogen	H	1	1.0080	
Helium	He	2	4.0026	
Lithium	Li	3	6.941	
Beryllium	Be	4	9.0122	
Boron	B	5	10.811	
Carbon	C	6	12.0111	
Nitrogen	N	7	14.0067	
Oxygen	O	8	15.9994	
Fluorine	F	9	18.9984	
Neon	Ne	10	20.179	
Sodium	Na	11	22.9898	
Magnesium	Mg	12	24.305	
Aluminum	Al	13	26.9815	
Silicon	Si	14	28.086	
Phosphorus	P	15	30.9738	
Sulfur	S	16	32.06	
Chlorine	Cl	17	35.453	
Argon	Ar	18	39.948	
Potassium	K	19	39.102	
Calcium	Ca	20	40.08	
Scandium	Sc	21	44.956	
Titanium	Ti	22	47.90	
Vanadium	V	23	50.9414	
Chromium	Cr	24	51.996	
Manganese	Mn	25	54.9380	
Iron	Fe	26	55.847	

The **atomic number** of an element is the number of protons in the nucleus and also the number of electrons around the nucleus of the neutral atom. All atoms having the same atomic number belong to the same chemical element.

The **atomic weight** is the relative mass of the atom; the unit employed in the table is $\frac{1}{12}$ the mass of the carbon-12 atom, taken as 12.000. The weights given here are generally the averages for two or more different kinds of atoms, or **isotopes,** of the same element having different numbers of neutrons in their nuclei.

Hydrogen comprises about 90% of all the atoms and 70% of all the mass of material in the stars. Helium is second with about 9% of the atoms and 28% of the mass. All the other elements together contribute only about 2% to the mass of the stars.

12.9 Model of the hydrogen atom and its spectrum

The ordinary hydrogen atom consists of one proton attended by a single electron. A pictorial model proposed by Niels Bohr in 1913 helps in an understanding of the atomic spectrum of hydrogen. In this model the electron revolves around the proton in one of many permitted orbits or energy levels. The atom can absorb energy from a radiation field, for example from a continuum of light that strikes it, only in the precise amount required to raise the electron from a lower orbit or energy level to a higher one. This **absorption** produces a **dark line** at the appropriate wavelength in the spectrum of the light. If the electron then falls back to the original lower level the energy is reemitted at the same wavelength. These bundles of light energy are called photons or particles of light.

At first one might think that this process should cancel out and leave the spectrum as it was. However, the light abstracted in the first process is removed from the line of sight to the observer, but the reemitted light is sent out in all directions, including the direction back to the source. Thus, only a small percentage of the light is reemitted toward the observer and the dark line is observed.

Although one hydrogen atom with its single electron absorbs energy corresponding to one dark line at a time, the many atoms in the atmosphere of a star together can produce the series of all permitted lines. The particular series where the electrons are raised from the second orbit of the Bohr model (Fig. 12.10) is known as the **Balmer series.** Its first line, Hα, is in the red region of the spectrum and its second line, Hβ, is in the blue-green region. From there the lines continue with diminishing spaces between them to a head in the ultraviolet. More than 30 lines of the Balmer series are recognized in the spectra of blue stars.

The **Lyman series,** based on the innermost Bohr orbit, is a similar array of lines in the far ultraviolet; it is recorded as a series of bright lines in the spectrum of the sun photographed from rockets and satellites. The **Paschen** series and others are in the infrared.

It is apparent that transitions from the third energy level to the second

Figure 12.10 *Conventional representation of the hydrogen atom. Possible orbits of the electron around the nucleus are shown as circles and represent the various energy levels the electron can occupy. Changes to and from the levels give rise to the emission and absorption series of hydrogen.*

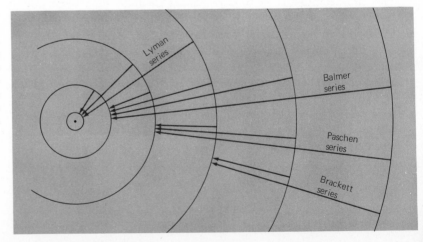

(Hα) are less energetic than those from the second to the first (Lα). Similarly, those from the fourth to the third (Pα) are less energetic than Hα, and so on. Transitions from the very high levels to levels nearby are consequently much less energetic and may fall in the radio region of the spectrum. The transition from level 110 to level 109 has been observed at λ 5.99 centimeters from the hydrogen atom in the low-density interstellar space. Many such **recombination lines** have been observed.

Another energy transition of hydrogen that has had enormous consequences for astronomy occurs in the lowest level or ground state. In our model we may picture the nucleus and electron as spinning, or rotating, as the electron revolves about the nucleus. Quantum theory tells us that the electron must have constant spin but the spin may be in the same direction as that of the nucleus (parallel) or opposite to the direction of the nucleus (antiparallel) as shown in Fig. 12.11. E. Fermi first pointed out that the energy in these two conditions is different. H. van de Hulst noted that when the atom flips from parallel to antiparallel spin the radiated energy has a wavelength of 21 centimeters. When the electrons of excited hydrogen atoms drop to the ground state, or lowest orbit, the results will be half of the electrons having parallel spin and half having antiparallel spin. Those having parallel spin will flip over to the antiparallel spin and this should be detectable if the amount of neutral hydrogen is sufficient. This single transition has been the most powerful tool, over the past decade, in determining the structure of the Galaxy, which we shall discuss later.

12.10 Atomic models of other elements

Each step in the succession of elements heavier than hydrogen adds one proton and usually one neutron to the nucleus of the atom and one electron in the outer structure. Thus the normal helium atom has two protons, two neutrons, and two electrons, and lithium has three of each. The electrons of the growing population take their places systematically in **shells,** which are the same as the Bohr orbits. The shells are filled in

Figure 12.11 *The λ 21-centimeter line results because the energy level of the spin of the electron in the antiparallel case is less than that of the parallel case.*

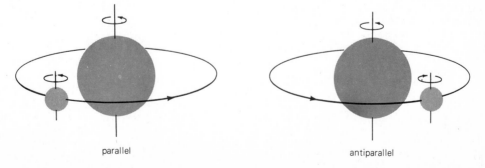

parallel   antiparallel

Figure 12.12 *The electron structures of a few atoms. The two electrons of helium are all that can occupy the first orbit, thus helium is tightly bound. Two electrons may occupy the lowest orbit of hydrogen giving rise to the H⁻ ion.*

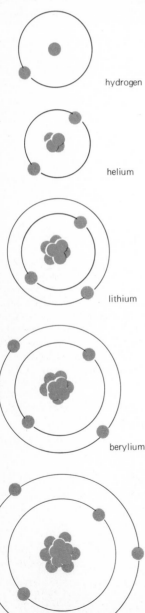

hydrogen

helium

lithium

berylium

boron

order of distance from the nucleus by $2n^2$ electrons in each, where n is the number of the shell. Thus the first shell is filled by two electrons, the second by eight electrons, and so on. A filled shell will not receive more electrons and is reluctant to release any that it possesses.

The electron structures of a few normal atoms are shown in Fig. 12.12. The helium atom has its two electrons locked in its filled innermost shell, which is not easily broken. The neon atom has two filled shells and is also relatively inactive. The lithium and sodium atoms, which are very active, each have in an outer shell a single electron that can be readily removed. These few examples will suggest that atoms of the different chemical elements in the atmospheres of stars in the same conditions may not be equally effective in producing lines in their spectra. We see presently that the different patterns of lines in stellar spectra are determined mainly by the different temperatures of the stars.

12.11 Neutral and ionized atoms

The **neutral atom** has its full quota of electrons. The **normal atom** has its electrons at the lowest possible levels, whereas the **excited atom** has absorbed energy to raise one or more electrons to higher levels. We designate the neutral atom by its symbol plus the Roman numeral I, e.g., Ca I, O I, H I, etc.

The **ionized atom** has lost one or more electrons. It has absorbed enough energy to expel these electrons beyond its outermost orbit. The process of expelling electrons is called **ionization.** They have become **free electrons,** free to move independently until they are captured by other ionized atoms. The singly ionized atom has lost a single electron and thereby carries a single unit positive charge of electricity. The doubly ionized atom has an excess of two positive charges. Each successive ionization makes the atom more difficult to excite or to ionize further. We designate a singly ionized atom by its symbol and the Roman number II, a doubly ionized atom by its symbol and the Roman numeral III, etc., e.g., Ca II, Ca III, etc.

We have seen that the sun is a glowing gaseous globe with a surface temperature of nearly 6000 K. The energy distribution approximates very nearly that derived by M. Planck for black bodies, that is, bodies that absorb and reemit all energy falling upon them and the energy distribution depends only upon temperature. Thus we refer to the sun and stars as thermal sources. The total energy emitted *(F)* per square centimeter per second is proportional to the fourth power of the temperature.

$$F = \sigma T^4$$

$$\sigma = 5.67 \times 10^{-5}$$

This is known as the Stephen–Boltzmann relation. The wavelength (λ_{max}) of the maximum of the distribution in centimeters is inversely proportional to the temperature.

$$\lambda_{max} = \frac{0.2898}{T}$$

This is known as the Wien displacement law. Note that as T becomes large λ_{max} shifts toward shorter and shorter wavelengths. Hence, for $T = 3$ K, λ_{max} is close to 1 mm; for $T = 3000$ K, λ_{max} is close to 9600 Å; and for $T = 6000$ K, λ_{max} is close to 4800 Å.

As the temperature rises the thermal agitation becomes increasingly severe. This is because the increase in temperature is proportional to the square of the velocity of the particles. This agitation excites greater numbers of atoms and eventually ionization begins and increases. The effect of thermal agitation is greater (1) as the temperature of the gas increases and thus the violence of the atomic collisions is greater; and (2) as the electrons are more loosely held by the atom for a particular temperature.

The energy required to excite or to ionize an atom is usually given in electron volts (eV).

$$1 \text{ eV} = 1.6 \times 10^{-12} \text{ ergs}$$

For hydrogen an energy of 10.2 eV is required to excite the atom from the ground state to the first excited level, and further excitation to the next level requires only an additional 1.9 eV. The energy required to ionize the hydrogen atom is 13.6 eV, hence the level to and from which all of the Balmer lines arise is only 3.4 eV from ionization. The first level of excitation for sodium, calcium, and iron requires only 3 eV or less, whereas that of helium is about 20 eV. These are a few of the elements prominent in the visible spectrum. The values, in electron volts, for the first stage of ionization are the following: sodium, 5.1 eV; calcium, 6.1 eV; iron, 7.9 eV; and helium, 24.6 eV. For all elements the corresponding numbers are smaller for higher degrees of excitation than the first and larger for higher stages of ionization than the first. For example, to remove the first electron from helium requires 24.6 eV as we have noted. To remove the second electron from helium requires 54.5 eV.

12.12 Molecular spectra

In a somewhat analogous manner to the hydrogen atom we may think of a simple molecule as being made up of two atoms revolving about each other. Such a molecule may vibrate as it rotates (Fig. 12.13). The atoms may revolve about each other in certain orbits, much as the

Figure 12.13 *Molecular spectra may arise from pure rotation of the atomic components (top), pure vibration of the atomic components (middle), or a combination of rotation and vibration (bottom).*

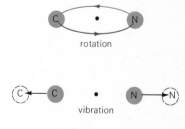

rotation

vibration

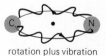

rotation plus vibration

electron about an atomic nucleus, and the vibration may occur only at certain frequencies. We say these motions are quantized, and they give rise to discrete lines in the spectrum. The energy radiated or absorbed when the molecule changes levels is referred to as the **rotation spectrum** if no vibration occurs or as the **vibration–rotation spectrum** of the molecule if both are occurring. Also, changes in the combined electron configurations may occur and these are referred to as **electronic transitions** of the molecule. The first two spectra have the transition energies so close together that they give the appearance of bands in the star's continuum. The electronic transitions may occur close together to form a band. Just as often, however, they will give rise to widely spaced lines and be mistaken for some unidentified atomic transition.

Many molecules have been identified in stellar spectra as we have already noted in the sun. The most interesting as far as stars are concerned are the two-atom, or **diatomic,** molecules of CN (cyanogen), TiO (titanium oxide), VO (vanadium oxide) and H_2 (molecular hydrogen). More complex molecules have been detected in interstellar clouds in the low-pressure conditions that exist there; we will note these later (Chapter 14) and call attention to their impact on our changing ideas on the origin and evolution of stars and "life" itself.

This introduction to atomic and molecular spectroscopy is necessary in order to describe and interpret the classification of stellar spectra.

12.13 Photographs of stellar spectra

Photographs of stellar spectra are taken generally in two different ways. One method employs the **slit spectroscope** at the focus of the telescope. It permits wider separation of the spectrum lines and the recording of a laboratory spectrum adjacent to the star's spectrum (Fig. 12.14). This method is accordingly useful for measuring wavelengths and Doppler shifts of the lines, but it is time consuming because it gives the spectrum of only one star at a time.

In the second method, a large prism of small angle is placed in front of the telescope objective, so that the whole apparatus becomes a spectroscope without slit or collimator. A single photograph shows the spectra of all stars of suitable brightness in the field of view of the telescope

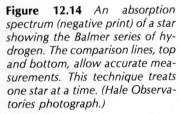

Figure 12.14 *An absorption spectrum (negative print) of a star showing the Balmer series of hydrogen. The comparison lines, top and bottom, allow accurate measurements. This technique treats one star at a time. (Hale Observatories photograph.)*

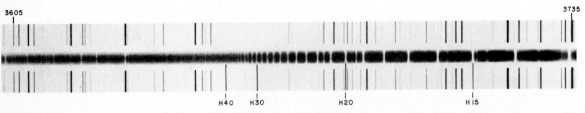

3605

3735

H 40   H 30   H 20   H 15

Figure 12.15 *An objective prism spectrogram. This technique records many stellar spectra at one time, but in much less detail than that of Fig. 12.14. (Warner and Swasey Observatory photograph.)*

(Fig. 12.15). This **objective prism** type is useful in qualitative studies of many stars, as in the classification of stellar spectra.

Studies of stellar spectra in objective prism photographs were in progress as early as 1885. A result of these studies is the Henry Draper Catalogue, mainly the work of A. J. Cannon, which with its extensions lists the positions, magnitudes, and spectral types of 400,000 stars. The Harvard types, together with the more recent luminosity classes (Section 12.20), remain the criteria in the present classification of stellar spectra.

12.14 Types of stellar spectra

The Harvard classification arrays the majority of the stars in a single continuous sequence with respect to the patterns of lines in their spectra. Seven divisions of the sequence are the principal spectral types and are designated by the letters O, B, A, F, G, K, and M. (A simple mnemonic for this sequence is Oh Be A Fine Girl, Kiss Me.) The various divisions are further subdivided on the decimal system. Thus an A5 star is halfway between A0 and F0. Some features of the principal types, as they appear in the violet and blue regions of the spectrum, are the following.

Type O. Lines of ionized helium, oxygen, and nitrogen are prominent in the spectra of these very hot stars along with lines of hydrogen. The O stars begin at type O5. An example is ξ Persei.

Type B. Lines of neutral helium are most intense at B2 and then fade, until at B9 they have practically vanished. Hydrogen lines increase in strength through the subdivisions. Examples are Spica and Rigel.

Type A. Hydrogen lines attain their greatest strength at A2 and then decline through the remainder of the sequence. Examples are Sirius and Vega. Thus far the stars are blue.

Type F. Lines of metals are increasing in strength, notably the Fraunhofer *H* and *K* of ionized calcium. These are yellowish stars. Examples are Canopus and Procyon.

Type G. Metallic lines are numerous and conspicuous in the spectra of these yellow stars. The sun and Capella belong to this type.

Type K. Lines of metals surpass the hydrogen lines in strength. Bands of cyanogen and other molecules are becoming conspicuous. These cooler stars are reddish. Examples are Arcturus and Aldebaran.

Type M. Bands of titanium oxide become strong up to their maximum at M7. Vanadium oxide bands strengthen in the still cooler divisions of these red stars. Examples are Betelgeuse and Antares.

Four additional and less populous types branch off near the ends of the sequence. **Type W** near the blue end comprises the Wolf–Rayet stars, having broad bright lines in their spectra. Near the red end, **types R and N** show molecular bands of carbon and carbon compounds, and **type S** has conspicuous bands of zirconium oxide.

Occasionally the spectrum of a star will show bright emission lines. If this is the case a lower case e is added to its spectral designation. Thus M5e means an M5-type star with emission lines. An example is Proxima.

The reader may suspect that such an apparently haphazard arrangement of letters for spectral types is not the way the system began. Originally the letters were arranged alphabetically and the principal criterion was the strength of the hydrogen lines. Thus the strongest hydrogen lines are seen in A stars. When it was realized that the general alignment was one of temperature, it became apparent that the O stars, which were very blue and hence very hot, had to be moved to the beginning of the classification. Similarly, the B stars had to be placed between the O and A stars. Many of the other classes were found to consist of so few stars that they were either small numbers of peculiar stars or minor mistakes that were corrected upon reexamination.

12.15 Sequence of stellar spectra

The sequence of stellar spectra (Fig. 12.16) is in order of increasing redness, and therefore of diminishing unit surface brightness of the stars. In general the progression in the line patterns is not caused by differences in chemical composition; helium is not more abundant in B stars, nor is

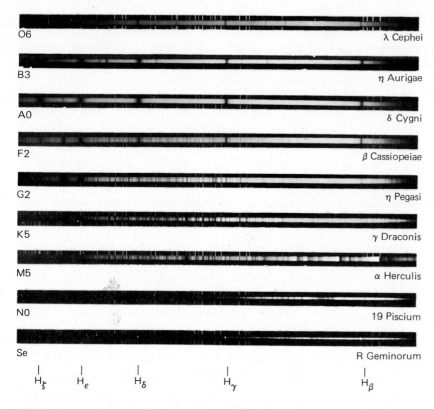

Figure 12.16 *The principal spectral types as recorded in the deep-blue region of the spectrum. Note the changing hydrogen line strengths with spectral type. (Hale Observatories photograph.)*

hydrogen in A stars. The lines of the different chemical elements become strongest at temperatures where their atoms in the stellar atmospheres are most active in absorbing the starlight.

The compact neutral-helium atoms imprint their dark lines at the high temperatures of B stars. Hydrogen is most active at the more moderate temperatures of A stars, although the great abundance of this element makes its lines visible throughout the sequence. Neutral atoms of the metals are more easily excited, so that their lines are strongest in the spectra of the cooler yellow and red stars. At higher temperatures the atoms of metals are ionized and have their stronger lines in the ultraviolet and x-ray regions of the spectrum. Atoms can assemble into molecules in the relatively cool atmospheres of red stars; bands of titanium oxide are conspicuous in the spectra of M stars as the accompanying photographs show.

Magnitudes of the Stars

In his star catalog compiled 18 centuries ago, Ptolemy gave estimates of brightness of the stars to aid in their identification. He based his estimates upon a system proposed by Hipparchus in 130 B.C. For these estimates the stars were divided into six classes, or magnitudes, in order of brightness. The brightest stars were assigned to the first magnitude. Somewhat less bright stars, such as the pole star, were of the second magnitude. Each succeeding class contained fainter stars than the one before, until the sixth magnitude remained for stars barely visible to the unaided eye.

The original magnitudes of the lucid stars continued in use until recent times, and the plan was extended somewhat arbitrarily to include fainter stars brought into view with telescopes. The present plan of magnitudes was proposed in 1856 by Norman Pogson when the brightness of stars had begun to be determined more accurately and was becoming an important factor in astronomical inquiries.

12.16 Apparent magnitudes and colors

Apparent magnitudes are the magnitudes that we assign stars as we see them according to the modified Ptolemaic scale. The apparent magnitude is a function of the distance from which we view a given object. Thus a car's headlights near to us are apparently brighter than the same car's lights far away. How much brighter or fainter?

Let us conduct an experiment in a long tunnel with perfectly nonreflecting walls. We place 100 candles at one end and we take away candles until we agree that the remaining candles are one-half as bright or one magnitude fainter. Let the fraction of the candles taken away be a. Now we take away candles again until we agree that the remaining candles are another magnitude fainter. Again we note that the fraction of candles taken away is a. And so we proceed for five times when we find that we have one candle remaining. The brightness changed by 100 while the magnitude changed by only five. In our example $100 = a^5$, or $a = 2.512$. In order to make a star appear one magnitude brighter to the eye you must increase its actual brightness by 2.512 times.

The astronomical magnitude scale runs contrary to what we expect for historical reasons, thus the star with the larger positive number is the fainter to our senses. In radio astronomy we give "brightnesses" in terms of flux units or the amount of energy received. Thus the smaller the number of flux units the fainter the object.

The brightness that we assign to a star depends upon our receiver and the temperature of the star. Suppose our eye responds to certain wavelengths (V), and another sensor, say, the photographic plate, responds at a different wavelength (P). The two sensors will assign different magnitudes to the same star. The blue-sensitive plate, properly calibrated, will see a 10,000 K star as brighter than our eye sees it. On the other hand the eye will assign a brighter magnitude to a 3000 K star than the photographic plate will.

These apparent magnitudes refer to the observed brightness of stars.

They may be determined to tenths of a magnitude by the trained eye, to hundredths of a magnitude in photographs, and to thousandths by more precise photoelectric methods.

The **visual magnitude** of a star refers to its brightness as observed with the eye, which is most sensitive to yellow light. The **photographic magnitude** refers to the star's brightness as recorded in a blue-sensitive photograph. The magnitudes differ in the two cases by an amount that depends on the color of the star. Note in Fig. 12.17 how much fainter, as indicated by its smaller image, a red star appears as photographed in blue than in yellow light.

Present studies often require the magnitudes of a star in several colors. The magnitudes are generally measured around specified wavelengths in the ultraviolet, blue, yellow or visual, red, and infrared, and are usually denoted by the letters $U, B, V, R,$ and L. The magnitudes of standard stars whatever the system used are being carefully determined in the different colors, so that all investigators may keep to the same color systems.

The **color index** (C.I.) of a star is the difference of its photographic magnitude (m_{pg}) and its visual magnitude (m_{pv}). Its value is often given as the difference, $B - V$, between the blue and visual magnitudes, but strictly speaking this is a measure of color, not the color index. The color index is then a numerical expression of the star's color. Taken as zero for stars of spectral type A0, the color is accordingly negative for the bluest hotter stars and attains a positive value for the reddest cooler ones. The observed color index of a star may exceed the normal index for the particular spectral type because of reddening of the starlight by intervening cosmic dust. Dust scatters the blue light, hence the transmitted light is redder than it should be. In any case, except for the effect of dust, the measured color of a star is independent of distance. If we are dealing only with thermal sources, in the simplest case the color assigns the star's temperature and hence its spectral type.

Figure 12.17 *A red star appears brighter on a red-sensitive plate (right) than it does on a blue-sensitive plate (left). (Yerkes Observatory photograph.)*

In our brief discussion of thermal energy (Section 12.11) we noted that the total energy *(F)* emitted was proportional to the fourth power of the temperature. In making comparisons between real stars and stars based upon theory it is often more convenient to refer to the total energy, that is, the sum of the energy from the very shortest wavelengths to the very longest wavelength radio radiation. The total energy can be given in magnitudes and such a magnitude is called a star's **bolometric magnitude.**

12.17 The brightest stars

The 23 stars listed in Table 12.3 are brighter than apparent visual magnitude +1.5 and are sometimes called "stars of the first magnitude," although they range through nearly three magnitudes. The visual *(V)* and blue *(B)* magnitudes were measured photoelectrically with the appropriate filters. The colors are shown by the differences $B - V$, which can be read from the table. The bluest stars, having the largest negative colors are alpha and beta Crucis and Spica. The reddest are Betelgeuse and Antares, which have the largest positive colors. The spectral types and luminosity classes are on the Morgan–Keenan system (Section 12.20); where the star is double, they refer to the brighter star of the pair.

The brightest stars have a considerable range in distance from us, as the parallaxes in the table show. Alpha Centauri, Sirius, and Procyon are among the nearest stars. Rigel, Canopus, and Deneb are the most remote in the list; they are evidently very luminous to shine so brightly in our skies. The significance of the last column of the table is explained in the next section. *The Yale Catalogue of Bright Stars* lists all of the stars brighter than 6.5 apparent magnitude and selected stars down to 7.5 magnitude. It lists a total of 9110 stars. Sixteen of the 22 brightest stars are visible in their seasons throughout the United States.

12.18 Absolute magnitudes

The apparent magnitude of a star relates to its observed brightness. This depends on the star's intrinsic brightness, or luminosity, and its distance from us. One star may appear brighter than another only because it is the nearer of the two; thus Sirius appears brighter than Betelgeuse, although the latter is actually many times the more luminous. In order to rank the stars fairly with respect to luminosity, it is necessary to calculate how bright they would appear if they were all at the same distance. By agreement the standard distance is 10 parsecs, or the distance at which the parallax is 0".1. Recall that the perceived brightness decreases with the distance squared. Let the absolute brightness be L and the perceived

Table 12.3
The Brightest Stars

Name	Spectrum	Apparent Visual Magnitude (V)	Apparent Photographic Magnitude (B)	Parallax	Distance in Parsecs	Absolute Visual Magnitude
Sirius[vb]	A1 V	−1.44	−1.45	0″.377	2.7	+1.4
Canopus	F0 II	−0.72	0.88	.018*	55.6*	−4.4*
α Centauri[vb]	G2 + K1	−0.27	0.44	.760	1.3	+4.2
Arcturus	K1 III	−0.05	1.19	.091	11.0	−0.2
Vega	A0 V	0.03	0.03	.123	8.1	+0.5
Capella[sb]	G8 + G0	0.09	0.90	.071	14.1	−0.6
Rigel[sb]	B8 Ia	0.11	0.06	.004*	250.0*	−7.0*
Procyon[vb]	F5 IV-V	0.36	0.77	.287	3.5	+2.7
Achernar	B5 IV	0.49	−0.66	.028	35.7	−2.2
β Centauri	B1 II	0.63	0.39	.008*	125.0*	−5.0*
Betelgeuse[v]	M2 I	0.4	1.89	.006*	166.7*	−5.9*
Altair	A7 IV-V	0.77	0.99	.204	4.9	+2.3
Aldebaran	K5 III	0.80	2.35	.048	20.8	−0.8
Acrux[vb]	B1 + B3	0.83	0.37	.012	83.3	−3.7
Antares[vb,v]	M1 Ib + B	0.94	2.77	.008*	125.0*	−4.7*
Spica[sb]	B1 V	0.97	0.74	.015	66.7	−3.1
Fomalhaut	A3 V	1.16	1.25	.143	7.0	+1.9
Pollux	K0 III	1.15	2.16	.093	10.7	+1.0
Deneb	A2 Ia	1.25	1.33	.002	500.0	−7.2
Mimosa	B0 III	1.29	1.04	.008*	125.0*	−4.3*
Regulus	B7 V	1.34	1.23	.038	26.3	−0.8
ε Canis Majoris[vb]	B2 II	1.48	1.31	.005*	200.0*	−5.0*

vb, visual binary.

sb, spectroscopic binary.

v, variable star.

*, uncertain values.

brightness be *b*. Then we can write $b = L/4\pi d^2$. Here *b* corresponds to *m*, *L* corresponds to *M*, and, of course, *d* corresponds to distance *d*. By arranging *L* to be the absolute brightness at 10 parsecs we can then derive the relation used by astronomers.

The **absolute magnitude** is the magnitude a star would have at the distance of 10 parsecs, or 32.6 light years. The relation is

$$M = m + 5 - 5 \log d,$$

where *M* is the absolute magnitude and *m* is the apparent magnitude at the distance *d* in parsecs. The absolute magnitude has the same character as the apparent magnitude from which it is derived, whether visual, photographic, or some other kind. Rearranging the relation slightly we can write

$$m - M = 5 \log d - 5,$$

where $m - M$ is known as the **distance modulus.** This term and its significance will be discussed later.

The sun's absolute visual magnitude is +4.8. At the moderate distance of 10 parsecs the sun would appear as a star only faintly visible to the unaided eye. Note in the last column of Table 12.3 that the apparently brightest stars are all intrinsically brighter than the sun.

12.19 Stars of the main sequence

When the spectral types of stars around us are plotted with respect to the absolute magnitudes of these stars, as in Fig. 12.18, the majority of the points are arrayed in a band running diagonally across the diagram. This band is known as the **main sequence** and the graph is known as the **H-R** diagram, after E. Hertzsprung and H. N. Russell who first introduced it. The middle line of the band drops from about absolute magnitude −3 for blue stars to fainter than +10 for red stars.

Figure 12.18   *The absolute magnitudes of relatively nearby stars plotted according to spectral types. This is a true H-R diagram. Note the tendency to align in columns.*

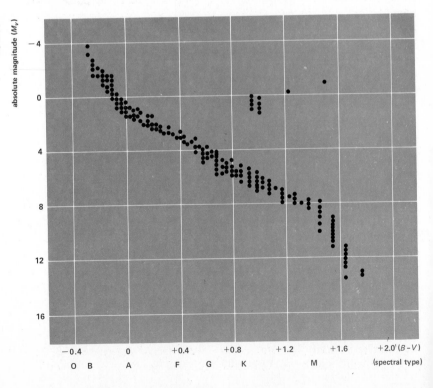

The sun, a yellow star of type G2 and absolute visual magnitude +4.8, is a main-sequence star, as we see. Blue stars of the sequence are more luminous than the sun because they are hotter and larger. Red stars of the sequence are less luminous than the sun because they are cooler and smaller. Remembering that the sun at the standard distance of 10 parsecs (32.6 light years) would be only faintly visible to the unaided eye, we understand why the red stars of the sequence are generally invisible without a telescope.

Since its introduction, the spectrum–absolute magnitude diagram or the similar color–absolute magnitude diagram has played a leading part in directing the studies of the stars. E. Hertzsprung had previously drawn attention to a sharp distinction between red stars of high and low luminosity, and had named them giant and dwarf stars, respectively. The term **dwarf star** is commonly used today to denote main-sequence stars fainter than about absolute magnitude +1, but this distinction is not generally applied to the white dwarf stars.

Giant stars, such as Capella and Arcturus, are considerably more luminous than main-sequence stars of the same spectral type. They are giants in size as well as in brightness. All stars of the same spectral type have about the same surface temperature, and therefore surface brightness per square meter. If one type-M star, for example, greatly surpasses another type-M star in luminosity at the same distance, its surface must contain many more square meters and its diameter must be much greater. Giants in the sun's neighborhood are mainly red stars.

The H-R diagram is a diagram of stellar evolution. We discuss the diagram further and make considerable use of it in the remaining chapters of this book.

Figure 12.18 shows the H-R diagram for the nearby stars. Strictly speaking, the H-R diagram is a plot of spectral type against magnitude, but in Fig. 12.19 we are substituting color for the spectral type. Since spectral types are arbitrarily divided into ten subclasses for every major class, the H-R diagram, as in Fig. 12.18, gives the impression of bunching in columns. This bunching effect is removed when we use color. The general relationship between spectral type and color is given in Table 12.4. The main sequence running from the upper left to the lower right in either Fig. 12.18 or 12.19 is obvious. Above the main sequence and on the red side of the diagram lay the giant stars. The dots well below the main sequence and generally to the left of the diagram represent the white dwarfs (Section 15.8), some of which are yellow and even red. These very faint stars are smaller than the larger planets. White dwarfs are occasionally designated by the letters wd. More recently they have been designated by the capital letter D followed by a capital letter of the Harvard sequence to indicate the color. Thus a white dwarf designated DG has a yellowish color similar to that of the sun.

Figure 12.19 *An absolute mag-*
nitude–color diagram for the
same stars as shown in Fig. 12.18.
The column effect is removed be-
cause color is not restricted to the
ten subdivisions of each spectral
type.

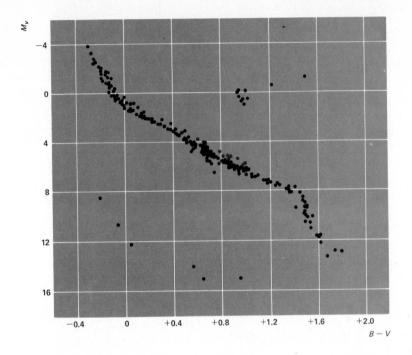

12.20 Luminosity classes of stars

The surface temperatures and normal colors of stars at intervals along the spectral sequence are shown in Table 12.4. The temperatures, as assigned by P. C. Keenan and W. W. Morgan, are based on an adopted temperature of 5730 K for the sun and spectral type G2. A few adjustments have been made in the table to bring it into agreement with later work by N. G. Roman. Diminishing temperature is mainly responsible for the succession of spectral patterns, as has been said. Note that the temperatures of yellow and red giants are lower than those of main-sequence stars of the same spectral type. The reason is that the giant stars are less dense, so that their atoms attain a particular degree of ionization at a lower temperature. Although the spectra of giants and main-sequence stars of type K0, for example, are generally similar, particular features may be considerably different. Some lines become stronger and others weaker as the stars are more luminous. Thus a complete description of a star's spectrum is obtained by adding its **luminosity class** to its spectral type.

The luminosity classes of Morgan and Keenan are numbered in order beginning with the most luminous stars, which have the least dense atmosphere. The numeral I refers to supergiant stars, Ia for the more luminous and Ib for the less luminous supergiants; II refers to bright giants; III to normal giants; IV to subgiants; V to main-sequence stars; and VI to

| Spectrum | MAIN SEQUENCE | | GIANTS | | |
	Temperature (K)	Color $B - V$	Temperature (K)	Color $B - V$	
O5	35000	−0.45			
B0	21000	−0.31			
B5	13500	−0.17			
A0	9700	0.00			
A5	8100	0.16			
F0	7200	0.30			
F5	6500	0.45			
G0	6000	0.57	5400	+0.65	
G5	5400	0.70	4700	0.84	
K0	4700	0.84	4100	1.06	
K5	4000	1.11	3500	1.40	
M0	3500	1.39	2800	1.65	
M5	2600	1.61	2200	1.85	

Table 12.4
Effective Temperatures and Colors

subdwarfs. Thus the two-dimensional designation in Table 12.3 for Deneb is A2 Ia; and Vega, A0 V. An H-R diagram showing the regions of the various luminosity classes is shown in Fig. 12.20.

12.21 Unusual stars

Many stars do not fit neatly into the spectral sequence and luminosity classes that we have just described. Most conspicuous are the different types of variable stars that we discuss in Chapter 13. The remaining nonconforming stars usually have circumstellar shells, envelopes of one sort or another, conspicuous radiation of an unusual nature, or peculiar abundances of elements. Among these stars are the planetary nebulae, T Tauri stars, x-ray stars, and infrared stars.

One of the most conspicuous of unusual stars is a group called **shell stars** whose prototype is P Cygni. These stars appear to be early B-type stars of absolute magnitude about −4, with a shell giving rise to very bright emission lines. Apparently related are the **Wolf–Rayet** stars, which are late O-type stars or early B-type stars, with absolute magnitudes around −5. The Wolf–Rayet stars divide into two classes, WN and WC depending upon whether nitrogen or carbon is the more obvious in their emission-line spectra. The Wolf–Rayet stars are explained as rapidly rotating stars that suddenly contract, leaving an extensive shell that expands outward at large velocities, 200 kilometers/sec or larger. The nuclei of planetary nebulae (Section 13.18) resemble Wolf–Rayet stars, but must be in a later stage of evolution because their absolute magnitudes are around +1. The shells of planetary nebulae expand much more slowly also.

Figure 12.20 *A schematic H-R diagram showing the locations of the principal luminosity classes: I, supergiants; II, bright giants; III, giants; IV, subgiants; V, main sequence; and VI, subdwarfs.*

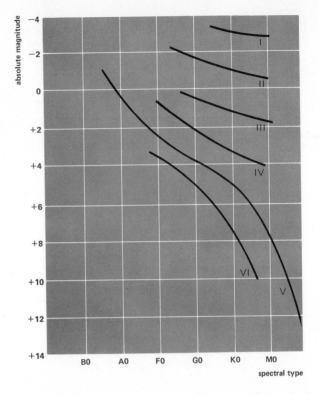

Another group of unusual stars are peculiar stars of type A, designated Ap stars and referred to as **A peculiar stars.** About 10% of all A-type stars show any of the following peculiarities; abnormal abundances of heavier elements, strong magnetic fields, variability in their light and magnetic fields, and spectral variability. It is believed that these stars have nuclear reactions taking place on their surfaces—a process referred to as **spalation.** This would explain the abnormal abundances of heavy elements because they are being formed near the surface. Normally the heavy elements are formed in the core and cannot come to the surface.

In the giant and supergiant region of the H-R diagram lie the carbon stars that, as we have seen, divide into classes R, N, and S. The carbon stars of class N, which are irregular variables, always seem to show an abnormal abundance of technetium and hence have been called **technetium stars.**

We have only touched upon a few of the more obvious unusual stars. We remind the students that no matter how normal a particular star may be, as soon as you begin studying it it reveals differences and peculiarities that cause it to stand out from all other stars and make it more interesting. Stars are like people in that each one is different; it is only their very gross properties that allow us to classify them.

Questions

1 What is meant by the term annual heliocentric parallax?
2 What do we mean by the term absolute parallax?
3 Of the nearest stars, how many are truly single?
4 Define the term radial velocity. How is it measured?
5 What is the most abundant element in the stars?

6 What is a neutral atom? An ionized atom?
7 How does the maximum of the black-body radiation curve shift with increasing temperature?
8 Describe the spectral sequence of stars.
9 Is an A star that appears to be of first magnitude, brighter or fainter at 10 parsecs?
10 What is meant by the term "the main sequence?"

Other Readings

Ashbrook, J., "How Far Away are the Stars?" *Sky & Telescope*, **47**, 165–167 (1974).

Carruthers G. R., and C. B. Opal, "A Far-Ultraviolet Rocket Survey of Orion," *Sky & Telescope*, **53**, 270–275 (1977).

Gieseking, F., "Measuring Radial Velocities with an Objective Prism," *Sky & Telescope*, **57**, 142–145 (1979).

Mihalas, D., "Interpreting Early Type Stellar Spectra," *Sky & Telescope*, **46**, 79–83 (1973).

Sitterly, B. W., "Changing Interpretations of the Hertzsprung-Russell Diagram, 1910–1940," *Vistas in Astronomy*, **12**, 357–371 (1970).

chapter 13

Double and Variable Stars

Binary stars and variable stars change in appearance with time. Binary stars are gravitationally bound stars which revolve around one another in fixed orbits. The motions of these stars about their common center give rise to a regular change in their positions in the sky as well as to the characteristics of their light curves and spectra. Variable stars change in aspect because of instabilities inherent in the stars themselves. In this chapter we will see that the systematic variations displayed by these stars not only tell us about the nature of individual stars, but also about the distances to the stars, and therefore about the scale of the universe.

Binary Stars

Binary stars are physically associated pairs of stars, and their study yields a direct approach to determining the masses of these stars. In some pairs the components are far enough apart to appear separated through a telescope; in other pairs the presence of the companions can only be detected by minute periodic changes in their positions; in still others the members can only be detected with a spectroscope. Some spectroscopic binaries mutually eclipse as they revolve, so that they vary periodically in brightness.

13.1 Visual double stars

Visual double stars appear as single stars to the unaided eye but separate into pairs when observed with the telescope. Mizar, at the bend in the Big Dipper's handle (Fig. 11.4), was the first of these, in 1650, to be reported casually by several observers; the only connection between the two stars of such pairs was believed to be that they chanced to have nearly the same direction from us (Fig. 13.1). Castor, beta Cygni, and alpha Centauri were other early known examples. By 1803 W. Herschel had determined from repeated observations that the two stars of Castor were revolving around a common center between them. He then made the distinction between "optical doubles" or "optical pairs" (stars that seem to be doubles because of their proximity to each other) and "real doubles" (stars that revolve around a common center). The latter, now called **visual binary stars,** are in the great majority.

A total of 65,000 visual binaries are recognized, mainly by the common proper motions of the component stars. They are tabulated in detail in the *Double Star Index and Observation Catalogue* maintained by the U.S. Naval Observatory. About 2500 binaries have had time since their discoveries to show evidence of orbital motion, and 20% of these have progressed far enough in their revolutions to permit reliable determinations of their orbits. Pairs having the smallest separations, which are likely to go around most rapidly, can be observed satisfactorily only with long-focus telescopes.

The presence of additional stars in binary systems, forming **multiple systems,** is not exceptional. Alpha Centauri is representative of a common type of triple systems. The double-double epsilon Lyrae is a familiar example of quadruple systems.

13.2 The apparent and the true orbit

The position of the fainter star of a visual binary relative to the brighter star is denoted by the position angle and distance (Fig. 13.2).

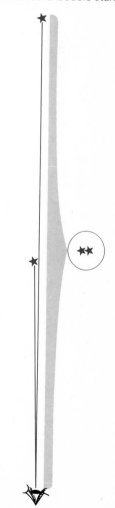

Figure 13.1 *Two stars far removed from each other may appear to be close enough (inset) to be called a double star.*

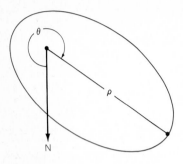

Figure 13.2 *Binary-star measures are the separation ρ and the position angle θ, measured from north through east and on around to south and west.*

The **position angle** θ is the angle at the brighter star between the directions of the fainter star and the north celestial pole; it is reckoned in degrees from the north around through the east. The **distance** ρ is the angular separation of the two stars. The position is measured micrometrically either directly at the telescope or in a photograph if the separation is wide enough.

When the recorded positions have extended through a considerable part of the period of revolution they are plotted after making corrections for precession. The relative **apparent orbit** of the fainter star is an ellipse that best represents the observed positions and in which Kepler's law of areas (Section 6.2) is maintained in the motion around the brighter star (Fig. 13.3). The apparent ellipse is projected on the plane tangent to the sky, and from it the true ellipse having the brighter star at one focus is readily calculated.

The **elements** of the relative **true orbit** are the specifications of the true ellipse, such as its eccentricity, period, time of periastron passage (nearest approach of the two stars), longitude of periastron, position angle of the node, inclination to the plane of the sky, and angular size, or the linear size, if the distance from us is also known. When the revolutions of both stars have been observed with reference to stars nearby in the photographs, the separate orbits can be determined; these differ from the relative orbit only in size. This then allows the determination of the individual masses.

Some characteristics of a few visual binaries are shown in Table 13.1. The first two binaries in the list have unusually short periods of revolution for visual binaries. The angular semimajor axis of an orbit may be converted to astronomical units by dividing the tabular value by the corresponding parallax. Thus the mean distance between the stars of the binary BD −8° 4352 is slightly more than the earth's distance from the sun, and that of the components of Castor is twice Pluto's distance from the sun. The separate masses of the stars are given in terms of the sun's mass, except that the value in parentheses is the sum of the two masses.

A study of the table leads to some interesting general conclusions. Visual binaries do not differ too greatly in brightness; this is an observational selection effect because we cannot easily see stars of great difference in brightness close together. The orbital eccentricities, except for very short-period stars, tend to be larger than 0.3. The masses of the various stars taken singly do not differ by a factor of 10 from that of the sun.

One of the nearest stars, the brilliant Dog Star, Sirius, drifts in the heavens three-fourths the apparent width of the moon in 1000 years. As early as 1844, F. W. Bessel announced that it is pursuing a wavy course instead of having the uniform motion expected of a single star. He concluded that Sirius is mutually revolving with a traveling companion; its orbit with the still unseen companion was derived in 1851. This was the first **astrometric binary**.

Figure 13.3 *The apparent orbit of the binary star Krüger 60. The apparent orbit can be studied to reveal the true orbit which, if the distance to the pair is known, gives the masses of the stars.*

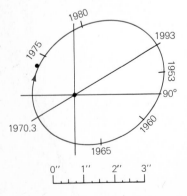

Table 13.1
Parameters for Selected Visual Binaries

Name	VISUAL MAGNITUDES m_1	VISUAL MAGNITUDES m_2	PERIOD (YEAR) P	SEMI-MAJOR AXES (α)	ECCEN-TRICITY (e)	PARALLAX (p)	SOLAR MASSES m_1	SOLAR MASSES m_2
BD −8° 4352	9.7	9.8	1.7	0″.22	0.05	0″.157	0.48	0.47
δ Equulei	5.2	5.3	5.7	0″.26	0.42	0″.056	1.57	1.51
42 Comae	5.0	5.1	25.8	0″.67	0.49	0″.054	1.41	1.46
Procyon	0.4	10.7	40.6	4″.55	0.40	0″.287	1.77	0.65
Kruger 60	9.8	11.4	44.5	2″.39	0.42	0″.254	0.26	0.16
Sirius	−1.4	8.5	50.1	7″.50	0.59	0″.376	2.12	1.04
α Centauri	0.0	1.2	79.9	17″.58	0.52	0″.760	1.05	0.89
ξ Boötes	4.7	6.8	150[a]	4″.88	0.51	0″.149	0.84	0.72
Castor	2.0	2.8	420[a]	6″.30	0.33	0″.074	(3.50)	

[a] Uncertain values.

The companion was first seen by the optician A. Clark in 1862, who was testing a new lens. The companion revolves in a period of 50 years and in an orbit of rather high eccentricity (Fig. 13.4). Despite the glare of Sirius, the companion is clearly visible with large telescopes except when the two stars are least separated. The latest periastron passage occurred in 1944 and the widest separation was reached in 1974, although apastron occurred in 1969. This happens because the widest separation in the apparent orbit does not occur at the same time as the widest separation in the true orbit. The companion of Sirius (Fig. 13.5) was among the first white dwarf stars to be known (Sections 12.19 and 15.8).

The companion of Procyon was likewise discovered from its gravitational effect on the proper motion of the bright star; it was first observed with the telescope in 1896. A similar case is the companion of the faint red star Ross 614, which was detected in 1936 at McCormick Observatory and later photographed using the 5-meter Hale telescope in 1955. Unseen companions of other stars have been studied at several observatories by their gravitational effects (perturbations) upon the motion of the visible star. An especially remarkable example is the companion of Barnard's star (Section 12.3), detected by P. van de Kamp. The companion's estimated mass is only 1.5 times the mass of Jupiter. It revolves with its primary in a period of 11.1 years. This companion is tentatively called a planet because, according to our present knowledge, it does not shine by its own light. Van de Kamp even advances the idea that the perturbation may result from two smaller objects.

Figure 13.4 *The apparent orbit of the famous binary Sirius. Note that the greatest separation in the apparent orbit does not correspond to apastron. This is due to the fact that the orbit is not perpendicular to the line of sight.*

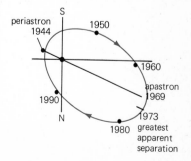

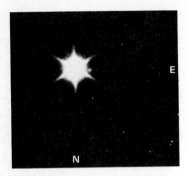

Figure 13.5 *A photograph of Sirius showing both components A and B. Sirius B is nine magnitudes fainter than Sirius A, hence it is hard to see and difficult to photograph. A special device was used to make this photograph, hence the "spikes" around Sirius A. (Sproul Observatory photograph by S. L. Lippencott and J. K. Wooley.)*

13.3 Masses of binary stars

The combined mass of a visual binary can be evaluated when the orbit of the binary has been determined. The method is another application of the general statement of Kepler's harmonic law, which we have already used to find (Section 6.5) the combined mass of a planet and its satellite. Applying the same law and choosing our units carefully, we can write the law such that the combined mass of the pair of stars equals the cube of their linear mean distance apart *(a)* in astronomical units divided by the square of the period *(P)* of revolution in years,

$$m_1 + m_2 = \left(\frac{\alpha}{p}\right)^3 \frac{1}{P^2}.$$

We have converted the angular separation (α) in the true orbit into astronomical units by dividing by the parallax *(p)* of the system. By selecting the earth–sun system to establish our units (we can neglect the mass of the earth because it is almost a factor of 10^6 less than the sun) we have established one solar mass as our unit of mass, which is much more convenient than grams, kilograms, or metric tons in this case.

As an example of the use of the above relation, the sum of the masses of the binary alpha Centauri is calculated from the data of Table 13.1 as follows:

$$m_1 + m_2 = \left(\frac{17.58}{0.760}\right)^3 \frac{1}{(79.9)^2} = 1.94.$$

Thus the combined mass of alpha Centauri is almost twice the sun's mass. The third component, Proxima, is so far removed it does not affect the calculation.

The sum of the masses is all that can be determined from the relative orbit. When, however, the revolutions of both stars have been observed, the individual masses become known, because the ratio of the masses is inversely proportional to the ratio of the distances of the two stars at any time from the common center around which they revolve:

$$\frac{m_1}{m_2} = \frac{a_2}{a_1} = \frac{\alpha_2}{\alpha_1}.$$

Solving this relation together with the harmonic law will yield the individual masses.

In Table 13.1 a few interesting anomalies occur. We recall from Table 12.1 that the companions of Sirius and Procyon are very faint. We also recall that the companions have colors (surface temperatures) very similar to the primaries. The only way that we can explain these facts is if the companions are very small, much smaller than the sun yet with masses similar to that of the sun. The fact that two white dwarfs are in our small sample indicates that white dwarfs must be fairly common.

13.4 Spectroscopic binaries

Spectroscopic binaries are mutually revolving pairs that often appear as single stars with the telescope. They are recognized by the periodic oscillations of the lines in their spectra caused by the Doppler effect as the two stars alternately approach and recede from us in their revolutions. When the two stars of a binary differ in brightness by as much as one magnitude, only the spectrum of the brighter star is likely to be observed. Capella and Spica are examples of spectroscopic binaries among the brightest stars.

The brighter star of Mizar's visual pair was the first spectroscopic binary to be detected. In 1889, during the early studies of stellar spectra, it was noticed that the dark lines in the spectrum of this star contained double lines in some photographs and single lines in others (Fig. 13.6).

When the displacements of the lines in the spectrum of a binary have been measured at intervals throughout the period of revolution, the **velocity curve** can be drawn to show the variation of the radial velocity

Figure 13.6 *Doppler shifts in the spectrum of the binary Mizar. When the components are moving toward and away from us the lines shift to the blue and red, respectively, and hence appear as double. When the components are moving across the line of sight the lines appear single. (Yerkes Observatory photograph.)*

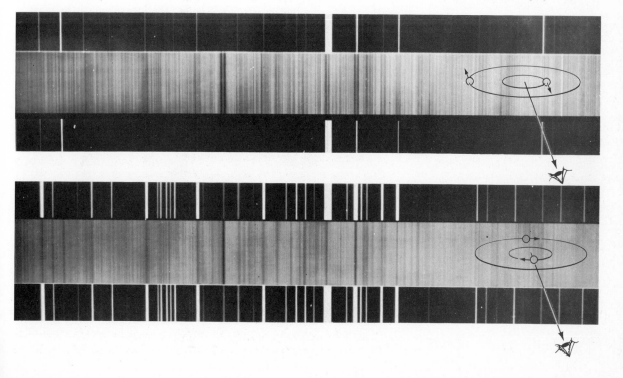

during the period. From this curve it is possible to determine the orbit projected on a plane perpendicular to the line of sight. The inclination of the true orbit to this plane is not determined from the spectrum; it may be derived from the light curve when the pair is also an eclipsing binary.

Generally, the orbital velocities of spectroscopic binaries are quite large even when we observe only the projected component. This is an instrumental selection effect because we determine the velocity component by measuring the Doppler shifts of the lines.

Spectroscopic binaries have one major advantage over visual binaries. So long as we can record a good spectrum we can determine the velocity curve for binaries at any distance. Then if we can measure a separation by some other technique we can ascertain the true separation in kilometers, convert this to astronomical units, and hence determine the parallax. This is a very powerful method for increasing the number of stars for which we determine masses.

13.5 The mass–luminosity relation

Studies of binary stars led to the discovery of a relation between the masses and luminosities of main-sequence stars in general. The more massive the star, the greater is its absolute brightness. The relation is shown by the purple line in Fig. 13.7, where the logarithm of the mass in terms of the sun's mass is plotted with respect to the star's absolute bolometric magnitude. **Bolometric magnitude** refers to the star's radiation in all

Figure 13.7 *The mass–luminosity diagram. Visual binaries are plotted as filled circles and the spectroscopic binaries as open circles. This nice empirical relation does not hold for non-main-sequence stars nor does it hold for white dwarfs—three of which are plotted below the curve. (Diagram adapted from one by K. A. Strand.)*

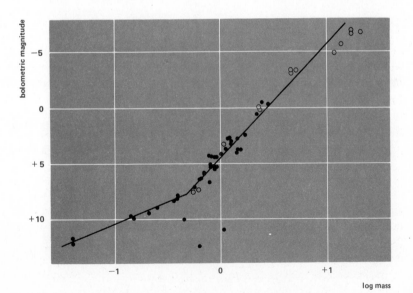

wavelengths with allowance for its absorption by the earth's atmosphere. The bolometric magnitude is used because the relationship is between the mass of a star and the total energy emitted by it. In general we will confine our use of magnitudes to the apparent and absolute visual magnitude.

The mass–luminosity relation, the harmonic law, and the relation $M = m + 5 + \log p$ (Section 12.18), allows us to obtain the parallax (p) for binary stars where the separation, period, and apparent magnitudes are known. Such a parallax is called a **dynamical parallax.** Estimating the sum of the masses and using the harmonic law we find a provisional parallax. Using this value of the parallax and the apparent magnitudes, corrected to apparent bolometric magnitudes, we determine the absolute magnitudes, which can now be entered in the mass–luminosity relation to give us a better estimate of the masses. Using this better estimate of the masses we go through the procedure again and again until the value of the parallax does not change.

Variable Stars

Variable stars are stars that vary in brightness. Their fluctuations in light are studied by comparing their magnitudes repeatedly with those of stars having constant brightness. The comparisons are often made in photographs taken at different times, or with the photoelectric tube when precision is required.

13.6 Light curve

The **light curve** shows the variation in magnitude of a star with respect to time. If the same variation is repeated periodically, we can determine its period. The curve for a single cycle may then be defined more precisely by plotting all the observed magnitudes with respect to **phase,** the interval of time expressed in fractions of the period since the epoch preceding the time of each observation.

13.7 Designation and classes of variable stars

The designation of variable stars follows a plan that started simply enough but became complicated when the discoveries of these stars ran into thousands. Unless the star already has a number in the Bayer system (Section 11.2), it is assigned a capital letter, or two, in the order in which its variability is recognized. For each constellation the letters are used in the order R,S, . . . , Z; RR,RS, . . . , RZ; SS, . . . , SZ; and so on until ZZ is reached. Subsequent variables are AA,AB, . . . , AZ; BB,

. . . , BZ; etc. By the time QZ is reached (the letter J is not employed), 334 variable stars have been named in the constellation. Examples are R Leonis, SZ Herculis, and AC Cygni. Following QZ the designations are V 335, V 336, and so on; an example is V 335 Sagittarii. The *General Catalogue of Variable Stars,* prepared by the Sternberg Astronomical Institute gives the names, and other information about recognized variable stars in our galaxy. The latest edition lists more than 16,000 variable stars. The *Catalogue* also lists more than 10,000 suspected variables. All variable stars, according to the *Catalogue,* may be divided into three main classes: eclipsing, pulsating, and eruptive variables. Each of these is further subdivided into several types.

Eclipsing Variables

Eclipsing variable stars are binaries having the plane of their orbits so nearly in the line of sight to the earth that the two stars mutually eclipse twice in the course of each revolution. They appear with the telescope as single stars that become fainter while the eclipses are in progress. The periods in which they revolve and fluctuate in brightness range from 82 minutes for WZ Sagittae to 27 years in the very exceptional case of epsilon Aurigae. AM Canum Venaticorum is believed to be an eclipsing binary with a period of only 17.5 minutes. (Some newly discovered stars have even shorter periods.) The average period lies between 2 to 3 days for this class of stars. This is the result mainly of observational selection because we tend to notice rapidly blinking stars more easily than those that blink only at long intervals. The very-short-period eclipsing binaries show a flickering in their light curves identical to that seen in the remnant stars of novae supporting the idea that the nova event is somehow related to double stars. HZ Herculis is such a binary, one component of which is a pulsar (Section 13.15).

Algol (beta Persei) is representative of eclipsing pairs in which the stars are nearly spherical. In pairs such as beta Lyrae the stars are elongated one toward the other by mutual tidal action, so that additional variations in brightness are produced by the different presentations of the ellipsoidal stars to us, from end-on at the eclipses to broadside between eclipses. In short-period pairs of the W Ursae Majoris type the components are so nearly in contact that the tidal effect is extreme. The spectra of some eclipsing pairs reveal gas streams that issue from the two stars and swirl in the directions of their revolutions. This is the case with beta Lyrae.

Combined photometric and spectroscopic observations of eclipsing stars lead to evaluations of their linear dimensions and masses—important data in studies of the constitution of stars. The photometry gives us the inclination of the orbit and the relative sizes of the stars in terms of the size of the orbit. The spectroscopy gives us the orbital velocity. The orbital velocity and the period give us the true size of the orbit. Hence, we get the true diameters of the stars.

In some systems the time of minimum (i.e., the period) changes in a periodic manner. This is often the result of the system rotating around a common center with a third unseen star. In some cases the variation results from a rotation of the lines of the apsides due to an eccentric orbit. When this is the case, it is possible to determine the distribution of the density in the stars involved. In some other systems the periods may change abruptly and erratically. Such changes are probably associated with mass loss from one or both components and/or mass being exchanged between the components. The study of these changes is still in its early stages and may provide some insight into the problems of stellar evolution.

Finally, from photometric observations of eclipsing binaries we can observe the limb darkening of stars. Only eclipsing-binary observations yield this important information excepting, of course, the case of the sun where limb darkening is directly observable. The limb darkening for both components of Algol is well determined.

Algol is a familiar example of eclipsing variable stars and was the first of these to be recognized, in 1783. In the severed head of Medusa, which Perseus carried in the old picture book of the skies, this "Demon Star" winks in a way that might have seemed mysterious until the reason for its winking came to be understood. The brighter star of Algol revolves in a period of 2^d21^h (Fig. 13.8) with a companion that is 20% greater in diameter but three magnitudes fainter. Once in each revolution the companion passes in front of the brighter star, partially eclipsing it for nearly 10 hours. At the middle of this eclipse the light of the system is reduced to one-third of its normal brightness. This is the deeper minimum and is called primary eclipse or primary minimum.

The small drop in the light midway between the primary eclipses

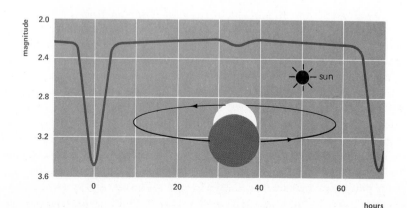

Figure 13.8 *The observed light curve and schematic drawing of the eclipsing binary Algol. The size of the sun on the same scale is shown. (Light curve and orbit by J. Stebbins.)*

Figure 13.9 *The light curve of the binary HR 6611. Note the difference in the relative depth of secondary eclipse as compared with that of Algol. This indicates that these two stars are very nearly equal in size, shape, and brightness. (Diagram by R. Zissell.)*

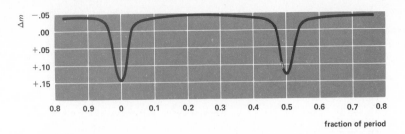

occurs when the companion is eclipsed by the brighter star. The light of the system rises slightly toward the secondary eclipse, because the hemisphere of the companion that is then turned toward us is made the more luminous by the light of the brighter star. This is called the **reflection effect.**

A binary similar to Algol, but where both stars are about equal in size, mass, etc., to each other is HR 6611. This system has been thoroughly studied. The larger and brighter star is 3 times the size of the sun. They revolve in a period of 3.89498 days at a distance of 0.08 AU. The light curve is shown in Fig. 13.9 and the relative system is shown to scale in Fig. 13.10.

Pulsating Variables

Intrinsic variable stars fluctuate in brightness from causes inherent in the stars themselves and not because of eclipses. Some of these stars are variable because they are alternately contracting and expanding, becoming hotter and cooler in turn. Pulsating stars include prominently the cepheid, RR Lyrae, and Mira-type variables.

13.8 Cepheid and RR Lyrae variable stars

Figure 13.10 *The apparent orbit of the binary HR 6611. The sun is shown to scale for comparison. Mercury would be 41 solar diameters away from the sun. (Diagram by R. Zissell.)*

Cepheid variable stars take their name from one of their earliest known examples, delta Cephei. This star fluctuates regularly in cycles of 5^d9^h, brightening more rapidly than it fades (Fig. 13.11). The velocity curve is nearly the mirror image of the light curve. Near maximum brightness the spectrum lines are displaced farthest to the violet, showing that the gases in front of the star are approaching us in the pulsation at the greatest speed. Near minimum light the lines are displaced farthest to the red, showing that these gases are receding from us at the greatest speed; the star is then redder than at greatest brightness and its spectrum has changed to the pattern of a cooler star. Cepheids generally show the greatest amplitude variation in the ultraviolet and the least in the red and infrared.

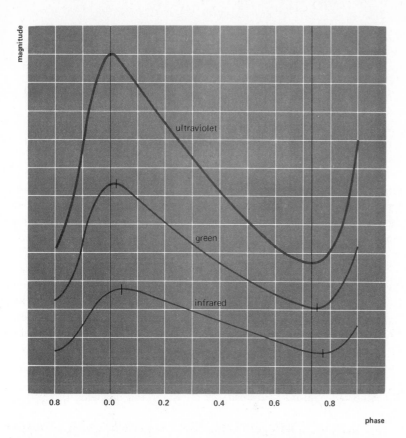

Figure 13.11   *The light curve of delta Cephei. At longer wavelengths (infrared) the light curve lags behind that of shorter wavelengths. Also, the maximum is not as high in the infrared. (Diagram from observations by J. Stebbins.)*

Also the light-curve maximum and minimum lag in time in the red as compared to the ultraviolet.

Cepheid variables are yellow supergiants; the more numerous **classical cepheids,** also referred to as **type-I cepheids,** resemble the prototype. Their periods range from a day to many weeks and are most numerous around 5 days. The visual range of the light variation is often around one magnitude. Those in our own galaxy congregate toward the Milky Way. Polaris is a cepheid having an exceptionally small range of variation.

The **type-II cepheids,** also called **W Virginis stars,** have been recognized more frequently in the globular clusters and toward the center of the Galaxy. Their periods are generally between 12 and 20 days. The light curves have broader maxima and are more nearly symmetrical. An example is W Virginis. Both types of cepheids are observed in other galaxies.

RR Lyrae variables are named after one of their brightest examples; they are also known as **cluster variables** because they were first observed in large numbers in the globular clusters. They are sometimes referred to as **short-period** variables because their periods are less than one day. They are blue giants of spectral class A varying in brightness in periods around one-half day and with ranges up to 1.5 magnitudes or more. The light curves of these stars with the shorter periods are likely to be almost symmetrical; at one-half day the curves become abruptly asymmetrical with very steep up-slopes and extreme ranges (Fig. 13.12), effects that moderate as the periods are longer.

RR Lyrae variables are pulsating stars with some characteristics like those of the cepheids. Their spectrum lines oscillate in the period of the light variation and the stars are bluer at maximum than at minimum brightness. These stars are recognized in greater numbers than are the cepheids even though they are much the less luminous. Because of this, the distance at which they can be detected is less than that for cepheids. Not one is visible to the unaided eye.

13.9 The period–luminosity relation

There is an interesting relation between the periods of pulsation of cepheids and their absolute brightness or luminosity: the longer the period of pulsation the brighter the star is. Originally established by H. Leavitt for stars in the Small Magellanic Cloud between 1908 and 1912, the relation was calibrated first by H. Shapley in 1913 and later by W. Baade in 1952.

Since the period–luminosity relation for the cepheids in the Small

Figure 13.12   *The light curve of RR Lyra as observed by T. Walraven. The rapid rise to maximum light is typical of this type of variable star. (Reproduced by permission of Astronomische Bulletin Nederlands.)*

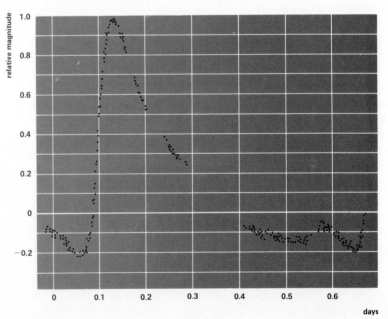

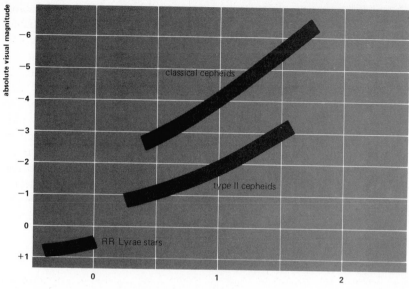

Figure 13.13   *The period–luminosity relationship for pulsating variable stars. Note the spread around the average at any given period and that the classical cepheids are about 1.5 magnitudes brighter than the type-II cepheids. (Drawn from data by H. C. Arp and M. K. Hemenway.)*

Magellanic Cloud was a simple straight line, Shapley and others assumed the same was so for stars in the Milky Way. Actually the cepheids are divided into two classes and the classical cepheids are 1.5 magnitudes brighter than type-II cepheids. Therefore, the classical cepheid line had to be raised above the existing period–luminosity line (Fig. 13.13). The confusion arose because the Small Magellanic Cloud contained only classical cepheids and Shapley calibrated the relation using type-II cepheids of the Milky Way. When Baade realized this, it had the effect of moving all extragalactic objects twice as far away.

The way for this revision of the period–luminosity relation was prepared by measures of cepheids in the Small Magellanic Cloud. It was found that the relation must be represented by a band at least one magnitude wide rather than by a simple line. The next step began with the discovery of two cepheids in galactic clusters and then later finding a few additional cases. Accurate distances and colors of cepheids that can be obtained in these clusters seem to sustain the revised relation as well as calibrate the relation.

When the period and mean color of a cepheid of either type is observed, the star's median absolute magnitude can be read from the period–luminosity relation. When the median apparent magnitude is also observed, the star's distance d in parsecs may be calculated by the formula (Section 12.18)

$$\log d = \frac{m - M + 5}{5}.$$

For the RR Lyrae variables the median magnitudes are independent of the period, and the value of M, formerly taken as zero, has now become about $+0.7$ on the average. All such distances require correction where cosmic dust intervenes (Section 14.7).

Because of their high luminosity, cepheid variables are useful as distance indicators for the nearer galaxies, whereas the less luminous RR Lyrae variables are employed generally as indicators for distances of objects in our own galaxy.

It is interesting to note that there is a relation between the pulsation period of a star and its mean density. The period of pulsation in days times the square root of the mean density is constant. The constant in this relation is 0.05. This relation, developed first from the theory of stellar interiors, tells us that the pulsation of stars is somehow related to the evolution of stars. This is supported by noting that the cepheids occupy a very well defined strip in the H-R diagram (Fig. 13.14).

13.10 Mira-type variables

Many red supergiant and giant stars are variable in brightness in a roughly periodic manner. The periods range from a few months to more

Figure 13.14 *The region of the H-R diagram occupied by the cepheid variables and the RR Lyrae stars.*

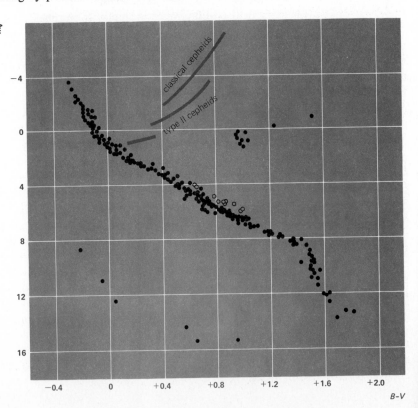

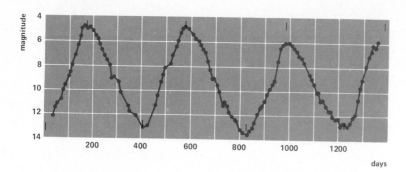

Figure 13.15 *Light curve of the Mira-type variable chi Cygni. Note that the variation is roughly periodic but that the maxima and minima vary. (American Association of Variable Star Observers.)*

than 2 years. The visual variation averages five magnitudes and may exceed ten magnitudes in the extreme case of chi Cygni (Fig. 13.15); yet the total radiation, as measured by the heating of a thermocouple at the focus of a telescope, varies only about one magnitude. These **Mira-type,** or **long-period variables,** are often regarded as pulsating stars and are so classified in the *General Catalogue.*

In addition to the dark lines and bands that characterize the spectra of red stars, these spectra show bright lines at certain phases, particularly lines of hydrogen. They are likely to be somewhat displaced to the violet at the maximum brightness of the stars, but a one-way shifting of the lines and not an oscillation is suggested.

The long-period variables lie in the upper right of the H-R diagram (Fig. 13.16). In general, as the periods of these stars increase the stars are redder and have fainter absolute magnitudes. Mira (omicron Ceti) is the best known and at times the brightest of these variables. It was in fact the first variable star to be recognized, aside from a few novae (Section 13.12), and was accordingly called **stella mira** (wonderful star). This red supergiant is at least 10 times as massive as the sun, and its diameter is 300 times the sun's diameter. The average period of the light variation is 330 days. The greatest brightness in the different cycles ranges generally from the third to the fifth visual magnitude and the least brightness from the eighth to the tenth magnitude, where the star is accordingly invisible to the unaided eye.

There is considerable doubt that the fluctuation of Mira is caused mainly or even at all by simple pulsation. The present interpretation is one of successive "hot fronts" that move outward from below the photosphere and disappear at the uppermost levels of the atmosphere. Here the dissipating waves may cause the gases to condense into droplets, which veil the photosphere until they finally evaporate.

The total energy change of Mira is not as great as the visual brightness range would indicate. A change of five magnitudes would suggest an energy change of a factor of 100. However, the bolometric magnitude of Mira changes by only slightly more than one magnitude or a total energy change

Figure 13.16 *The region of the H-R diagram occupied by the long-period variables.*

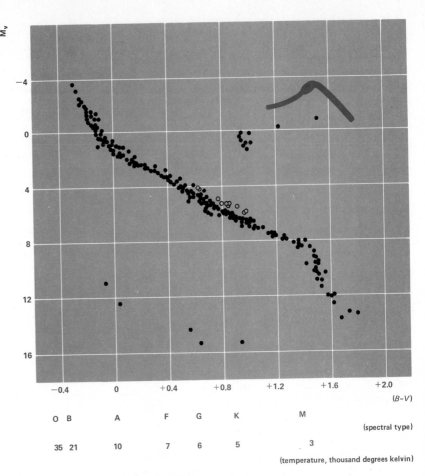

of a factor of about 3. This is because Mira has a low surface temperature—somewhat less than 2000 K. At this temperature the wavelength (Section 12.11) of the energy-distribution maximum is at 14,500 Å, which is in the infrared region of the spectrum at the wavelength twice that observable by the human eye.

13.11 Irregular and semiregular variables

Many red supergiants and giants vary irregularly in narrower limits, often not exceeding one-half magnitude. Betelgeuse is the brightest of the irregular variables. The red supergiant alpha Herculis is another example. This star varies between visual magnitudes 3.0 and 4.0 in irregular cycles of several months' duration. Its distance is 150 parsecs, and its diameter is several hundred times the sun's diameter. This star is surrounded by an expanding shell of patchy clouds which rise to heights above the surface of at least 700 times the earth's distance from the sun. At very high levels they become clouds of solid particles, which disappear by dilution as they move outward and are replaced by other clouds. This is an example of mass being lost from a star. Their partial veiling of the photosphere is believed to contribute to the variability of alpha Herculis. Similar phenomena occur in other large red stars.

RV Tauri stars are semiregular yellow and reddish supergiants that form a sort of connecting link between the cepheids and the Mira-type variables. Other types of semiregular variables having small memberships are recognized.

Eruptive Variables

Eruptive, or explosive, variable stars include novae, supernovae, flare stars, and T Tauri stars (Section 15.1). Some astronomers prefer to refer to the latter two types of stars as spasmotic variables. **Novae** are stars that rise abruptly from relative obscurity and gradually decline to their former faintness. They are designated either by the word nova followed by the possessive of the constellation name and the year of the outburst, or more recently by letters along with other variable stars. Thus Nova Herculis 1934 is also known as DQ Herculis. In addition to typical novae, recurrent novae, and dwarf novae, there are the even more spectacular supernovae, which are however unrelated to novae.

13.12 Typical novae

Prior to its one recorded outburst, a typical nova is smaller and denser than the sun. It is often pictured as a semidegenerate star that is collapsing to become a white dwarf. When more energy is liberated by the contraction than the small surface can radiate, the star blows off the excess energy together with a small fraction of its gaseous material in a succession of violent explosions. The total amount of material ejected in a single explosion is about $\frac{1}{10,000}$ of the star's mass. Clouds of gas emerge with speeds of about 1000 kilometers/second, as shown by the Doppler shifts of the spectrum lines. With the growing volume of these hot gases the star may rise 12 magnitudes, or about 60,000 times in brightness and attain an absolute magnitude of about -7.7.

The emerging gas is opaque at first. The dark spectrum lines are displaced strongly to the violet, giving the illusion that the whole star is swelling enormously. Soon after maximum brightness the expanding gas becomes more nearly transparent. Bright undisplaced lines then appear in the spectrum (Fig. 13.17), much broadened because the light comes from parts of the gas that are approaching and other parts that are receding

13 Oct 1967

2 Jul 1968

10 Dec 1968

Hα

Figure 13.17 *Three spectra of Nova Delphini 1967. In October the spectrum resembled that of P Cygni, a well-known shell star. By July 1968 the spectrum shows very broad, violet-displaced emission lines. As the ejected material cools the lines narrow as seen in the spectrum of December 1968. The emission line of the extreme right is Hα. (Courtesy of J. Gregar, Ondrejov Observatory.)*

from us in the expansion. The broad bright lines are bordered at the violet edges by narrow dark lines absorbed as before by the gas immediately in front of the star. When the envelope has become still more tenuous, the spectrum of the nova changes to the bright-line pattern of an emission nebula (Section 14.1) except that the lines are wider. Meanwhile, the brightness fades as the envelope is dissipated by expansion, until after 20 to 40 years the star returns to about its original status.

The typical nova event may be divided into premaximum and postmaximum phases (Fig. 13.18). The premaximum phase consists of the **prenova star,** which many astronomers suspect shows a flickering light variability prior to the event, the **initial rise,** which is very rapid, probably lasting only a few hours to a day at most, a **premaximum halt** in the rise about two magnitudes below its greatest brightness, and then a rather slow **final rise** to maximum. These phases are not very well studied because novae seldom are noticed before they reach maximum light. The next stage is the **principal maximum.** After maximum light the nova begins an exponential decline, finally ending with the star's return to its original appearance prior to the nova event. The postmaximum **early decline** is characterized

Figure 13.18 *The principal phases in a nova outburst from observations of Nova Persei 1901.*

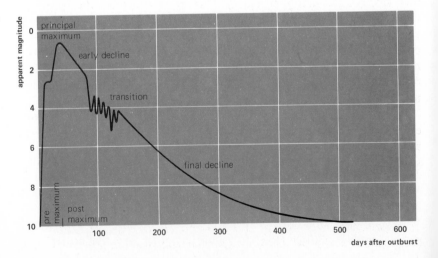

by the spectral changes discussed above. After declining about three or four magnitudes in brightness a **transition phase** sets in where the light curve may suddenly drop to the prenova brightness, or it may fluctuate violently, or it may show a smooth decline with only slight fluctuations. After this the nova makes its **final decline** to the **postnova** star. The postnova star, when studied intensively, proves to be a binary star in almost all cases, and may show violent flickering. In any event, the typical nova phases scale with unusual precision. For example, the transition phase takes about 20% of the duration of the nova event whether it is 4 months, in which case the transition phase lasts about 24 days, or 4 years, in which case the transition phase last about 290 days.

Five typical novae in the present century became stars of the first magnitude or brighter. It is estimated that 25 novae burst out yearly in the Milky Way; the majority escape detection and most of the observed ones are invisible to the unaided eye.

The envelopes produced by eruptions of novae have sometimes become large enough to be observed directly with the telescope. Nova Aquilae 1918 had a spherical envelope; the envelope began to be visible 4 months after the outburst and increased in radius at the rate of 1″ a year. Spectroscopic observations showed a velocity for the material of 1600 kilometers/second, which yields a distance of 336 parsecs to the nova. This is obtained by assuming the expansion rate is the same across the line of sight as it is along the line of sight (i.e., the radial velocity v) (Fig. 13.19). If the shell expands at μ seconds of arc per year due to a velocity of v kilometers/second, our problem is to determine its parallax or the angle subtended at that distance by one astronomical unit.

At a velocity of v kilometers/second we can calculate how many kilometers are covered in one year by multiplying by the number of seconds in a year (60 sec/min $\times$ 60 min/h $\times$ 24 h/day $\times$ 365.25 day/yr). Now if we divide this by the number of kilometers in an astronomical unit (1.496 $\times$ 10^{10} km) we convert the motion to astronomical units per year, or 0.21 $\times$ v astronomical units per year. The angle subtended by 1 astronomical unit is then just $\mu/0.21v$ which is therefore the parallax p. Then from our distance formula, $d = 0.21v/\mu$. There are several other ways to arrive at this formula, but all are equivalent.

In 1940, the vanishing envelope of Nova Aquilae 1918 had a radius exceeding 5000 times the earth's distance from the sun. The envelopes around typical novae have generally disappeared after a few years. Their short durations contrast with the longer lives of supernova envelopes.

13.13   Recurrent and dwarf novae

Recurrent novae, which have two or more recorded outbursts, also differ from typical novae in their more moderate rise in brightness (around

Figure 13.19   *The distance of a nova can be determined if the rate of expansion (v) of the shell is known from the spectrum and if the angular expansion (μ) can be measured.*

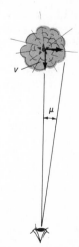

seven magnitudes). An example is RS Ophiuchi, normally around the 12th magnitude; it rose abruptly to the fourth magnitude in 1898 and 1933, and to the fifth magnitude in 1958.

Dwarf novae have some characteristics of typical novae. An example is SS Cygni (Fig. 13.20). Normally around apparent magnitude 12, this star brightens abruptly about four magnitudes at irregular intervals and declines in a few days. The discovery in 1943 that SS Cygni is a spectroscopic binary star may have prepared the way for the eventual understanding of all novae. Other typical and dwarf novae have since proved to be members of binary systems. It seems likely that membership in a particular type of binary system may be a necessary condition for a star to become a nova.

13.14 Supernovae in our galaxy

Supernovae are stars considerably more massive than the sun, which explode once in the course of their lifetimes. During their explosions they become many million times more luminous than the sun and blow into space gaseous material amounting to at least one solar mass. Allowing for the many that are too remote to be conspicuous or which are hidden from our view, we estimate that supernovae are exploding in the system of the Milky Way at an average rate of one in 30 to 60 years. Novae and supernovae in other galaxies are discussed in Chapter 18.

The apparently brightest supernova on record became as bright in the sky as the planet Venus. It flared out in Cassiopeia in 1572 and was observed by Tycho Brahe; after 18 months it was no longer visible to

Figure 13.20 *Light curve of SS Cygni, a recurrent nova, during 1965. The numbers along the top are Julian days beginning with J.D. 2,435,100. The numbers at the bottom are the tabulation of eruptions and the letter–number combination indicates the type of eruption. (American Association of Variable Star Observers as plotted by M. W. Mayall.)*

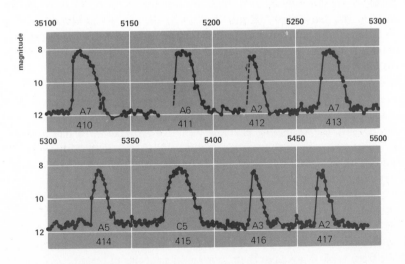

the unaided eye. "Kepler's star" in Ophiuchus in 1604 rivaled Jupiter in brightness. A supernova in Taurus in the year 1054 became as bright as Jupiter and remained visible about two years. Supernovae are divided into two groups: group I attains an absolute magnitude of −19, while group II reaches an absolute magnitude of −17.

Distances to supernovae can be obtained independently from their expanding shells as explained above for ordinary novae. Another method for determining distance makes use of the velocity of light if the event happens to illuminate interstellar material in the vicinity. In this case the radial velocity is not necessary so long as the expansion rate of the illuminated ring (Fig. 13.21) is measured in seconds of arc per year (μ). Since we know the velocity of light, we know the rate at which the ring is expanding in kilometers/second; that is, we already know v in the formula given in Section 13.12. This gives us the distance d in parsecs:

$$d = 63000/\mu.$$

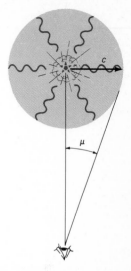

Figure 13.21 *The distance to a supernova (or nova) can be determined if its light illuminates interstellar material. The principle is the same as that shown in Figure 13.19 except that the velocity of light (c) replaces v.*

Supernova events are quite different from that of the ordinary nova discussed above. It is clear that some cataclysmic event has occurred, and the star involved is shedding a large portion of its mass. We will theorize later that any star possessing too much mass as it reaches the end point of its evolution is forced to release the excess mass through just such a violent process.

About 24 remnants of supernova envelopes are recognized or suspected in the Milky Way. The first known of these is the Crab nebula (Messier 1), the counterpart of a source of strong radio radiation called Taurus A, which is shown in Fig. 13.22. This conspicuous nebula is still increasing in radius at the rate of 1100 km/sec, around the site of the supernova of 1054. At a distance of 1000 parsecs, its present linear diameter is about 2 parsecs. The Crab nebula consists of a homogeneous central region surrounded by an intricate system of filaments. Fragments of the envelopes of the supernovae of 1572 and 1604 have been photographed and are still spreading.

The familiar loop of nebulosity in Cygnus (Fig. 13.23) is an example of the remnants of more ancient supernovae in the Galaxy. These appear in photographs as complex patterns of filamentary nebulosity. Other supernova remnants are observed beyond the relatively nearby obscuring material by their radio radiation. Pictures or radiograms of their form can be constructed by synthesis techniques. The largest supernova remnant has a diameter of some three degrees on the sky. The emissions from supernova remnants in the radio wavelengths are mainly nonthermal, like those produced by the synchrotron type of accelerator in radiation laboratories.

Figure 13.22 *The Crab nebula in Taurus. This is the remnant of the supernova event observed by the Chinese and North American Indians in A.D. 1054. The immediate impression is one of a violent event. (Lick Observatory photograph.)*

Pulsars

In late 1967, J. Bell and A. Hewish, together with their colleagues at Cambridge, discovered a new type of variable which emitted short, rapid pulses of radio waves at regularly timed intervals. Once they knew what to look for, three other pulsing radio sources, or **pulsars,** were quickly found. Six months later, they announced their discovery which proved to be one of the most profound and unexpected astronomical discoveries of the decade.

The timing of the radio pulses from these mysterious sources was remarkably precise, showing a regularity of one part in 100 million. A clock with that precision would keep time to better than one-hundredth of a second per year. Equally amazing was the extreme sharpness of the individual pulses and the short periods between them. The duration of each pulse, known as the pulse width, was only about 20 milliseconds ($\frac{1}{50}$ of a second), while the periods of the first discovered pulsars (the length of time between the pulses) were only 1.38, 0.25, 1.19, and 1.27 seconds.

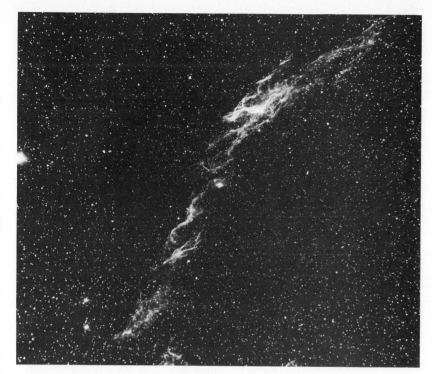

Figure 13.23 *Filamentary nebula in Cygnus referred to as the
Loop nebula. This is the remnant
of a supernova event which occurred 70,000 years ago. For scale
see Fig. 11.9. (Hale Observatories
photograph.)*

13.15 Nature of pulsars

A flurry of papers was published in an attempt to explain pulsars.
One of the first suggestions was that extraterrestrial life forms were sending
the signals in an attempt to communicate with other intelligent beings.
However, it was soon discovered that the radiation was spread over a
large band of wavelenghts. If intelligent beings were sending out signals
to announce their presence, they would surely restrict their broadcasts
to only one frequency, just as we use only one channel in our radio and
television broadcasts. To transmit a signal over a broad band is an extremely
inefficient use of energy and would not be a very intelligent thing to do
at all. Scientists began searching for a natural explanation for the pulsars.

The extremely narrow pulse widths place a limit upon the size of
the objects emitting them. The radius cannot be larger than the light
time equal to the pulse width or about 6000 kilometers. Also, the objects
must be very faint visually. Only white dwarfs and stars predicted by
theory—neutron stars—fit this description. **Neutron stars** are expected
to be about 15 kilometers in diameter and extremely dense, packing more
than a solar mass into their small volume (Section 15.9).

Three explanations for the regular periods are possible; they are eclipsing binaries, they are pulsating stars, or they are rapidly rotating stars with a narrow beam of radiation much like the beacon of a lighthouse. Various arguements led to the adoption of the latter explanation. The discovery of a very short period pulsar in the Crab nebula (Fig. 13.25) eliminated the white dwarfs from consideration and thus it was concluded that the pulsars were rotating neutron stars.

Almost 1000 pulsars have been cataloged now, and more are being found with great regularity. However, only a small percentage of the neutron stars which actually exist are expected to be observable. This is because the highly directional beam of radio waves which they emit only traverses a limited portion of the sky. If the earth does not lie in the path of the pulsar's beam, it will remain undetected. What is more, after a few million years, pulsars stop beaming radiation altogether. It must be concluded that many more neutron stars exist in our galaxy than have been detected as pulsars. It has been estimated that one out of every thousand stars is a neutron star.

Neutron stars are expected to have extremely strong magnetic fields, billions of times stronger than the sun's. This is because neutron stars collapse from stars larger than our sun down to compact objects only 15 kilometers in diameter. As the star collapses, the magnetic field lines which are frozen into the gas in the star are crowded into a smaller and smaller surface area. The more field lines emerging from a given surface area of the star, the stronger the magnetic field becomes. If a star the size of the sun were to collapse to the size of a neutron star, it would have a magnetic field ten-billion times stronger.

It is the extreme strength of the magnetic dipole field around neutron stars which gives rise to the highly directional beam of radiation which they emit. Charged particles in a magnetic field find it much easier to move along magnetic field lines than across them. Around a neutron star, this causes charged particles which have been accelerated by the magnetic field to follow the magnetic field lines away from the magnetic poles of the star. Oscillations of these particles around the magnetic field lines cause them to emit synchrotron radiation, and a beam of radio waves is generated in the direction of the neutron star's magnetic axis (Fig. 13.24). If the rotational axis of the neutron star is not aligned with the magnetic axis, the beam of radiation will trace out a cone across the heavens. When the earth lies on this cone, the neutron star is detected as a pulsar by the pulse of light each time the beam of radiation flashes past the earth.

For a long time it was a puzzle to astronomers how the Crab nebula could shine so brightly (Fig. 13.22). Some source of energy must be continually supplied to the glowing gases to keep them alight, and yet no bright star in the vicinity is hot enough to power such a profuse outpouring of radiation. With the discovery of the Crab pulsar, the mystery was solved. Observations show that its period is increasing at a rate of 3×10^{-8}

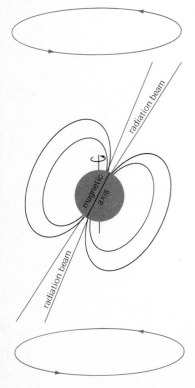

Figure 13.24   *A schematic drawing of a pulsar whose magnetic axis is tilted with respect to its spin axis. The observer must be located on the cone swept out by either beam in order to see the radiation.*

seconds/day. This is equivalent to an increase of only one part in a trillion in the interval between each successive pulse. Although this is an extremely small change, it represents a vast amount of energy, because the neutron star is so incredibly dense. The rotational energy which is converted to kinetic energy in the electrons in the nebula not only powers the sychrotron radiation in the diffuse gases, but also is speeding up the expansion rate of the nebula.

The Crab pulsar is the most carefully studied of all the known pulsars. It has a number of exceptional properties which distinguish it from other pulsars. Most notable is the fact that it emits radiation at all observable wavelengths, from radio wavelengths, through visible light, to the highest-energy x rays. All these wavelengths produce the same pulse period and shape, although they arrive at different times on earth due to the dispersion of radiation in the interstellar medium. In fact, the visible component of the pulsar's radiation had been carefully studied long before it was realized that the radiation was arriving in bursts. Because the long-exposure photographs taken with telescopes accumulate the light of many pulses on their emulsions, the Crab pulsar had been photographed and its spectrum studied without the least suspicion that it was a pulsating source. Once the radio bursts from the pulsar were discovered, special equipment was devised to detect any short-period optical variability and it was quickly found. The light curve and the corresponding images of the star are shown in Fig. 13.25. The Crab pulsar and a pulsar in Vela are the only pulsars so far discovered which emit radiation in the visible region of the spectrum.

All pulsars are believed to have formed in supernova explosions, as did the Crab pulsar. However, the resulting nebula surrounding the pulsar, known as the supernova remnant, is extremely short lived. Most supernova remnants disperse within 100,000 years after they have formed, which is a short time on an astronomical scale. Only four pulsars are still seen in association with their supernova remnants. The Crab pulsar is the youngest and therefore the most prominent. It was formed a little over 900 years ago in a bright supernova event observed in A.D. 1054. The extreme youth of the pulsar is revealed by its rapid rotation (30 revolutions per second) and the wide range of wavelengths over which it emits radiation. As it ages, the Crab pulsar will slow down and its radiation will become limited to the radio region of the spectrum. Eventually, the pulsar will no longer be able to emit its highly directional beam of radiation, and it will switch off. The Vela pulsar, which is about 10,000 years old, is found in a partially dispersed supernova remnant, and its rotation rate has slowed to 11 revolutions per second. Most pulsars rotate much slower than this, and the supernova remnants that formed with them have all dispersed long ago.

One supernova remnant known as W 50 presents circumstantial evidence for having a neutron star at its center although pulsing has not been observed. The central object, SS 433, exhibits a bizarre behavior. Its spectrum is that of synchrotron radiation and emission lines showing

Figure 13.25 *Pulsar NP 0531 in the Crab nebula observed at high speed. The star's period is 0.033 second and the light curve is shown below for reference. (Kitt Peak National Observatory photograph by H-Y. Chin, R. Lynds, and S. P. Maran.)*

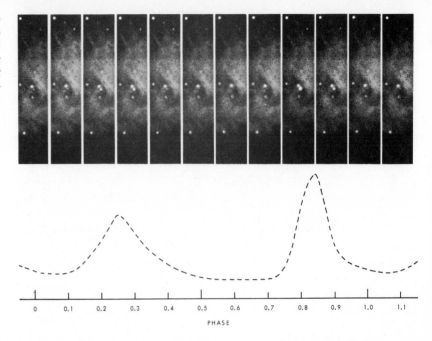

Doppler displacements of ±30,000 km/sec—one-tenth the speed of light! The most plausible explanation is that a neutron star in a binary system is beaming high energy particles and radiation into the remnant shell from the inside.

Another pulsar, not associated with an obvious supernova remnant, is part of a binary system. Its orbital period is shortening with time. The simple explanation for this is that the system emits gravitational waves, i.e., energy, and this lost energy is compensated for by a shortening of the orbital period as predicted by the theory of general relativity. If this proves to be the case, it will be an elegant proof of the validity of the theory.

13.16 X-ray pulsars

Another type of pulsar has been discovered which emits radiation solely at x-ray wavelengths. Known as x-ray pulsars, they differ from radio pulsars in that they always occur in binary systems. The source of the x-ray bursts is believed to be neutron stars with extremely strong magnetic fields, just as in the case of radio pulsars. Two x-ray pulsars have been discovered in our galaxy and one is known to be located outside our galaxy.

Hercules X-1, the first x-ray source discovered in the constellation Hercules, is a companion star to the spectroscopic binary HZ Herculis. Every 1.7 days, the visible star occults the x-ray source. In addition to the 1.7-day period of the binary-star system, Hercules X-1 shows a 1.24-

second period which is due to a rotating beam of x rays emanating from the pulsar's surface. The short period and precise timing of the x-ray bursts indicate that the source is a rapidly rotating neutron star. The mass of Hercules X-1, calculated from the Doppler shift in its period and that of its visible companion, is about 1.3 solar masses, which is consistent with the expected mass of a neutron star.

The x-ray pulses from Centaurus X-1 are similar in character to those of Hercules X-1. The pulse period of Centaurus X-1 is 4.84 seconds. This period shifts slightly due to the Doppler effect. The x-ray source is also periodically occulted in a 2.087-day cycle. It is therefore assumed that Centaurus X-1 is a rotating neutron star in orbit around a visible star, just like Hercules X-1, although in this case the visible companion has not been directly observed.

SMC X-1 is found in the Small Magellanic Cloud, a small companion galaxy to our own. (It was the first x-ray source discovered there, hence the abbreviation SMC X-1.) This x-ray pulsar is a companion to a supergiant star. Analysis of the spectroscopic Doppler shifts indicates that the supergiant is about 16 times more massive than the sun, while SMC X-1 has only one solar mass, again in keeping with the expected mass of a neutron star. The x-ray pulsar has a pulse period of 0.716 seconds and an orbital period of 3.9 days.

The fact that all three x-ray pulsars are in binary systems suggests that the x rays are generated on the neutron stars by the transfer of mass from their companions. In order for x rays to be emitted by thermal processes in a gas, it is necessary for the gas to attain a temperature of at least 1 million degrees Kelvin. Because of the extremely strong gravity at the surface of a neutron star, any matter accreting onto its surface would easily be heated to this temperature. The transfer of mass from the giant gaseous star to a compact companion neutron star could be accomplished by outward streaming as in the stellar wind of the giant star. The strong magnetic field of the neutron star would then funnel this accreting gas onto the magnetic poles of the star. When the gas approached the surface, it would be heated to millions of degrees, and x rays would be generated at the polar caps of the star. It is estimated that the size of the patch of hot atmosphere on the polar caps of the neutron star where the x rays are generated is no larger than the area of Central Park in New York City. The rotation of these small areas of x-ray emission towards and away from the earth results in the regular pulsation of the x-ray sources.

Flare Stars and Planetary Nebulae

We discuss flare stars and planetary nebulae here for completeness. Flare stars exhibit an eruptive flare phenomena. The planetary nebulae are ejecting material more strongly than the shell stars, but not nearly as violently as novae or supernovae.

13.17 Flare stars

Some red main-sequence stars are subject to repeated intense outbursts of very short duration reminiscent of the solar flares (Section 10.10) and hence are called **flare stars.** At least 20 stars of this type have been listed. An example was the sudden brightening by 1.5 magnitudes of the normally fainter component of the visual binary Krüger 60 (Fig. 13.26).

Flare stars are designated in the *General Catalogue* as UV Ceti-type variables after a typical representative first recorded in 1948. This star is the fainter component of the binary Luyten 726-8. The main outbursts of UV Ceti occur at average intervals of 1.5 days, when the rise in brightness of the star is generally from one to two magnitudes. On one occasion in 1952, however, an increase of six magnitudes was observed, the greatest flare on record for any star. Only a small fraction of the star's surface is affected, as in the case of a solar flare. Between the main outbursts the light of the flare star varies continuously and irregularly in small amplitude.

Recently, observations show that the flare stars, particularly UV Ceti, flare in the radio region of the spectrum as well. In this way, the flare star's outbursts are different from even the most violent flare on the sun since the power emitted by UV Ceti in the radio wavelengths is enormous compared to a solar flare. Even more powerful flares have been detected from flare stars in the Orion nebula. These stars are considerably farther away and hence must radiate prodigious amounts of energy in order to be detected at all here in our solar system.

The location of flare stars in the H-R diagram is shown in Fig. 13.27. In this figure we summarize the locations of the various variable stars discussed in this chapter.

13.18 Features of planetary nebulae

About 1000 **planetary nebulae** have been recognized (Fig. 13.28). They range in size from the ringlike NGC 7293 in Aquarius, having half

Figure 13.26 *Krüger 60B flaring on 26 July 1939. The last of four successive exposures (left) shows that the faint companion became as bright as the primary in only a few minutes time. (Sproul Observatory photograph.)*

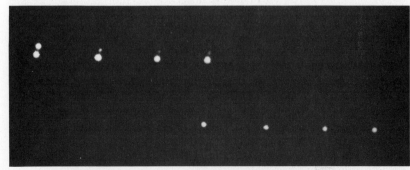

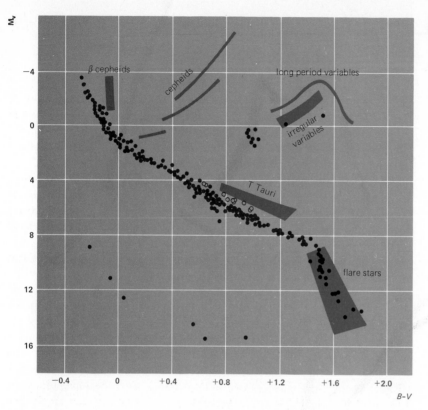

Figure 13.27 *A schematic H-R diagram showing the domains of various variable stars discussed in the text.*

the moon's apparent diameter, to objects so reduced in diameter by distance that they are distinguished from ordinary stars only by their peculiar bright-line spectra. Because of their spectra they are sometimes erroneously classified as Wolf–Rayet stars. Their linear diameters range from 20,000 to more than 100,000 times the earth's distance from the sun.

The central stars of planetary nebulae are about as massive as the sun, but are much smaller and denser than the sun. Having surface tempera-

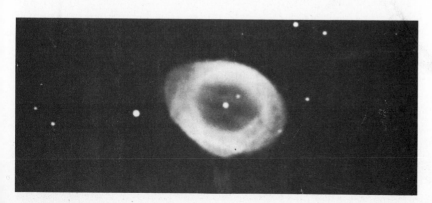

Figure 13.28 *The Ring nebula in Lyra. This beautiful planetary nebula is an excellent example of a star returning material to the interstellar medium. See color plates. (Dominion Astrophysical Observatory photograph.)*

tures of 50,000 K or more, they furnish a rich supply of ultraviolet and x-ray radiation causing the illumination of these emission nebulae. Because their radiations are mainly in the ultraviolet, the central stars are less easy to see than are the nebulae themselves; the stars come out clearly in the photographs taken in blue light, however.

Planetary nebulae having a rich supply of high-energy photons show the forbidden lines typical of emission nebulae, particularly the oxygen and nitrogen lines noted above. Again, although these lines are prominent, the abundances of these elements are quite normal. In fact the various elements appear in about the same amounts as we find in other nebulae, stars, etc. The distribution of planetary nebulae is interesting. They seem to concentrate in one general region of the sky that corresponds to the direction of the center of the Milky Way. Other types of stars and objects have similar distributions and are very old. We conclude that the central stars of planetary nebulae are old stars also. From their small numbers we can conclude that the planetary-nebula phase in the lifetime of a star is quite short, lasting only about 10^5 years.

Planetary nebulae are expanding around their central stars, as the Doppler effect in their spectra shows. Although they resemble the envelopes around novae in this respect, the planetaries are expanding more slowly and their lifetimes are much longer. Their moderate speeds of expansion in radius range from 10 to 50 kilometers/second. After lifetimes as long as 20,000 years they begin to disintegrate by breaking into separate clouds of gas, whereas the envelopes of normal novae have disappeared only a few years after emerging from the stars.

There is also a pronounced difference in the amount of material involved. The mass of a planetary nebula is about $\frac{1}{10}$ the sun's mass; but the mass of a normal nova envelope does not exceed $\frac{1}{10,000}$ the sun's mass.

Questions

1 What is the distinction between the apparent orbit and true orbit of a binary star?

2 What is meant by the term astrometric binary?

3 Explain how we determine a dynamical parallax.

4 Why are short-period eclipsing stars more numerous than long-period eclipsing stars?

5 Distinguish between classical cepheids, type-II cepheids, and RR Lyrae stars.

6 Describe a typical nova event.

7 Pulsars are the product of what type of celestial event?

8 How does a flare-star flare differ from a solar flare?

9 Why do we say pulsars are neutron stars?

10 Sketch an H-R diagram and locate the domains of the various variable stars.

Other Readings

Helfand, D. J., "Recent Observations of Pulsars," *American Scientist* **66(3),** 332–339 (1978).

Kirshner, R. P., "Supernovas in Other Galaxies," *Scientific American* **235(6),** 88–101 (1976).

Lewin, W. H. G., and J. van Paradijs, "What are X-ray Bursters?," *Sky & Telescope* **57,** 446–451 (1979).

Percy, J. L., "Pulsating Stars," *Scientific American* **232(6),** 66–75 (1975).

Stephenson, F. R., and D. H. Clark, "Historical Supernovas," *Scientific American* **234(6),** 100–107 (1976).

chapter 14

Cosmic Gas and Dust

Nebulae in general are clouds of cosmic gas and dust. Diffuse nebulae are condensations of the interstellar material that is abundant in our Milky Way and in the arms of spiral galaxies. Such nebulae supply the material from which stars are formed.

Diffuse Nebulae

Diffuse nebulae have irregular forms and often large angular dimensions. Some resemble the cumulus clouds of our atmosphere, whereas others have a filamentary structure that is reminiscent of our high cirrus clouds. Shocks and compressions of the colliding turbulent material and effects of magnetic fields account for the complex structures and also for the light of nebulae in certain cases.

Some diffuse nebulae are made luminous by the radiations of stars in their vicinity. The quality of the nebular light depends on the temperature of the associated stars. Where the star is as blue as or bluer than type B1, the nebular spectrum differs from that of the star, being mainly a pattern of bright emission lines. Where the star is cooler than B1, the light is mainly reflected starlight, so that the spectrum resembles that of the star. Thus there are two types of bright diffuse nebulae with respect to the quality of their light: emission nebulae and reflection nebulae. The dark nebulae, in the absence of a nearby star, reveal themselves by dimming and/or reddening the light of stars located beyond them, by imposing their absorption lines upon that of a more distant star, or by emitting radiation in the infrared or radio region of the spectrum.

14.1 Emission nebulae

The extreme-ultraviolet radiations of very hot stars contain enough energy to remove electrons abundantly from atoms of certain elements in the gases of the surrounding nebulae. As the ionized atoms capture other electrons, the nebulae emit light that may differ from that of the stimulating stars and are referred to as **emission nebulae.** The spectrum of the emitted light, as seen in Fig. 14.1, is more conspicuous than that of the starlight reflected by the same nebula for two reasons: (1) the emitted light is concentrated in a few bright lines of the spectrum, whereas the reflected light is dispersed over the entire spectrum; (2) much of the reflecting dust may have been blown away from the vicinity of the star by the star's radiations, hence reducing the amount of reflected light. Because diffuse emission nebulae and the associated blue stars in our galaxy are

[O II]   Hγ

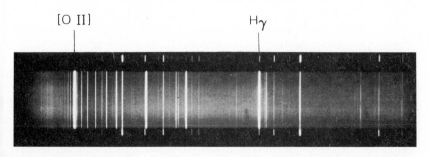

Figure 14.1 *A low-dispersion spectrogram of the Orion nebula by D. E. Osterbrock showing the characteristic bright emission lines of an emission nebula. The forbidden doublet of oxygen (O II) and the Hγ line are indicated. The comparison lines are hydrogen and helium.*

Figure 14.2 *(Opposite page) The Great nebula in Orion (NGC 1976, top, and NGC 1977, bottom). These nebulae are faintly visible to the unaided eye and are superb examples of emission nebulae. (Copyright by Akademie der Wissenschaften der DDR.)*

features of the large-scale pattern (spiral arms), they have been useful in the optical tracing of the arms (Section 17.6).

Prominent in the spectra of emission nebulae are "forbidden lines," so called because they are not likely to be observed under ordinary laboratory conditions. These bright lines of oxygen, nitrogen, and some other elements remained unidentified and were thought to be due to a hypothetical element called *nebulium* until they were explained theoretically in 1927. In our discussion of atoms and atomic spectra in Chapter 12, we pointed out that only a photon of the proper energy can excite an atom from a given level to a higher level. The forbidden oxygen and nitrogen lines observed in the nebulae are high-excitation lines, that is, they require high-energy photons to excite them. High-energy photons of the proper energy did not seem to exist. Then it was noted that an emission line from singly ionized helium (He II) in the far untraviolet (λ 304 Å) consisted of photons with an energy that agreed precisely with the energy required to raise a photon from the unexcited state of doubly ionized oxygen (O III) to a higher or excited level. The following transitions to lower levels resulted in emission of the observed lines. Helium is quite abundant, hence a source of the high-energy photons was found. A little further work showed that the photons from one of the O III emission lines had just the energy required to raise doubly ionized nitrogen (N III) from the unexcited state to an excited state. The resulting decay to lower levels explains the observed N III emission lines. Thus a long standing mystery was explained.

This line of reasoning has been applied to explain unusual lines, not only in interstellar clouds, but in planetary nebulae, circumstellar rings and shells, novae and supernovae, etc. A similar line of reasoning may explain some lines observed from interstellar clouds in the radio region of the spectrum. Despite their stronger showing in the spectra, oxygen and nitrogen are less abundant in these nebulae than are hydrogen and helium; in collisions with other atoms, however, they are able to utilize greater quantities of energy provided by the exciting starlight.

A strong pair of oxygen lines at wavelengths 4950 and 5007 Å in the green region of the spectrum imparts the characteristic greenish hue to emission nebulae (see color plate). In another oxygen pair at λ 3726 Å and λ 3729 Å in the ultraviolet, the strength of one line relative to the other depends on the density of the gas. This relation has been employed to determine the densities of these nebulae.

The Great Nebula in Orion, Fig. 14.2, is the brightest diffuse emission nebula in the direct view with the telescope. Scarcely visible to the unaided eye, it is the middle "star" of the three in Orion's sword. With the telescope the nebula appears as a greenish cloud around a star, which itself is resolved into the familiar **Trapezium** of type-O stars. In photographs the Orion nebula is spread over an area of the sky having twice the apparent diameter

of the moon. At its distance of 500 parsecs the corresponding linear diameter is 8 parsecs, or about the distance of Vega from the sun.

By the relation mentioned above, we find that the gas of the Orion nebula has an unusually high density, about 20,000 atoms/cm³, in one of the brightest central regions. The density is reduced to 300 atoms/cm³ at a point halfway from the center to the edge. While these numbers sound large they really represent very low densities. If we assume that all of the material is hydrogen then we are considering densities of 3×10^{-20} and 5×10^{-22} g/cm³, respectively. We should compare this with the atmosphere of the sun which is 10^{-7} g/cm³ and the atmosphere of the earth which is 10^{-3} g/cm³ at sea level.

Among other conspicuous diffuse emission nebulae are M8, the Lagoon nebula in Sagittarius, and the nebula surrounding the star eta Carinae, the latter being a most remarkable nebula.

14.2 Reflection nebulae

Where stars involved in interstellar material are cooler than type B1, the nebulae around them glow with the starlight reflected by their dust. The bright nebulae surrounding stars of the Pleiades are examples of **reflection nebulae** (Fig. 14.3). These have the same spectra as the associated stars. We have already noted that we occasionally see nebular material by the reflection of light from novae and supernovae. The densities of reflection nebulae are of the order of 10^{-22} g/cm³.

The similarity in color of reflection nebulae and of the stars responsible for their shining is shown in photographs taken with filters of a dusty region of Scorpius. Here the reflection nebulae surround stars like the glow around street lamps on a foggy night. The nebular light around the red star Antares is scarcely noticeable with blue filters but becomes conspicuous with red ones. The opposite is true for the nebulae around blue stars.

It is important that the reader not be misled by the previous paragraph. A reflection nebula always shows the characteristic spectrum of the star illuminating it. However, its actual color is somewhat bluer than that of the star rendering it visible. The reason for this is that some of the red light passes through and is not reflected, causing the reflected light to be a little bluer than the star itself.

14.3 Dark nebulae

Dark nebulae are clouds of gas and dust that are sufficiently dense to obscure the stars behind them and have no stars near enough to light them effectively. Their faint illumination by the general star fields can be detected only by measurements of high precision. Dark nebulae make

Figure 14.3 *The reflection neb-ula around the star Merope in the Pleiades.   (Hale   Observatories photograph.)*

their presence known optically by obscuring whatever lies behind them as we see in Fig. 14.4. The rifts they imprint on the bright background of the Milky Way are conspicuous in the photographs. Some rifts, such as the Coalsack near the Southern Cross, are easily visible to the naked eye and have accordingly been known for a long time; but their interpretation as dark clouds rather than as vacancies is fairly recent.

The darkest clouds in the Milky Way are relatively near us at distances of 90 to 500 parsecs. At greater distances their contrast with the bright background is diluted by increasing numbers of stars in front of them. Few dark nebulae are recognized optically in our galaxy at distances exceeding 1500 parsecs. Even so, these dark nebulae are of great interest as we will see later in this chapter because they show evidence of complex molecules in interstellar space.

14.4 Dark globules

A feature of many emission nebulae is small, almost spherical globules seen as **dark globules,** or **Bok globules** (Fig. 14.5), against the glowing nebulae. These globules are also seen as dark patches against a uniformly

Figure 14.4 *The Horsehead nebula in Orion. The illuminated nebula is quite thin, as evidenced by counting stars in the light and dark portions. (Hale Observatories photograph.)*

Figure 14.5 *The Lagoon nebula in Sagittarius. Note the irregular dark globules between us and the emission nebula. (Lick Observatory photograph.)*

Figure 14.6 *Dark globules in Sagittarius called Barnard 68 and Barnard 69. (Photograph by B. J. Bok.)*

distributed star field (Fig. 14.6). Because of their small size they are not detectible at distances much greater than 500 parsecs.

The Bok globules are extremely dense, some having at least 20 magnitudes of absorption. There is one case where the absorption is more than 25 magnitudes. An absorption of 25 magnitudes means that if an A0 star were only 10 parsecs away, but behind such a cloud we might just detect the A0 star on a very-long-exposure plate taken with the 5-meter telescope. Normally such a star would have 0.0 apparent magnitude and be one of the brightest stars in our night sky. The amount of absorption tells us the amount of dust present. The average globule contains some 20 times the mass of the sun in dust alone. Estimates of the gas and molecular content of such clouds range from 50 to 100 times the amount of dust. Thus the globules contain some 1000 to 2000 times the mass of the sun. Further evidence of the extreme density is demonstrated by globules holding together in strong radiation fields.

This evidence that the globules are strongly bound suggests that they are near collapse and hence initiating star formation. One globule shows positive evidence of star formation. On one edge of this globule are two stars of the T Tauri type, very young stars that have not yet become stable. Photographs give the definite impression of two stars shining out through the nebula.

The Interstellar Medium

In addition to the more obvious bright and dark nebulae, certain regions of our galaxy contain an abundance of tenuous gas and dust. The interstellar dust dims and reddens the light of more distant stars. The gas imprints dark lines in the spectra of stars beyond it. The presence of optically invisible neutral hydrogen in the gas is recorded with radio telescopes. The average density of the interstellar gas is around 10^{-24} g/cm^3 and interstellar dust is 100 times less.

14.5 Interstellar spectra

Interstellar lines in the spectra of stars, first noted and correctly interpreted by J. Hartmann in 1904, are abstracted from the starlight by the gas through which the light passes. These dark lines are generally much narrower than the lines in the spectra of the stars themselves and have different Doppler displacements than those of corresponding star lines. Among the chemical constituents of the interstellar gas indicated by these lines are atoms of sodium, potassium, calcium, iron, and titanium, and molecules of cyanogen and various hydrocarbons. Hydrogen atoms are doubtless very abundant, but in these conditions their lines would appear only in the generally unobservable extreme-ultraviolet region of the spectrum and at λ 21 centimeters when observed with radio telescopes.

The division of interstellar lines into two or more components was recognized some time ago (Fig. 14.7). The division is given important interpretation by more recent studies which show that the principal components of the interstellar lines are absorbed by gas in two intervening arms of our galaxy. Very precise measures of the interstellar hydrogen lines at λ 21 centimeters confirm this (Fig. 14.8).

14.6 Interstellar hydrogen

Where a cloud of interstellar gas surrounds a hot star, the gas is set glowing within a radius around the star that depends on the temperature of the star and the density of the gas. This sphere is referred to as the **Strömgren sphere.** Outside this region the gas is normally dark. Because hydrogen is the most abundant chemical element, it has been the custom to speak of the two parts of such a cloud as the H-II and H-I regions, respectively.

In the part of the cloud that is nearer the star, hydrogen atoms are ionized by the star's ultraviolet radiation. Because the atoms that have lost electrons soon capture others reemitting the energy, the gas is thereby made luminous. In addition to the conspicuous emission nebulae, many

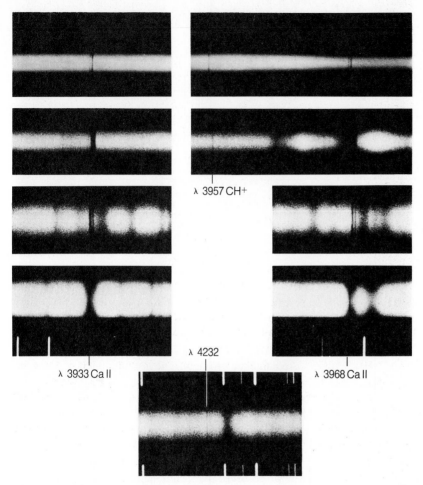

λ 3957 CH+

λ 4232

λ 3933 Ca II   λ 3968 Ca II

Figure 14.7   *Interstellar lines in the spectra of various stars. The interstellar lines are extremely sharp compared to those of the stars and are often split into several lines, indicating many thin clouds between us and the star. (Hale Observatories photograph.)*

faint H-II regions are detected in photographic surveys with wide-angle cameras and plates especially sensitive to the spectral region of the red line of hydrogen. Such surveys have been useful in tracing the spiral arms of our galaxy and other nearby galaxies.

The bright material in the inner region of a cosmic cloud is likely to expand and to collide with the gas in the dark region outside. Such collisions may be vigorous enough to set the contact areas glowing conspicuously. This reason has been assigned for the familiar loop of bright nebulosity near epsilon Cygni (Fig. 14.9).

In the part of a gas cloud that is too remote from a hot star to be much stimulated by it, the hydrogen is generally dark, cold, and optically unobservable. Radiation by neutral atoms of cosmic hydrogen at a wavelength of 21 centimeters (Section 12.9) previously predicted by H. C. van

Figure 14.8   *The hydrogen emission profile in the direction of the star HD 14134. Velocities measured for the interstellar calcium lines are shown by the vertical bars and the radial velocity of the star is indicated by the asterisk. The radial-velocity agreement between the hydrogen and calcium indicates that they are in the same clouds.*

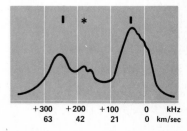

| +300 | +200 | +100 | 0 | kHz |
| 63 | 42 | 21 | 0 | km/sec |

Figure 14.9 *Hydrogen clouds and wreaths in Cygnus. In this mosaic of negative prints from the Mt. Palomar Schmidt telescope emission nebulas and stars appear black. Astronomers often work with negatives to avoid loss of detail in reproduction. The familiar loop nebula is seen at the bottom. (Composite photograph by J. L. Greenstein, California Institute of Technology.)*

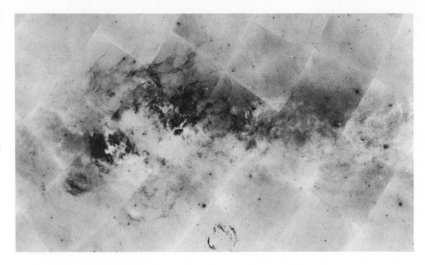

de Hulst, was first recorded with a radio telescope by H. I. Ewen and E. M. Purcell in 1951. This emission line has been employed effectively in tracing the spiral arms of the Galaxy (Section 17.6) to great distances from the sun. It would be difficult to overstress the prediction and observation of the λ 21-centimeter line too strongly. The last equivalent instrumental advance was the introduction of the photographic plate in the 1850s.

The emission by neutral hydrogen has also allowed us a detailed look at the center of the Galaxy. It has been used to trace the extent of nearby galaxies as well. Where a bright source is located behind cooler neutral hydrogen the corresponding absorption line at λ 21 centimeters is observed.

At λ 18 centimeters we observe the rotational lines of the OH molecule. Again, depending upon circumstances, it may be observed in emission or in absorption. Many other molecules have likewise been observed (Section 14.9).

Our general impression so far is that the interstellar medium is cold. Actually there is evidence that the interstellar medium is made up of a cold and hot component, hot in the same sense that the corona of the Sun is hot. Indeed, some investigators believe that the hot component is the dominant component and is composed of high-energy particles released by supernova events.

14.7 Dimming of starlight by dust

Interstellar dust dims the light of stars behind it (Fig. 14.10). The dust also reddens the light of those stars, because dimming in the shorter,

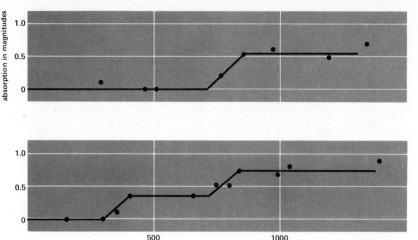

Figure 14.10 *Photographic absorption by dust in two regions of Aurigae. Star counts reveal one dust cloud in region F and two in region G.*

blue wavelengths is greater than in the longer wavelengths of the light. A corresponding effect is observed in the reddening of the light of the setting sun by particles of our atmosphere. Careful observations show that the absorption *(A)* by the interstellar material is inversely proportional to the wavelength of the observation $(A \propto 1/\lambda)$. A typical absorption curve as a function of wavelength is shown in Fig. 14.11. The wavelength dependence is just that expected for particles having sizes of 3×10^{-5} centimeters or slightly larger. If the particles were atoms or molecules the absorption should be inversely proportional to the fourth power of the wavelength $(A \propto 1/\lambda^4)$ which is shown by the light purple line in Fig. 14.11. We also know that if the particles were several times the wavelength in size we would have a neutral absorption, that is, all starlight would be dimmed by the same amount in all wavelengths. Thus our observations tell us that the absorption is due to dust and tell us the size of the dust. We cannot rule out the possibility that larger particles are dimming all light accordingly, or that gas is not present. Indeed, we shall see that the gas accounts for 100 times more of the interstellar mass than does the dust.

The **color excess** of a star is the difference in magnitudes by which the observed color, blue minus visual *(B − V)*, exceeds the accepted value for a star of its spectral type; it is a measure of the reddening of the star by the dust. When the color excess is multiplied by an appropriate factor, we have the **photographic absorption,** that is, the degree to which the star is dimmed by the dust when photographed with a blue-sensitive plate.

The calculated distance of a reddened star is greater than its true

Figure 14.11 *Interstellar reddening as observed (black line) approximates a 1/λ law. If atoms or molecules were causing the reddening the curve would follow a 1/λ⁴ law (purple line).*

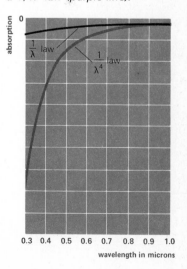

distance and thus requires correction for absorption. The corrected distance d is calculated by the formula

$$5 \log d = m - M + 5 - K,$$

where m is the observed photographic magnitude, M is the corresponding absolute magnitude for a star of this particular spectral type, and K is the photographic absorption.

Since dimming and reddening of starlight is caused by the interstellar dust, these quantities must vary depending upon where one looks in the Galaxy. The determination of dimming and reddening is not very difficult in principle, but in practice it is a most perplexing task. Thus, any method promising to yield distances, magnitudes, or colors independent of dimming and reddening is studied to the utmost.

14.8 The dust grains

The dimming of stars by intervening cosmic dust is attributed to particles of the order of a thirty-thousandth of a centimeter in diameter. Dust grains of this size would scatter the light in inverse proportion to the wavelength, which is not far from the observed relation for the reddened stars. Considerably larger particles would obstruct but not redden the starlight. The origin of the dust grains is not clearly understood; whether they build up from atoms of the interstellar gas or are particles transferred to the medium from the upper atmospheres of stars, or both, is conjectural. Present theory favors the latter idea. It is suggested that long linear chains of carbon (10 or more carbon atoms in a string) are stable and form the basic ingredient of the grains.

Starlight becomes polarized in its passage through clouds of interstellar dust. A possible explanation is that the polarization is produced by dust grains shaped like needles and rotating on their short axes, resulting in a concerted effect by setting themselves along magnetic lines of force of the Galaxy; a magnetic field of 10^{-5} gauss is indicated. This and other indications that our galaxy has a general magnetic field have added other items to the growing list of problems in celestial hydromagnetics, not the least of which is the Zeeman splitting observed in the OH lines—if that is what it really is. Local galactic magnetic fields of about 10^{-2} gauss are implied by the OH lines and this is some 100 times larger than fields previously observed. Other radio observations, such as pulse delay times as a function of wavelength from pulsars, indicate field strengths of 10^{-5} gauss in agreement with the earlier resutls.

14.9 Interstellar molecules

We have seen that the interstellar material seems to bunch into nebulae. Within these clouds, shielded from the high-energy radiation of the ultraviolet and x-ray photons that would destroy them, interstellar molecules could form. The first interstellar molecule, CH, was discovered in 1937. A few additional molecules were detected in the visual region of the spectrum, CN and CH^+ (singly ionized CH), but if our reasoning is correct we cannot detect the more complex molecules, which are also more fragile, simply because they must be hidden deep in the interior of the great interstellar clouds.

Fortunately, radiation at radio wavelengths passes through the interstellar clouds, hence the clouds can be probed for emission from molecules. One other factor works in favor of detection by radio techniques. We recall that the strength of any atomic spectral line is proportional to the number of atoms involved, and this is also true for molecular lines. Thus, if the abundances of molecules are in a constant ratio, the very densest clouds will have the highest concentrations rendering them more readily detectable.

In 1963 emission from the OH molecule was discovered at λ 18 centimeters. This was followed in 1968 by the discovery of ammonia (NH_4) and water vapor (H_2O). These discoveries served to show that complex molecules do exist in the interstellar medium. Then, in early 1969 came the remarkable discovery at λ 6.2 centimeters of formaldehyde (H_2CO), a **polyatomic organic molecule.** An organic molecule is one where hydrogen is attached to carbon; the term originates from the fact that such molecules are found in living organisms. This discovery has been followed by the detection of some 40 additional molecules, most of which are organic (Table 14.1). It is tempting to think that a carbon chemistry is possible if not common throughout the Milky Way. This thought has great consequences on our ideas of the origin and evolution of life. We will return to this in Chapter 19.

Several of the interstellar molecules are not normally found on the earth; they must be carefully made in the laboratory. All of the rest are common ordinary molecules readily available on earth. Identification of interstellar molecular lines is carried out in several ways. One is to concoct the molecule in the laboratory and then observe its microwave and/or infrared spectrum. The laboratory lines can then be compared with the observed lines. When no laboratory spectrum is available the spectroscopist must resort to calculations.

The discovery of interstellar molecules proceeds on two fronts with part of the trick being in knowing where to look. We may discover a molecule by observing its lines and comparing it with laboratory lines

Table 14.1
Interstellar Molecules (not including isotopic species)

Year of Discovery	Molecule	Chemical Symbol	Part of Spectrum
1937	methylidyne	CH	visible
1940	cyanogen radical	CN	visible
1941	methylidyne ion	CH$^+$	visible
1963	hydroxyl radical	OH	radio
1968	ammonia	NH$_3$	radio
1968	water	H$_2$O	radio
1969	formaldehyde	HCO	radio
1970	carbon monoxide	CO	radio
1970	hydrogen cyanide	HCN	radio
1970	cyanoacetylene	HC$_3$N	radio
1970	hydrogen	H$_2$	ultraviolet
1970	methyl (wood) alcohol	CH$_3$OH	radio
1970	formic acid	HCOOH	radio
1970	formyl ion	HCO$^+$	radio
1971	formamide	HCONH$_2$	radio
1971	carbon monosulfide	CS	radio
1971	silicon monoxide	SiO	radio
1971	carbonyl sulfide	OCS	radio
1971	methyl cyanide, acetonitrile	CH$_3$CN	radio
1971	isocyanic acid	HNCO	radio
1971	methylacetylene	CH$_3$C$_2$H	radio
1971	acetaldehyde	CH$_3$CHO	radio
1971	thioformaldehyde	H$_2$CS	radio
1971	hydrogen isocyanide	HNC	radio
1972	hydrogen sulfide	H$_2$S	radio
1972	methanimine	H$_2$CNH	radio
1973	sulfur monoxide	SO	radio
1974	diazenylium	N$_2$H$^+$	radio
1974	ethynyl radical	C$_2$H	radio
1974	methylamine	CH$_3$NH$_2$	radio
1974	dimethyl ether	(CH$_3$)$_2$O	radio
1974	ethyl alcohol	CH$_3$CH$_2$OH	radio
1975	sulfur dioxide	SO$_2$	radio
1975	silicon sulfide	SiS	radio
1975	acrylonitrile, vinyl cyanide	H$_2$CCHCN	radio
1975	methyl formate	HCOOCH$_3$	radio
1975	nitrogen sulfide radical	NS	radio
1975	cyanamide	NH$_2$CN	radio
1976	cyanodiacetylene	HC$_5$N	radio
1976	formyl radical	HCO	radio
1976	acetylene	C$_2$H$_2$	infrared
1976	cyanoethynyl radical	C$_3$N	radio
1976	ketene	H$_2$C$_2$O	radio
1977	ethyl cyanide	CH$_3$CH$_2$CN	radio
1977	carbon	C$_2$	infrared
1977	nitroxyl radical	HNO	radio
1977	cyanotriacetylene	HC$_7$N	radio
1978	cyanotetracetylene	HC$_9$N	radio
1978	butadiynyl radical	C$_4$H	radio
1978	nitric oxide	NO	radio
1978	methane	CH$_4$	infrared
1979	methyl mercaptan	CH$_3$SH	radio
1979	isothiocyanic acid	HNCS	radio

as explained above. Or we may deduce from various molecules already observed in a given cloud that a certain molecule should be there also. We then look for lines of the suspected molecule in the same cloud.

The physics and chemistry of the interstellar molecules and medium is only now being unraveled. It will be several years, perhaps decades, before we will really begin to understand what we are observing. Many of the emissions observed are from levels that are not normally found on earth or in the laboratory, and in this sense they are similar to the forbidden lines that we have discussed for atoms. Molecules have several lines close together and, as often as not, the relative strengths of these lines are not those we predict on the basis of calculations or laboratory measurements. We do not know the excitation mechanism that causes these anomalies, but work is being done in this area and answers should be available soon. We do not even know what causes abundance anomalies. The strength of the HD molecule ("heavy" hydrogen) compared to H_2 (normal hydrogen) gives a ratio of hydrogen to deuterium, an isotope of hydrogen, of 300; this is several orders of magnitude smaller than we would predict. This result, which was obtained by the spacecraft OAO-C (Orbiting Astronomical Observatory-C), is quite unexpected and typical of the subject of interstellar molecules.

Complex molecules are found in interstellar clouds almost everywhere in the Milky Way. Occasionally, the clouds are releasing large amounts of energy, and they are sometimes variable in their energy output. At least one cloud, with water in great abundance, has solar system dimensions.

Our conceptual picture of an interstellar cloud containing organic molecules is that in the densest innermost regions are found the most complex molecules along with all the other molecules making up the cloud (Fig. 14.12). The molecules become less and less complex as we move outward from the center to regions where photons can penetrate and dissociate them. In the outermost regions only the tightly bound CO molecules can exist, and at the boundary even this molecule is photodissociated. In a cloud that is not very dense, perhaps the most complex molecule that can exist is water.

Studies of these clouds and the chemistry tell us about the density of the material in the clouds and the temperature as well. In one case, the material in the cloud is colder than 3 K. In general the temperatures in the dense clouds are around 10 K. The densities are deduced from the chemistry. For example, if ammonia is observed and hydrogen cyanide is not, the density of the cloud is between 10^3 and 10^5 molecules/cm³. If hydrogen cyanide is present, then the density is greater than 10^6 molecules/cm³. Of course, most of the 10^6 molecules are molecules of hydrogen (H_2), but these are not observable in the radio or optical region of the electromagnetic spectrum.

Figure 14.12 *Highly schematic representation of an interstellar cloud. Simple atoms and dust fill the entire cloud. More and more complex molecules are found deeper and deeper in the cloud as the density increases.*

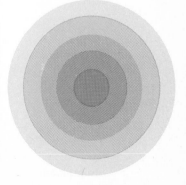

The large interstellar clouds with their great molecular concentrations account for a large fraction of the mass of the Milky Way Galaxy.

Our present picture of how molecules are formed involves the cold grains of ice and dust in a dense interstellar cloud. To form a molecule of OH, for example, an atom of hydrogen collides with and sticks to a grain. It is free to move around on the grain. Later the grain is struck by an oxygen atom that sticks and also moves around on the surface of the grain. During the migrations of the atoms the hydrogen and oxygen move close and bind, forming an OH molecule. This is a very simple picture because we are ignoring electric charges and how the molecules are released from the grains.

The grains, particularly the dust, appear to originate in the atmospheres of stars, probably giant stars, by processes not yet known. It is interesting that the ratio of dust to gas is almost constant, whether we are considering the tenuous interstellar medium or a very dense globule.

Interstellar chemistry is a new and developing subject. Since the chemistry takes place under a wide variety of conditions, reactions that take place in one region do not necessarily take place in another. Furthermore, we often must extrapolate from laboratory conditions to those of interstellar space.

Atoms and molecules in the thin interstellar medium undergo reactions which ultimately cease when a molecule is destroyed because of the flux of ultraviolet photons, x rays, and cosmic rays. The gases deep inside a cloud, shielded from destructive radiation, undergo other reactions. These latter reactions may be relatively rare, but being protected the reactions products accumulate into large concentrations.

Ions of molecules are found in the dense clouds, indicating that an energy source for ionization exists inside some clouds. In some cases the source may be very-high-energy cosmic rays, but in most cases it is a newly formed star not yet visible because the star has not had time to blow away the surrounding cloud. In a few cases we detect the new star by means of the infrared radiation from the grains in the cloud. The grains absorb the stellar radiation and reradiate it as heat.

Questions

1 List three ways that we detect dark nebulae.
2 Distinguish between emission nebulae and reflection nebulae.
3 Describe a Bok globule.

4 How do we differentiate between an interstellar absorption line in a stellar spectrum and the star's own absorption lines?

5 What is the effect of interstellar absorption upon the distance of a star?

6 Are most complex interstellar molecules organic or inorganic?

7 Why are complex interstellar molecules found inside interstellar clouds?

8 Describe one method we believe is responsible for forming molecules in interstellar space.

9 What is the ratio of dust to gas in interstellar space?

10 Can you explain why no discussion of helium appeared in this chapter?

Other Readings

Dickman, R. L., "Bok Globules," *Scientific American* **236(6),** 66–81 (1977).

Heiles, C., "The Structure of the Interstellar Medium," *Scientific American* **238(1),** 74–84 (1978).

Jura, M., "Interstellar Clouds and Molecular Hydrogen," *American Scientist* **65,** 446–454 (1977).

Wesson, P., "The Case of the Missing Cosmic Dust," *New Scientist* **73,** 207–209 (1977).

chapter 15

The Lives of the Stars

In studying stellar evolution we try to understand how stars are born and how they spend their lives until they cease to shine. The idea that stars condense from nebulae was pioneered by Kant and Laplace in the 18th century and has persisted to the present time. Stars of large mass evolve faster than less massive stars and, during the course of their evolution, return material to the interstellar medium. Finally all stars end their careers as black dwarfs, white dwarfs, neutron stars, or black holes.

Youthful Stars

New stars are continually forming from condensations in the interstellar gas and dust, according to our present theory of stellar evolution. The primitive stars are heated by contraction until they begin to shine. Thereafter, they go on shrinking until their central temperatures are hot enough to permit the building up of atoms of heavier chemical elements from lighter ones. Nucleosynthesis, the formation of helium from hydrogen, replaces contraction as the main source of stellar energy. The star then settles down to a long life on the main sequence.

15.1 The birth of stars

The account of stellar evolution begins conveniently with structureless interstellar gas. When a denser cloud develops in the gas and is not considerably heated in the process, the cloud is likely to condense further under its own gravity and to break into smaller clouds. The fragmented cloud would be the beginning of a cluster or an association of stars (Chapter 16).

How does a dense cloud form? How does the cloud ever become dense enough so that the self-gravitation is greater than the forces that would disrupt the cloud? The disrupting forces are the random motions in the cloud resulting from the heat caused by the contraction of the cloud. The densities required to make the cloud self-gravitating are of the order of 10^{-19} g/cm³, whereas the average interstellar cloud has a density of 10^{-23} g/cm³ or somewhat lower. Thus, a typical interstellar gas cloud must be compressed to 10,000 times its initial density in order to begin collapsing on its own.

A possible answer may be provided by clouds containing complex molecules. These clouds are generally quite cold; when formaldehyde is present, we can see it in absorption against the general cosmic background (Section 19.5). This proves that the temperature of the inner part of the cloud must be less than 3 K. If the temperature of the interstellar medium is 50 or 100 K, then the hotter interstellar medium exerts an inward pressure on the cooler cloud, raising its density. This temperature difference will cause a significant pressure difference, but not enough. We now think that the extra push needed to make an interstellar gas cloud collapse is provided by the shock wave sent out by supernova explosions in the vicinity of the gas cloud. The momentary compression caused by the explosion would create a self-gravitating region of gas around which the rest of the cloud could collapse. Stars may therefore be born as a consequence of the death of other stars.

Photographs of the Milky Way show small roundish "globules" of dark material against backgrounds of star-rich regions (Fig. 15.1) and of bright nebulosities, such as M 8 (Fig. 14.5). These dark spots, which continue in a sequence of diminishing sizes to the limit of the largest telescopes, are viewed with interest as perhaps representing the first visible stages in the formation of stars from nebulae.

An objection to the idea that such protostars can be observed is that stars are likely to form in the interiors of cosmic clouds, where they would be concealed by dust in the clouds. Exceptions are considered possible: One is that the star must begin to shine in the infrared first; infrared photons may then escape the cloud. Another is that the most massive stars of a very young group have already become hot blue stars. In the first case we can observe the stars as infrared objects. In the second, the intense radiations of the hot stars may have dispersed enough of the dust around the group so that the redder members can be seen still forming within the cloud. This may be the case with the T Tauri stars.

The T Tauri stars, named after the prototype, were first recognized in a clouded area of Taurus. Now observed in other heavily obscured

Figure 15.1 *The small dark globule, Barnard 335. This globule may be collapsing to form stars. (Courtesy of B. J. Bok.)*

regions of the Milky Way as well, these yellow and red stars are irregularly variable in brightness. Their spectra are characterized by strong bright lines of various elements superposed on ordinary dark-line patterns. In some cases the stars illuminate portions of the surrounding clouds, producing fan-shaped reflection nebulae that may be variable in brightness and appearance.

Young stars very recently formed in the clouds, the T Tauri stars, are believed to vary in brightness and to have bright lines in their spectra because of the instability of extreme youth rather more than by their interaction with the dust grains of the clouds. A few such stars appear in the photographs in places where no stars were previously observed; these may have been revealed lately by rapid thinning of the dust in front of them.

15.2 Arrival at the main sequence

When the contracting protostars of a cluster or an association have begun to shine, they are enormous distended clouds of gas and dust and are quite cool. They radiate principally in the infrared and radio regions of the spectrum. Such stars have been observed and are referred to as **infrared stars.** They are always found in highly obscured regions, and intense radiation from water vapor is observed. When they appear in the visible region of the spectrum they lie rather high up in the H-R diagram. The gravitational energy released in the deep interior of the cloud is carried outward by convective currents. This causes the cloud to collapse rapidly to the point where a radiative core develops. During the collapse the energy radiated per square centimeter remains constant, but the surface area decreases rapidly, therefore the protostar moves almost vertically down the H-R diagram and reaches thermal equilibrium (emitting as much energy as it generates) on the main sequence (Fig. 15.2). The stars are brighter and attain higher surface temperatures as their masses are greater. Thus they array themselves along the diagonal band of the main sequence in order of mass, the most massive stars at the blue end and the least massive at the red end.

The time required for the development of the stars of a group depends on the masses. The more massive blue stars may arrive at the main sequence only a few hundred thousand years after their birth. The least massive stars may require a few million years to arrive at the main sequence. Thus in some very young clusters the blue stars are already settled on the main sequence, while the other stars are still approaching it.

This recently recognized effect is shown in Fig. 15.3 for the very young cluster NGC 2264 in Monoceros, near the central line of the Milky Way about 15° east of Betelgeuse and 700 parsecs distant from us. The points representing the blue stars appear on the main sequence, whereas

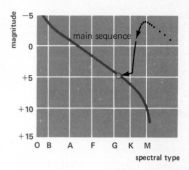

Figure 15.2 *The Hayashi evolutionary track for a solar-type star. The star becomes visible as a very distended object (top of dashed line), then rapidly shrinks in size with little change in temperature until thermal equilibrium is established. At this point the star moves slowly to the left until it reaches the main sequence (purple line).*

Figure 15.3 *Hertzsprung–Russell diagram of the young cluster NGC 2264. Most of the stars have not reached the main sequence. The dots represent photoelectric measures and the circles photographic measures. Vertical lines denote variable stars. Horizontal lines denote stars having hydrogen emission lines. The solid curve is the standard main sequence. (Diagram by M. F. Walker.)*

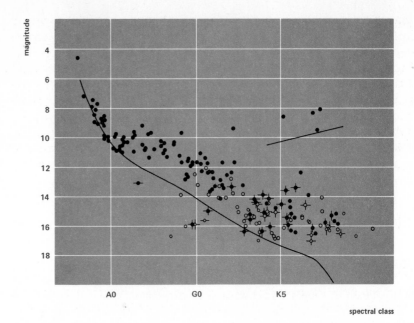

those for the yellow and red stars depart abruptly to the right from the standard sequence at type A0 and tend to lie about two magnitudes above the curve. Many of the latter are variable stars of the T Tauri type.

Some condensations may not have sufficient mass for thermal equilibrium to set in. In this case they will continue to contract into degenerate objects called **black dwarfs.** Presumably, if such condensations are gravitationally attached to a larger condensation, they might become planets.

15.3 The stable main-sequence stars

Contraction is halted when the protostars reach the main sequence. This process, regarded in earlier theories as the main source of stellar energy, is inadequate to supply the stars during their long lives. Contraction suffices in the present account to raise the central temperature high enough to initiate the fusion of lighter chemical elements into heavier ones. Atomic syntheses then becomes the primary source of stellar energy and the object is then a bona fide star.

Main-sequence stars are stable for a time, neither contracting nor expanding. Therefore, they vary little in size, temperature, and brightness. Consider the sun as an example. At any level in its interior the sun is in equilibrium. The weight of overlying gas is just supported by the upward push of the gas pressure at that level. Because the gas pressure depends on the temperature of the gas, among other factors, the temperature becomes known when the other necessary information is available. Obtained in this way, the sun's present central temperature is about 15 million degrees Kelvin. Another form of balance in the sun concerns its shining. The rate of radiation of energy from the sun's surface is 3.8×10^{33} ergs/second. In order to preserve the equilibrium, this must also be the rate at which energy is provided in the hot core of the sun. This constant rate for a normal star is very constant indeed. There is fossil evidence on the earth that the sun's total energy output has been constant to within 1% over a period of three billion years. There is reason to believe that this period has been even longer, perhaps five billion years, and will continue for another five billion to ten billion years.

The Stars in Middle Age

The stars stay on the main sequence as long as they remain practically homogeneous in composition. The energy released by the fusing atoms keeps the stars shining steadily. As the buildup of helium in their cores continues, the stars eventually lose stability; they then begin to expand and brighten, leaving the main sequence to become giant stars. The bluer stars evolve faster than the redder ones.

15.4 Fusion of hydrogen into helium

The relative weight of the nucleus of a hydrogen atom is 1.0080 and that of a helium atom is 4.003 in terms of the **atomic mass unit (amu).** One atomic mass unit is equal to 1.66×10^{-24} grams of matter. The electrons normally associated with atoms need not be taken into account here, because these atoms are stripped of their electrons in the hot interiors of stars. When four hydrogen atoms unite to form one helium nucleus, 0.7% of the original mass is left over. The excess mass is converted into energy. By the relativity theory, $E = mc^2$, i.e., the amount of energy E (in ergs) released equals the mass m (in grams) of the excess material multiplied by the square of the speed of light c (in centimeters per second). Using this formula we can calculate that enough energy is thereby made available to keep the sun shining at its present rate if 700 million metric tons of hydrogen gas are converted each second to 695 million metric tons of helium in the sun's interior. This consumption is trivial compared with the vast supply of hydrogen in the sun.

Taking as a lower limit that 70% of the present sun is hydrogen, then the sun has 1.4×10^{33} grams of hydrogen to convert. Each second, as noted above, 7×10^{11} grams is converted so it will take 2×10^{21} seconds to convert all of the hydrogen into helium. One year is about 3×10^7 seconds so the sun would last about 10^{14} more years. Of course,

as we will see shortly, the conversion process does not go up to 100%
completion so the life of the sun is considerably less than this.

Two possible processes by which hydrogen may be fused into helium
in the sun and stars were explained in 1938 by H. A. Bethe. One of
these, known as the **carbon cycle,** because carbon serves as a catalyst*
to promote it, becomes effective in stars hotter than the sun. The second
process, known as the **proton–proton reaction,** is considered more appropriate at the central temperatures of the sun and the main-sequence stars
that are cooler than the sun.

It is likely that the first **burning** process is that which transforms
deuterium into helium. In this process a deuterium nucleus, which we
have seen exists in space, interacts with a proton, or hydrogen nucleus,
releasing energy in the form of a gamma ray and forming an isotope of
helium (^{3}He). This process, made possible by the temperature increase
due to the collapse of the cloud, then raises the temperature to a point
where the proton–proton reaction can take place.

In the proton–proton reaction six protons (or hydrogen nuclei) are
required to form a helium nucleus. The reaction proceeds when two protons
unite to form deuterium, releasing energy and a neutrino. Deuterium and
a proton then react to form an isotope of helium, releasing a gamma
ray. When two helium isotopes combine, normal helium is formed and
two free protons are released. The two unused protons go back into circulation. The excess mass in the four united protons is released as energy,
which is carried up to the sun's surface to contribute to the sunlight.
The reaction may be written symbolically in the following three steps:

(1) $$_1^1\text{H} + {_1^1}\text{H} \rightarrow \quad _1^2\text{H} \quad + {^1}e + \nu\,,$$

(2) $$_1^2\text{H} + {_1^1}\text{H} \rightarrow {_2^3}\text{He} + \gamma\,,$$

(3) $$_2^3\text{He} + {_2^3}\text{He} \rightarrow {_2^4}\text{He} + {_1^1}\text{H} + {_1^1}\text{H}\,.$$

The notation is necessary but should not detract the reader from noting
that a gamma-ray (γ) photon is released each time step (2) takes place
and also that a neutrino *(ν)* is released when step (1) takes place. The
neutrino is a nuclear particle having energy but no electronic charge and
zero mass. The subscript indicates the number of protons and the superscript indicates the mass of the nucleus.

In stars somewhat more massive than the sun, perhaps of spectral
type F2 and earlier, the carbon cycle dominates. In it carbon (^{12}C) combines
with a proton, forming an isotope of nitrogen (^{13}N) that decays into an
isotope of carbon (^{13}C), releasing energy. This carbon isotope combines

* A catalyst is a substance that initiates a reaction. It remains unchanged after the reaction
  is complete.

with a proton to form stable nitrogen (^{14}N), releasing additional energy. A proton then interacts with the nitrogen to form an isotope of oxygen (^{15}O) that immediately decays into another isotope of nitrogen (^{15}N). This nitrogen isotope then combines with a proton, but instead of forming stable oxygen, which we might expect, it forms a carbon nucleus plus helium. The end result is the conversion of hydrogen into helium, with the initial amount of carbon unchanged. For reference we may write the carbon cycle symbolically in six steps:

(1) $$^{12}_{6}C + {}^{1}_{1}H \rightarrow {}^{13}_{7}N + \gamma \, ,$$

(2) $$^{13}_{7}N \rightarrow {}^{13}_{6}C + {}^{4}e + \nu \, ,$$

(3) $$^{13}_{6}C + {}^{1}_{1}H \rightarrow {}^{14}_{7}N + \gamma \, ,$$

(4) $$^{14}_{7}N + {}^{1}_{1}H \rightarrow {}^{15}_{8}O + \gamma \, ,$$

(5) $$^{15}_{8}O \rightarrow {}^{15}_{7}N + {}^{4}e + \nu \, ,$$

(6) $$^{15}_{7}N + {}^{1}_{1}H \rightarrow {}^{12}_{6}C + {}^{4}_{2}He \, .$$

Again the important points for the reader are the release of energy through the high-energy gamma-ray photons in steps (1), (3), and (4), and the release of neutrinos in steps (2) and (5).

Eventually the hydrogen in the center of a star is converted to helium. When the pressure becomes great enough, due to contraction, the helium **burns** to carbon by means of the **triple alpha process.** This process combines three helium nuclei (sometimes called alpha particles) directly into carbon. The triple alpha process may be written in two steps:

(1) $$^{4}_{2}He + {}^{4}_{2}He \rightleftharpoons {}^{8}_{4}Be + \gamma \, ,$$

(2) $$^{8}_{4}Be + {}^{4}_{2}He \rightarrow {}^{12}_{6}C + \gamma \, .$$

In step (1) the reverse arrow means that the beryllium decays back into two alpha particles spontaneously after a very short period of time. If, before this, another alpha particle interacts with the beryllium the second step takes place. Both steps yield high-energy gamma-ray photons.

15.5   Evolution from the main sequence

A star spends the majority of its life on the main sequence, where it gradually converts the supply of hydrogen in its core to helium. Since the nuclear reactions occur in the center of the star, the helium is concentrated there, and the percentage of hydrogen steadily decreases. Eventually the core is converted almost entirely to helium. At this point the nuclear fires in the core are extinguished. Unable to support itself against the immense weight of the overlying layers, the core contracts and heats up to an even higher temperature than before. Since the temperature in the thin shell around the core where hydrogen is still burning is raised to the temperature of the helium core, the hydrogen burning reactions proceed at an accelerated rate, greatly increasing the energy production of the star. As the core continues to contract it becomes hotter and hotter and the nuclear reactions in the surrounding shell run faster and faster. This continued increase in energy output causes the star to expand. Although the luminosity of the star increases, because the amount of energy generated in the star's interior increases, the surface temperature of the star drops. As a result, the star moves up and to the right on the H-R diagram (Fig. 15.4). The expansion continues until convection sets in in the gaseous envelope surrounding the core of the star. Energy transfer to the surface is then rapid enough to prevent the temperature of the core from rising any further. The star is now extremely large and very luminous; it is a red giant.

Figure 15.4 *Semiempirical evolution tracks for stars in the globular cluster M 3. Values of log T_e (effective temperature) correspond to the following spectral types: 3.9 to A5, 3.8 to F5, 3.7 to K0 for the main sequence and G0 for giants, 3.6 to K5 for the main sequence and K0 for giants. (Diagram by A. Sandage.)*

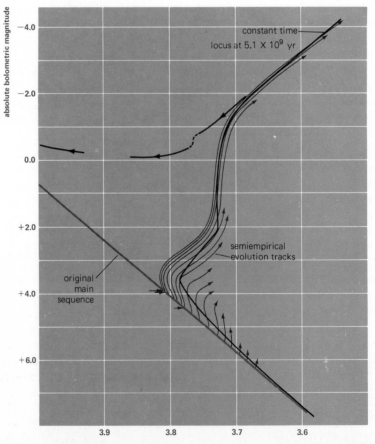

In a red giant, matter is in a very peculiar configuration. The core of a red giant is in an extremely compressed state, with a density equal to one metric ton per cubic centimeter. Surrounding the core is a shell a few thousand kilometers in thickness in which hydrogen fusion is taking place. The exterior of the star consists of a tenuous gas extending to a diameter which would encompass the orbit of the earth. Thus, the core of a red giant, which contains about one-quarter of the mass of the star, is about twice the size of the earth, but weighs nearly 100,000 times more. If the size of the red giant were considered to be the size of a beach ball, the core at its center would be no larger than a grain of sand.

The length of time which a star spends on the main sequence, and the speed with which it evolves into a red giant, depends upon the mass of the star. This is because the rate of nuclear fusion reactions is highly dependent upon the temperature at the center of the star, and more massive stars have higher central temperatures. The luminosity of a star increases with the 3.5 power of its mass ($L \propto M^{3.5}$). Thus, a star twice as massive as the sun is radiating more than 10 times as much energy and hence is using up its available fuel 10 times more rapidly. Although it has twice as much fuel to burn, it will live only about one-fifth as long as the sun.

The diagram of Fig. 15.4 shows the probable evolutionary tracks of stars of the globular cluster M3 since they began to evolve away from the main sequence more than 5 billion years ago. Stars originally a little bluer (more massive) than the sun are now red giants. For example, an F4 star identified in the diagram by the number 1 on the main sequence has evolved to the tip of the arrow marked with a 1. It is almost eight magnitudes brighter than it was on the main sequence. This star will continue to climb up the H-R diagram to become a fully evolved red giant. Those stars that were much bluer than the sun, say the A5 stars, have already passed through their red-giant stage. Meanwhile, the cluster stars that were similar to or redder than the sun have not yet risen far above the main sequence. The general evolutionary trace and the relative sizes of the stars at a given age of the cluster are shown in Fig. 15.5.

All stars eventually become red giants, but the particular course which a star follows after it reaches the red-giant stage is again dependent on its mass. The fate of stars which are as massive as the sun is discussed here. The evolution of more massive stars are considered in the next section.

In the red-giant stage, hydrogen burns in the shell surrounding the core, producing helium which is continually added to the core; this increases the core's mass and raises its temperature. Eventually a temperature of 100 million degrees Kelvin is reached in the core and the helium there begins to form carbon by the triple alpha process. This initiates a profound change in the internal configuration of the star.

In a star on the main sequence, the gas at its center is in equilibrium with its surroundings. Any departure from this equilibrium would be quickly restored by the self-regulating mechanism of nuclear reactions

Figure 15.5 *Relative sizes of stars on a logarithmic scale. The giant star in the upper right should actually be 200 times larger than the sun which is shown shaded in the lower center of the diagram. The cluster variable stars are shown at maximum and minimum size along the horizontal branch.*

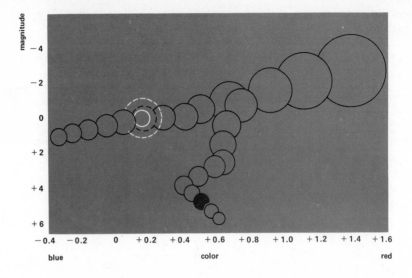

occurring in a gas. Imagine, for example, that the rate of nuclear fusion suddenly increased. This would heat up the gas, causing it to expand and cool. The decrease in temperature would lower the rate of nuclear fusion and the original conditions in the star would be restored. This self-regulating mechanism in stars on the main sequence is what accounts for their steady energy output over periods of billions of years.

In the core of a red giant, the safety-valve mechanism which controls the rate of nuclear fusion is no longer operative because the matter in the core no longer behaves like a gas. Instead, it is more like a rigid solid. When the helium in the core begins to burn, the rigid core cannot expand, and the temperature begins to rise rapidly. With the increase in temperature, the rate of helium fusion increases, causing a further increase in temperature. A vicious cycle ensues in which the rate of nuclear burning in the core of the star runs out of control. Within hours after helium fusion begins, the core of the star explodes. This event is called the **helium flash** because it occurs so quickly.

The helium flash does not result in an increase in the brightness of the star. In fact, what is observed is quite the opposite. When the hot core explodes, it expands and cools, so that the temperature in the surrounding shell where hydrogen was burning is reduced. As a result, the star's nuclear fires are temporarily extinguished. With its internal heat supply removed, the star can no longer support its large distended envelope and it begins to collapse. The temperature at the center of the star increases until it reaches 100 million degrees again. The helium is reignited, but now the core is in a gaseous state, so the reaction proceeds at a steady rate. In the course of this event, the star becomes smaller and hotter, and it moves from right to left across the H-R diagram. This evolutionary path is called the **horizontal branch.**

On the horizontal branch, helium is gradually converted to carbon in the center of the star. Some of the carbon also reacts with the helium to form oxygen. Eventually carbon and oxygen become so concentrated that helium fusion can no longer occur and a dense rigid core forms again. Two shells of nuclear fusion then exist in the star: a helium-burning shell around the carbon core and a hydrogen-burning shell around the helium shell. As the carbon core grows, the exterior shells are forced to occupy progressively thinner zones closer and closer to the surface of the star. When the helium-burning shell migrates so close to the surface that the temperature in this zone drops below the critical temperature needed for helium fusion, helium burning stops and the star shines solely by the energy released from hydrogen fusion. The star is nearing the end of its red-giant stage.

When the amount of hydrogen left in a star has been reduced to about one-tenth of a solar mass, the shell in which hydrogen fusion is occurring is so close to the surface that the outward force of radiation pressure exceeds the weight of the overlying layers and these outer layers are lifted off the star; the expelled layers form an advancing sphere of glowing gas known as a planetary nebula (Section 13.18). The hot core of the star left exposed at the center of the nebula has a surface temperature of at least 50,000 degrees Kelvin and is located to the far left in the H-R diagram. However, because the core has no more usuable fuel, its luminosity drops quickly and it moves down the H-R diagram to become a white dwarf.

The sun today is approaching middle age. It has brightened only slightly in the 5 billion years since it reached the main sequence and will continue to brighten moderately for 5 billion more years. Life on the earth should go on in reasonable comfort to the end of that period. Thereafter, the sun will begin to consume its remaining hydrogen so rapidly that in the following 500 million years the temperature on the earth will rise to 800 K. The oceans will boil away, lead will melt, and life will cease to exist. At its maximum the sun will appear as a red globe in the sky 30 times its present diameter. This is the sun's red-giant phase (Fig. 15.6).

With little available fuel remaining, the sun will then decline rapidly, moving to the left on the horizontal branch. Water on the earth will condense again to form oceans and these will soon freeze. The sun will end its visible life as a white dwarf star having a higher surface temperature for a longer time than that of the present sun but a diameter more nearly like that of the earth. The sunshine will be less than a thousandth as bright as it is now, and the temperature at the earth's surface will be more like the present low temperature at the surface of Neptune.

Figure 15.6 *Predicted variation in the earth's temperature as evolution alters the sun's radius and temperature. The temperature scale is in degrees Kelvin. (Diagram by A. Sandage.)*

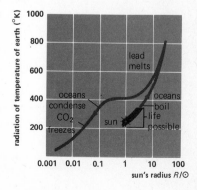

15.6 Evolution of more massive stars

Stars more massive than the sun evolve further than less massive stars because they are able to burn elements heavier than helium in their interiors. These stars spend a comparatively short amount of time on the main sequence and then move quickly through their final evolutionary stages. For example, a 15 solar mass star may only be on the main sequence for 10 million years.

Like smaller stars, a massive star accumulates a helium core and becomes a red giant. Because the helium core never becomes solid as it does in smaller stars, a helium flash does not occur. Instead, the helium core ignites gradually and burns steadily. Nonetheless, the effect is the same, and the star moves from right to left across the H-R diagram (Fig. 15.7) after the red-giant stage.

When all the helium in the core is consumed, the carbon core contracts and heats up and the outer envelope of the star expands. Again the star moves up and to the right on the H-R diagram, just as it did the first time it became a red giant. However, the rate of nuclear energy generation is greater the second time. The star therefore follows a higher path when it climbs up the H-R diagram. It is now an extremely luminous supergiant.

Eventually the temperature in the core of a supergiant reaches 600 to 700 million degrees Kelvin. The carbon in its core ignites and the carbon nuclei fuse to form magnesium, sodium, and neon. The carbon burns until it is totally consumed and the nuclear fires in the core are extinguished. Again, the core contracts and heats up. If the star is massive enough, the temperature in its interior will reach a billion degrees and oxygen in the core of the star will start to burn, producing sulfur, phosphorous, and silicon. With each new cycle of element production, the star crosses the H-R diagram, climbing higher and higher as its luminosity increases. However, this process cannot go on indefinitely. When a temperature of 3.5 billion degrees Kelvin is reached in the center of a star, iron is formed in the core.

Iron is fundamentally different than any of the elements that have formed in the star up to this point. When reactions involving iron occur, energy is absorbed rather than released. Thus, with the beginning of reactions involving iron, the core of the star suddenly behaves like a giant freezer unit, soaking up the heat in the center of the star. With the thermal gas pressure removed, the outer layers collapse towards the center of the star. In the earlier stages of the star's evolution, such a collapse would have heated the interior of the star and the collapse would have been halted. But with iron in the core, the heat is continuously removed as the iron nuclei react to form other elements and the outer layers of the star continue to implode. Pressures become so high in the interior of

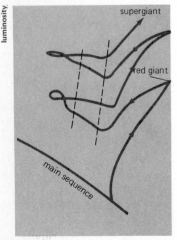

Figure 15.7 *The evolutionary track of a massive star. The successive crossings of the H-R diagram are due to the "burning" of helium and heavier elements. The dashed lines show the boundary of the instability strip where the star is a pulsating variable star.*

the star that electrons tunnel into the protons in the nuclei and free neutrons and neutrinos are produced.

Ordinarily, neutrinos pass right through matter, making them extremely difficult to detect. However, in the core of the collapsed star, the density is so great that even neutrinos are stopped. The neutrinos push away the infalling layers of the star and generate intense shock waves. As the outer layers of the star are blasted into space, the free neutrons are driven into the nuclei. The number of neutrons in each nucleus increases until the nucleus becomes unstable and one of the neutrons decays into a proton and an electron. This increases the atomic number of the nucleus by one, and a new element is formed. For example, neutrons can be added to ^{56}Fe (iron-56), forming ^{57}Fe, ^{58}Fe, and ^{59}Fe. When ^{59}Fe decays by emitting an electron ^{59}Co (cobalt-59) is formed. More neutrons are then driven into the nucleus and heavier and heavier elements are built up. By this process, called **neutron capture,** elements heavier than iron can be formed.

This violent explosion at the end of a star's supergiant phase is called a **supernova.** So much energy is released in this cataclysmic event that the star may temporarily outshine all the other stars in the galaxy combined (Fig. 18.13). The gaseous outer layers of the star which are blown off form a **supernova remnant,** such as the Crab nebula (Fig. 13.22), while the core of the star becomes a neutron star (Section 15.9).

While a star is still in the red giant stage, it crosses an area of the H-R diagram where it becomes unstable and begins pulsating. At this point in its life a massive star is a cepheid variable. (Less massive stars in this portion of the diagram become RR Lyra variables.) The gases in the outer layers of these stars continue to oscillate because they can not attain thermal equilibrium. When these stars contract, the gases in the outer layers are compressed to a higher density than they should have at that level in the star. As a result, the gases heat up and expand. The momentum of the expansion then carries the outer layers beyond their equilibrium point. Since the outer layers of the star are now too extended and diffuse to support themselves, they contract and again they overshoot the mark. These fluctuations would be stopped by friction in the expanding and contracting gases if some mechanism were not continually feeding energy into the pulsations.

Temperatures in the outer layers of a star in the instability strip are in the proper range to partially ionize helium at a critical level in the star. Because partially ionized helium becomes more opaque to radiation as it is compressed, the partially ionized helium prevents energy from escaping from the interior of the star. Energy is thereby trapped inside the star, causing the gas in its outer layers to heat up and expand. When the outer layers have expanded, the partially ionized helium becomes transparent, and the radiation can escape from the interior of the star, enabling it to collapse again. The partially ionized helium therefore acts like a heat valve which amplifies the fluctuations in the star's outer layers and keeps the star pulsating in spite of the energy lost to friction. It is this mechanism which accounts for the restriction of cepheid variables to the instability strip of the H-R diagram.

15.7 Return of material to cosmic clouds

All main-sequence stars are observed to have some heavier elements in their spectra. Since elements heavier than helium are not produced in stars until they leave the main sequence, all young stars must have formed from the remnants of other stars. How can stars return the heavier elements that they produce in their cores to the interstellar gas clouds so that other stars may form? We have already discussed some of the mechanisms. Supernovae blast a large portion of their mass into space when they explode, forming heavier elements in the process. The normal novae eject much less material each time they flare up, but they do so much more frequently. Stars shed mass when forming a planetary nebula. Red giants lose several tenths of a solar mass in the form of stellar wind. On the average a star jettisons one-half of its original mass during its lifetime.

In the exchange of material between the clouds and stars, the clouds are becoming more complex chemically as the galaxy grows older. The stars of successive generations formed in the clouds contain a greater percentage of metallic gases compared with those of the lightest elements.

The interstellar gas cloud from which the solar system condensed must have been cycled through several generations of stars. The heavier elements in the sun and planets are the products of nuclear fusion in stars. Investigations of the isotope abundances in meteorites have indicated that the last addition of material to the solar nebula may have occurred only shortly before the solar system formed. For example, an excess of an isotope of magnesium has been found in the Allende meteorite which fell in Mexico in 1969. The excess ^{26}Mg results from the decay of ^{26}Al (aluminum-26). ^{26}Al is a radioactive isotope not found on earth because it decays with a half-life of only one-million years which is much shorter than the age of the solar system. In order for the decay product of ^{26}Al to be found in detectable quantities in the Allende meteorite, it must have been added to the solar nebula within a few million years of the solar system's formation. It is currently thought that the ^{26}Al was formed in a supernova in our neighborhood and that it was injected into the solar nebula in dust grains which condensed from this explosion. Some scientists have suggested that the collapse of the interstellar gas cloud may have been initiated by the shock wave from this supernova event. We may owe our existence to the violent death of a star 5 billion years ago! Furthermore, the heavier elements in our bodies were formed in the incredible violence of these stellar furnaces. In a sense, we are all descendents of the stars.

The Declining Stars

After a star has exhausted its supply of nuclear fuel, it can no longer support itself against the pull of gravity towards its center and it begins to collapse. How far this collapse will go depends on the mass of the star. Stars comparable in size to our sun end their lives as white dwarfs. More massive stars undergo a supernova explosion and become neutron stars. The most massive stars continue to collapse right out of existence, becoming black holes. All stars must experience one of these three fates. Thus, in spite of the wide variety which stars display during their lives, they have only three alternatives as to their final state.

15.8 White dwarfs

Most stars end their lives as white dwarfs. The size of a white dwarf depends on its mass. Unlike stars on the main sequence, which increase in size with increasing mass, white dwarfs shrink in size as their mass increases. A white dwarf as massive as the sun will have a diameter comparable to the earth's. With greater mass, they get smaller and smaller as their density progressively increases, until at 1.4 solar masses they reach the smallest diameter which a white dwarf can possibly have. At this critical mass, called the **Chandrasekhar limit,** densities in a white dwarf would be so great that the electrons would be driven into the protons in the nuclei of the atoms, producing neutrons, and a neutron star would form. Thus, white dwarfs cannot be more than 1.4 times as massive as the sun.

A star on the main sequence whose mass exceeds the Chandrasekhar limit can still become a white dwarf if it loses its excess mass before collapsing. There are a number of ways this can happen. As a red giant, a star can shed about one-tenth of its mass in forming a planetary nebula. Also, red giants can eject a great deal of matter in their stellar winds. Thus, stars on the main sequence ranging up to about four solar masses can still end their lives as white dwarfs.

The matter inside a white dwarf is compressed to extraordinarily high densities, on the order of 100 million g/cm^3. A piece of this material the size of a pea would weigh more than a truck. In ordinary terrestrial matter, electrons orbit around the nucleus of atoms at large distances. To get some idea of the scale involved, imagine that the nucleus of the atom were the size of a marble. If it were then placed on the home plate of a baseball diamond, the nearest electron would be orbiting at the distance of the grandstand. Thus, ordinary matter is composed almost entirely of empty space. When matter is compressed, the electrons surrounding adjacent nuclei approach one another and are repelled by their like charges. However, at the pressure found in white dwarfs, the atoms become stripped of their electrons. Once the electrons are dissociated from their nuclei, the electron barrier separating the nuclei is removed, so that they can become much more closely spaced. The matter in a white dwarf therefore consists of tightly packed nuclei embedded in a sea of electrons.

It is the uniform electron gas which supports the mass of a white dwarf against further gravitational collapse. According to a principle of quantum mechanics, the **Pauli exclusion principle,** no more than two electrons with equal energy and opposite spins may occupy a given volume of space. In main-sequence stars, the density of the electrons is low, and the number of electrons per unit volume is far below the limit imposed by the exclusion principle. However, at the densities found in white dwarfs, the exclusion principle dominates the behavior of the electrons. Under these conditions, the electrons occupy all the lowest-energy levels available and the electron gas is said to be **degenerate.** To prevent more than two oppositely spinning electrons with the same energy from crowding into the volume permitted by the Pauli exclusion principle, some of the electrons must be promoted to higher-energy levels, which means that they have greater velocities. The greater the velocity of the electrons, the greater is the force with which they impinge upon their neighbors, and the greater is the pressure which they exert on their surroundings. The slightest increase in the density of a degenerate gas will therefore greatly increase the velocity of the electrons in the gas, which results in a tremendous increase in pressure. Consequently, a degenerate electron gas is virtually incompressible. The center of a white dwarf is 10^{20} times more rigid than normal steel. The degeneracy pressure in a white dwarf is not thermal, but is totally quantum mechanical in origin. Consequently, a white dwarf will remain at its equilibrium size, even if the temperature of the star drops to absolute zero.

Most of the nuclei in a white dwarf are heavier elements which form in the core of red giants, such as carbon and oxygen. These heavy nuclei are completely stripped of their electrons and consequently they have large positive charges. Although the strong electrical forces of these highly charged nuclei do not significantly affect the electrons because their behavior is dominated by the much greater force of electron degeneracy, they do influence the behavior of the nuclei themselves. First, the positively charged nuclei evenly distribute themselves throughout the white dwarf because of their electrical attraction for the negatively charged electrons which support the star. If it were not for this attraction, the nuclei, which contain most of the mass of the star, would sink to the center of the white dwarf. Second, the nuclei arrange themselves in an ordered geometrical pattern because this minimizes the mutual repulsion of their positive charges. The resulting crystalline lattice of the star is more like a solid than a gas. Since solids are good conductors of heat, the temperature of a white dwarf is relatively uniform throughout its interior.

White dwarfs are surrounded by an envelope of gas several hundred kilometers in thickness. It is from this layer that the light of the star is released. The spectra of more than 50 white dwarfs have been photographed. Some spectra show only helium, others only hydrogen, others only a few metallic lines, and still others no visible lines at all (Fig. 15.8). There is no relation between the spectral patterns and the colors of white dwarfs, as there is with main-sequence stars.

According to Einstein's general theory of relativity, time is slowed down as the force of gravity increases. At the surface of a white dwarf, gravity is very strong because of the high density of matter in its interior. Atoms can serve as tiny clocks because they emit radiation at specific frequencies. If the atomic clocks are slowed down on white dwarfs, their frequencies should decrease, resulting in a redshift of their spectral lines. This gravitational redshift, or **Einstein redshift** has been observed in the spectrum of 40 Eridani's companion white dwarf, and in Sirius B, providing yet another confirmation of Einstein's theory.

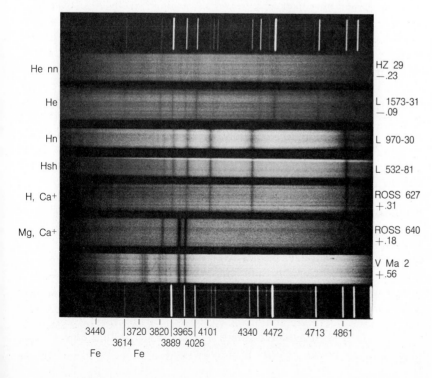

Figure 15.8 *Spectra of white dwarfs. Elements prominent in each spectrum are shown on the left. The wavelength scale is at the bottom. (Spectra by J. L. Greenstein, California Institute of Technology.)*

None of the stars visible to the unaided eye are white dwarf stars. This is because they are very faint, being well below the main sequence in the H-R diagram. In order to discover them, indirect methods of observation are often used. One method is to look for blue stars with large proper motions. If a faint star moves substantially with respect to the distant background stars in the course of a few years, it must be relatively nearby. Only white dwarfs could be so close to the earth and yet have such a low luminosity. Another method of discovering white dwarfs is to look for a slight wobble in the motion of brighter stars. Observations of the motions of Sirius, the brightest star in the sky, led to the suggestion in 1844 that Sirius has an invisible companion. Now called Sirius B, this faint white dwarf was first observed directly in 1862. At least three of the 50 stars within 5 parsecs of the sun are binaries with white dwarfs as companions. This suggests that white dwarfs are very common. So far a few hundred white dwarfs have been discovered. It is estimated that they account for 3% of the stars in our galaxy.

White dwarfs are faint because they have depleted their supply of nuclear fuel and can no longer generate energy the way normal stars do. They are also prevented from collapsing any further by the degeneracy pressure of their electrons so they can not release energy by gravitational contraction either. They shine solely by the release of thermal energy which has been stored in their crystal lattices since they collapsed from red giants. White dwarfs radiate light much like a hot iron poker that has been pulled from the fire. As their thermal energy is gradually released they cool and grow fainter, moving down and to the right in the H-R diagram. It requires a white dwarf about 8 billion years to move from the blue to the red end of the spectrum and then even longer before it ceases to shine altogether and become a black dwarf. This is the ultimate endpoint in the evolution of a star the size of our sun. However, it may be that our galaxy is not yet old enough to possess a single fully evolved star such as this.

15.9 Neutron stars

Stars which are more than 4 times as massive as the sun while on the main sequence experience a supernova explosion when they reach the end of their lives as red supergiants. The core of the star which remains becomes a neutron star. Neutron stars are only about 15 kilometers in diameter and are 100,000 to a billion times more dense than white dwarfs. If a white dwarf the size of a basketball collapsed to a neutron star, it would be smaller than the period at the end of this sentence.

Neutron stars were first predicted theoretically in 1934 by F. Zwicky, but it was not until 1967 that there was any evidence that they actually existed. In that year, pulsars were discovered, and it did not take long before it was realized that pulsars were rapidly rotating neutron stars. Some of these pulsars are found in supernova remnants, such as the Crab nebula, lending support to the idea that neutron stars are formed in supernova explosions. A most unusual neutron star is known as SS 433. This curious star is burning in a supernova remnant and appears to be beaming particles into the remnant, thus restoring energy to the expanding gases. This appears to be a more extreme case than the Crab nebula and its pulsar.

The interior structure of a neutron star consists of a solid crust surrounding a fluid core. Within the first meter of the crust, the density rises from zero at the surface to 10^8 g/cm^3. At these densities, electrons form a degenerate gas in which the nuclei arrange themselves in a rigid crystalline lattice, just as in white dwarf stars. At depths greater than 1 meter, the electrons are moving near the speed of light. These high velocities enable the electrons to tunnel into the protons in the nuclei, forming neutrons. With increasing depth in the star, more and more protons are converted to neutrons and unusually heavy neutron-rich nuclei are produced. In fact, equilibrium with the electrons controls the elemental composition of the nuclei, and elements with atomic numbers up to 140 can exist and be stable. (The heaviest elements above atomic number 90 yet produced in terrestrial accelerators only have atomic numbers up to 106, and they are highly unstable, decaying within a fraction of a second.) The control over the nuclear composition exercised by fast-moving electrons tends to make the composition of the star extremely pure at any one level, although the atomic number of the stable element changes with depth from 42 (molybdenum) on up to 140. Then, at a depth of about 1 kilometer below the surface of the star, the nuclei become so rich in neutrons that the neutrons begin to evaporate out of the nuclei to become uniformly distributed among the electrons.

When the density in the neutron star reaches 10^{14} g/cm^3, all the nuclei disintegrate. This marks the boundary between the crust and the core. In the lightest neutron stars, the crust may extend all the way to the center of the star. However, in most neutron stars, the crust is expected to be only a few kilometers in thickness.

The core of a neutron star consists almost entirely of neutrons. The few protons and electrons that remain are scattered uniformly throughout this neutron sea. Since nearly all the electrons have been absorbed, the pressure of electron degeneracy can no longer support the star. Instead, neutron degeneracy takes on the burden of the overlying layers. Like electrons, neutrons have spin and they obey Pauli's exclusion principle. (Neutron degenerency is therefore analogous to electron degeneracy except that matter is compressed to much higher densities.) The core of a neutron star may be thought of as a gigantic atomic nucleus with a diameter of 15 kilometers or less.

Neutrons have no charge, and so they can not arrange themselves in a rigid crystalline lattice like the nuclei in the crust of neutron stars. The neutrons in the core are expected to flow past one another without any friction at all. This gives the material in the core the properties of a superfluid. Superfluids exhibit very bizarre behaviors. For example, helium-3 when cooled to near absolute zero will spontaneously climb up the walls of its container, against the force of gravity!

Neutron stars rotate rapidly (up to 30 times per second) and their rigid crusts bulge at the equator much like the earth's. This rotational energy is gradually converted to radiant energy which powers the pulsar's beam of radiation and provides the energy needed to make the supernova remnant around it shine. This loss of rotational energy causes the rotation rate of the neutron star to gradually slow down, relaxing the equatorial bulge. The rigid crust of the neutron star responds to this change by rupturing, causing a **starquake.** Since angular momentum must be conserved, the rotational speed of the star will suddenly increase as the size of the bulge decreases. A readjustment as small as 1 millimeter in the crust of a neutron star is observable as a change in the period of the pulsar. By studying starquakes, scientists can test their theories about the interior structure of neutron stars and thereby learn more about the conditions of matter at ultrahigh densities.

15.10 Black holes

In the previous two sections we have seen the fate of stars which retain less than three or four solar masses after their collapse; they become either white dwarfs or neutron stars. What happens to larger stars that undergo supernova explosions and retain more than four solar masses? According to current theory, the core of a star which exceeded three solar masses would continue collapsing indefinitely, becoming a black hole. A **black hole** is an object whose gravitational pull is so great that anything that enters it is swallowed up, never to leave again. Not even light can escape from a black hole. They are totally dark objects which sweep up everything that has the misfortune of encountering them. Once something has entered a black hole, it has left our universe in the sense that nothing further can ever be learned about it. No messages could be sent from its interior and no explorer could ever return from its bowels with tales of its inner secrets. They are awesome cosmic prisons as well as the most efficient vacuum cleaners ever conceived.

The possibility of a black hole was first suggested by Laplace in 1796. Working with Newton's universal law of gravity, Laplace considered what would happen to the escape velocity of the earth as its mass increased. The escape velocity at the surface of the earth is 11 kilometers/second. If the earth were 25,000 times larger and its density remained the same, the escape velocity of the earth would then be 300,000 kilometers/second, which is the speed of light. Laplace reasoned that such a body would appear perfectly dark to an outside observer, no matter how much radiation it was emitting, because any light leaving its surface would be pulled back by gravity. The only problem with this conjecture is that light has no mass and so according to Newtonian physics, it would be unaffected by gravity. The possibility of black holes was therefore disregarded until Einstein reformulated our understanding of gravity in 1915.

The modern theoretical basis for understanding black holes is found in Einstein's general theory of relativity (Section 6.6). Rather than viewing gravity as a force, as Newton had, Einstein conceived of gravity in terms of the geometry of space and time. Ordinarily gravity is weak and space is flat, so that the shortest distance between two points is a straight line. But, according to general relativity, gravity warps space and time: the stronger the gravitational field, the more space and time are warped. In order to comprehend this, it is necessary to think of four-dimensional space–time as a two-dimensional surface (like a sheet of paper). A warp in space–time can then be pictured as a curved surface.

On a curved surface, the shortest distance between two points is not a straight line but a curve. For example, on the curved surface of the earth, the shortest distance between New York and Peking (assuming you are constrained to move on the surface of the earth) is a great circle that wraps halfway around the world. Similarly, in space–time which has been warped by gravity, the shortest distance between two points is a curve. Since light always follows the shortest distance between two points, the path of light rays through a strong gravitational field should be bent. General relativity therefore predicts that light will be influenced by gravity.

In order to test his theory, Einstein proposed that astronomers observe the positions of stars near the sun during a total eclipse. At this time, the stars around the sun can be seen and their light passes close enough to the sun to be noticeably deflected (Fig. 15.9). In 1919, astronomers photographed an eclipse of the sun and compared the positions of the stars around the sun with their positions on photographs taken many months earlier when the sun was in another part of the sky. They found that the positions of the stars closest to the sun were shifted by just what Einstein had predicted. Einstein's theory was confirmed: gravity warps space and time!

Usually the effect of gravity on space and time is so slight that it

Figure 15.9 *Deflection of starlight by the sun in a schematic drawing. Light rays passing close to the sun are bent by the sun's gravitational field. Consequently, a star close to the edge of the sun appears to be further away from it during a solar eclipse than it actually is. The dotted line indicates the apparent position of the star during eclipse.*

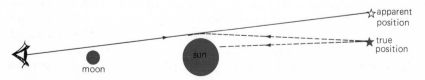

can be safely ignored. But when objects of huge mass are considered, the influence of gravity on time and space becomes significant. As the gravitational field around a body increases, the curvature of space–time increases. Eventually, if the gravitational field around a star becomes strong enough, space and time will fold in on themselves and the star will disappear from the universe—a black hole is formed.

The more mass is compressed into a given volume, the stronger is the gravitational field around that mass. Thus, any mass that is crushed into a small enough volume can become a black hole. The exact radius an object must achieve in order to become a black hole was first worked out by Karl Schwarzschild in 1916. Now called the **Schwarzschild radius,** or **event horizon,** it is equal to twice the gravitational constant G times the mass of the star M, divided by the velocity of light squared c^2:

$$R_s = \frac{2GM}{c^2}.$$

The Schwartzschild radius of a 3-solar-mass star is 3 kilometers. Since the Schwarzschild radius is directly proportional to the mass of the object, the greater the mass of a black hole, the *larger* its size becomes. Black holes therefore grow in size as they sweep up matter. There is no limit to the size which a black hole can achieve. In time, all the matter in the universe could be consumed by black holes. Perhaps this is the ultimate fate of the universe.

Black holes may already be common objects in our own galaxy. Any star which possesses more than 8 solar masses while on the main sequence is expected to become a black hole. If every star with more than 8 solar masses collapses to form a black hole, there may already be 100 million black holes in our galaxy. That is one out of every 1000 stars!

Theoretically, once a star collapses past the Schwarzschild radius, it is squeezed into an infinitely small volume with infinitely high density. This infinitesimally small point at which all matter in a black hole is crushed out of existence is called a **singularity.** If a star is massive enough it is possible that it could collapse to a black hole without ever experiencing a supernova explosion. Once the black hole was formed at the center of the star, nothing could oppose the inward rush of stellar matter. On the surface of a star collapsing towards a black hole the gravitational field would increase in strength, warping space and time. At first the effect would be slight and light could leave the surface of the star at any angle. However, as the collapse continued, the light rays which left the surface at a large angle to the vertical would be pulled back into the star. Only rays which pointed nearly perpendicularly away from the surface could escape. Separating the light rays that escape from those which are recaptured are light rays with just the right inclination to the surface to go into orbit around the collapsing star. These light rays trace out a cone, called the **exit cone,** whose apex is on the surface of the collapsing star (Fig. 15.10). Light which leaves on the exit cone orbits the star at $\frac{3}{2}$ times the Schwarzschild radius. This shell of orbiting light is called the **photon sphere.**

As the star continues to collapse towards the Schwarzschild radius, the exit cone gets narrower and narrower, so that less light is escaping from the star. To an outside observer, the star would be growing dimmer and dimmer. As the surface of the star got closer to the Schwarzschild radius, time would also be dilated by the increasing strength of the gravitational field. Thus, the atomic processes on the surface of the star would appear, to an outside observer, to slow down and the frequency of the light escaping from the star would decrease, resulting in a redshift of the star's spectral lines. This is the same gravitational redshift which has been observed in white dwarf stars, only the effect would be much more extreme in the case of a star collapsing into a black hole. To an outside observer, the star would become redder and redder until its light was redshifted out of the visible part of the spectrum altogether.

Since time would be slowing down on the surface of the star, its rate of collapse would appear to slow down as the collapse proceeded; the closer to the Schwarzschild radius the surface of the star got, the slower the collapse would seem to proceed. In fact, an outside observer would have to wait an infinitely long time to see the last few atoms on the surface of the star reach the Schwarzschild radius, because at this surface, time stops. However, an observer on the surface of the star would not notice this effect because everything around him would slow down; he would have no way of knowing that time was being so radically altered. From his perspective, the star would collapse into the black hole within a fraction of a second. Black holes are perfectly dark objects, not because their gravitational field pulls light back to their surfaces, but because the light that is emitted is infinitely redshifted, so it can not be observed. If light can not escape from a black hole, how can we ever expect to discover one? Actually, if a black hole has matter falling onto its surface, it may be detected by the effect it has on the accreting material. Stars in binary systems often exchange mass. For example, if a red giant flares up, its loosely held outer layers would be pulled off by its companion. Matter which is captured by a black hole in this way would form a disk resembling a greatly enlarged version of Saturn's rings. Friction between the rapidly orbiting gas close to the black hole and the slower-moving gas farther out (Kepler's second law) would heat up the gas to temperatures of hundreds of millions of degrees. At these temperatures, copious amounts of x-ray radiation would be produced.

A number of x-ray stars in binary systems are currently considered prime candidates for black holes. Cygnus X-1 is the leading contender.

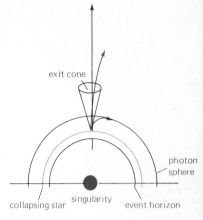

Figure   15.10 *Light   escaping from the surface of a star collapsing into a black hole. Light inside the exit cone escapes from the black hole, but light outside the exit cone is pulled back into the black hole. Light on the exit cone goes into orbit around the black hole, forming the photon sphere.*

Cygnus X-1 is in a binary star system with a blue supergiant which is about 20 times as massive as the sun. By examining the Doppler shift in the spectrum of this supergiant, it has been shown that Cygnus X-1 and the blue supergiant orbit each other with a period of 5.6 days. From this orbit, Cygnus X-1's mass is calculated to be about 10 solar masses. This invisible x-ray source must therefore be a black hole: it is too massive to be a neutron star and it would be plainly visible if it were a main-sequence star.

What we have presented so far is the picture of black holes predicted by relativity. Current work suggests that a more elegant theory may show that black holes do not exist for very long nor are they all consuming. Many scientists studying this type of object believe that a complete analysis will show that black holes evaporate in a sputtering type of process. The reason for our vagueness here is that this is a frontier area in scientific research.

15.11 Cosmic rays

We include cosmic rays at this point in our study because of their relation to star formation, stellar evolution, and the nature of the Milky Way Galaxy which we take up in Chapter 17. Cosmic rays were shown to be of cosmic origin in 1912 and came under intense study in the following decades. Interest in cosmic rays, a somewhat erroneous name, increased when it was deduced that they were particles striking the earth's atmosphere with extremely high energies (velocities). Many of the particles were protons traveling at almost the speed of light. At such velocities, they would shatter atomic nuclei when they collided with them. Thus, cosmic rays were of interest in experimental particle physics and theory and a stimulus to this interest. Studies of cosmic rays establish several significant facts: (a) the particles are nuclei of elements and have a positive charge; (b) there is an overabundance of heavy nuclei, especially carbon, among the cosmic rays; (c) the distribution of particles striking the atmosphere is isotropic. The sun is a source of cosmic rays, but for the most part the solar cosmic rays are low-energy (i.e., low-velocity) protons. The really high-energy particles come from other sources and must be accelerated to their high velocities by some mechanism.

The overabundance of heavy nuclei as compared to the normal solar or cosmic abundance of the elements suggests that the origin of the cosmic rays lies in the violent explosions of the stars; that is, supernova events. Such events inject the particles into interstellar space at high velocities. Even so the imparted velocities are nowhere near those observed.

The additional acceleration occurs as the ionized particles move along the interstellar magnetic field (Section 14.8) and are reflected back and forth by it. Each reflection adds energy and changes the direction in which the particle is traveling. Thus we can explain the very high energies observed and the isotropic distribution of the cosmic rays. Detailed arguments suggest that the particles have been traveling through space for something like 500,000 years up to about 5,000,000 years. The inferred number of particles is in agreement with the estimates of two or three supernova events per century in the Milky Way Galaxy.

The fact that the particles are positively charged ions is a piece of information useful to our discussion in Chapter 19 about the nature of the universe. For every atomic particle and subatomic particle there is an antiparticle, often referred to as matter and antimatter. For example, the counterpart to an electron is a positron. A positron has the same mass as an electron but it has unit positive charge instead of unit negative charge. When an electron and positron interact they annihilate each other and release energy. Since the cosmic rays are composed almost entirely of matter particles, our Milky Way Galaxy can not contain very much antimatter.

15.12 An interesting problem

Throughout this chapter we have drawn attention to the sun by using it as an example and explaining what the course of evolution is for the sun. Tacitly assumed is the fact that we take the sun to be a normal, well-behaved star evolving according to the processes described. We have seen that the various nuclear sources of energy release gamma rays and neutrinos. The highly energetic photons released in the same reactions are absorbed and reemitted many times before they escape as light from the sun's surface. The time for the escape of these photons is about 10^7 years. The neutrinos, however, escape directly from the sun and stars, thus offering us the opportunity to detect what happened deep in the solar interior only 8 minutes earlier (Fig. 15.11). In addition the various nuclear reactions produce neutrinos in different energy ranges so we can determine which of the reactions are taking place.

Neutrinos do not interact readily with matter, which makes them very difficult to detect. However, they can be detected and elaborate experiments have been undertaken to detect them. New experiments are detecting neutrinos 5 times less frequently than predicted. Something is wrong.

The neutrino flux from the sun is at least 5 times less than predicted, and this is cause for alarm. The neutrinos being sought and detected are not those given in the main reactions given in Section 15.4. There are several minor branches of the proton–proton process, one of which gives rise to fairly energetic neutrinos. It is these neutrinos that are being detected at too low a rate.

There are several possible explanations. (1) The interior of the sun is not undergoing nuclear reactions, which means the sun is shining on

Figure 15.11 *Neutrinos pass directly out of the sun and reach the earth only 8 minutes after they are released. The γ-ray photons are gradually degraded in a random-walk process into visible photons. This whole process takes some 10 million years so the photons we see today originated deep in the sun some 10 million years ago.*

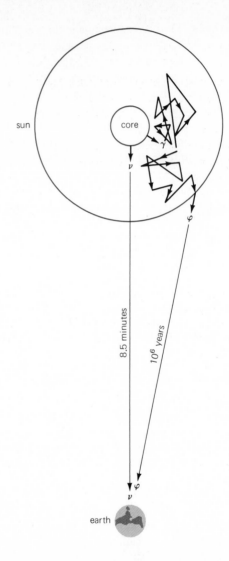

energy released previously. (2) The nuclear reactions turn off and on, and at present we are in an off cycle. (3) Our knowledge of nuclear energy generation is wrong, and the secondary neutrinos are not produced as predicted. (4) Perhaps there is a layer of material in the sun that is not transparent to neutrinos? All of these possibilities tax our credibility. In any case it is clear that we do not know as much about the workings of a "normal" star as we thought we did and we are left with an exciting problem on our hands. New experiments are being proposed using substances that will detect the primary reaction neutrinos. Hopefully, in a few years we will have answers to this problem.

Questions

1. What must happen to a normal interstellar cloud in order for it to form stars?
2. Trace the early history of a protostar from the nebular state to the main sequence.
3. What are the three initial nuclear reactions that cause the release of thermonuclear energy?
4. What are infrared stars?
5. Trace the eventual history of the sun from the main sequence.
6. How does the evolution of a massive star differ from that of the sun?
7. How is the interstellar medium enriched in metals at the expense of hydrogen?
8. What are the three end products of a supernova event?
9. What are the arguments that pulsars are neutron stars?
10. What is the origin of cosmic rays?

Other Readings

Chevalier, R. A., "Supernova Remnants," *American Scientist* **66**, 712–717 (1978).

Hawking, S. W., "The Quantum Mechanics of Black Holes," *Scientific American* **236(1)**, 34–40 (1977).

Herbst W., and G. E. Assousa, "Supernovas and Star Formation," *Scientific American* **241(2)**, 138–145 (1979).

Zeilik, M., "The Birth of Massive Stars," *Scientific American* **238(4)**, 110–118 (1978).

chapter 16

Star Clusters

Star clusters are physically related groups of stars having their members less widely spaced than are the stars around them. The stars of a cluster had a common origin and are moving along together through the star fields, so that the cluster will maintain its identity for a very long time. Although the members of a cluster are of about the same age, the more massive stars have shorter lives and are further developed in their evolution than the less massive ones. Together, they give important information about the course of stellar evolution. Star clusters are of two types: galactic clusters and globular clusters. A related grouping of stars is called an association.

The two types of clusters of stars associated with the Galaxy are called galactic, or open clusters, and globular clusters. They are found in other galaxies as well. The reason for studying star clusters at this point is that they are excellent tools for studying stellar evolution and their location in space leads to an idea of the size and shape of our Milky Way Galaxy.

The brighter open clusters are often called by special names, such as the Praesepe and Pleiades clusters. More generally, the clusters are designated by their running numbers in one of two catalogs where they are listed together with nebulae and galaxies. Thus the great globular cluster in Hercules is known as M 13 or NGC 6205. The first designation is by the number in the catalog of 103 bright objects, which C. Messier published in 1784. Messier compiled this catalog to assist people searching for comets, since, in many cases, they could be mistaken for comets by an amateur astronomer. The second catalog designation of the Hercules cluster is by its number in J. Dreyer's *New General Catalogue* (1887), which, with its extensions in the later *Index Catalogue* (IC), lists over 13,000 objects.

All of the stars in a given cluster may be assumed to have been born at the same time. Thus the H-R diagram of a cluster tells us the age of a cluster. Extremely young clusters might be expected that will still have stars moving onto the lower main sequence. Middle-aged clusters might be expected to have all of the stars on the main sequence, while older clusters might have their evolved early type stars moving off the main sequence. Very old clusters might be expected to exhibit a very late stage of evolution where all but the least massive stars have evolved away from the main sequence.

Clusters reveal something about the structure of the Milky Way by virtue of when and where they were born. Galactic clusters suffering frequent gravitational interaction with other stars will gradually lose members and hence blend into the general star field. Thus we should expect to see only relatively young or middle-aged galactic clusters and these will be located not far from their birthplace. Clusters formed far from the dense regions of the Milky Way should be expected to keep their identity for a very long period of time.

Galactic Clusters

Galactic clusters are so named because those in our galaxy lie near its principal plane. They accordingly appear close to the Milky Way except for a few of the nearest ones notably the Coma Berenices cluster, which is in the direction of the north pole of the Milky Way. They are also known as **open clusters,** because they are loosely assembled and are not greatly concentrated toward their centers.

16.1 Features of galactic clusters

The Pleiades in Taurus and the Hyades in the same constellation are familiar examples of galactic clusters. Their brighter stars are clearly visible to the unaided eye and those of the Coma cluster are faintly visible. The Praesepe cluster in Cancer, also known as the "Beehive," the double cluster in Perseus (Fig. 16.1), and some others are hazy spots to the unaided eye and are resolved into stars with binoculars. These are fine objects with small telescopes.

About 500 galactic clusters are recognized in our region of the galaxy. Their memberships generally range from two dozen to a few hundred stars, and even exceed 1000 stars in the rich Perseus clusters. The known galactic clusters are all within 6000 parsecs from the sun. More remote ones are too faint to be noticed in bright areas of the Milky Way or are concealed by dust in the dark areas. It is estimated that there are about 50,000 galactic clusters in the Milky Way.

Galactic clusters may or may not have gas and dust associated with them. Those clusters with bright blue stars present always have gas and dust nearby. We will see that this is an effect of the age of the cluster.

16.2 The Pleiades cluster

This beautiful cluster meaning "the seven sisters" is visible to the naked eye as six stars in a hazy compact region (Fig. 11.12 and color plate). The hazy appearance comes from the numerous faint stars of the cluster plus the nebula in which it is imbedded. Much speculation centers upon the question whether or not a seventh star was visible in antiquity when the cluster was named. If it was, it has either faded gradually until it was below detection by the unaided eye, or one of the stars became a supernova during a springtime period. There is no direct or indirect evidence for either supposition, and it is possible that the name was not meant to match the number of stars or that the poetry was written with another purpose in mind.

The Pleiades cluster has a membership of about 250 stars located in a tenuous cloudy region. The nebular material is visible around the brighter stars as reflection nebulae. The brightest stars are of type B5, indicating that the cluster is rather young. The Pleiades does not have a separate giant population in the H-R diagram. Unlike many open clusters the Pleiades is sufficiently compact and will retain its principal identity for the lifetime of its stars. This is because as a member is lost the cluster contracts and will do so on each successive loss until it is a tight cluster able to withstand dissolution forces from the outside. These forces arise in the tidal effects of the Milky Way and the shearing effect of Keplerian motion around the center of the Milky Way.

Figure 16.1 *A fine pair of open clusters called h and χ Persei. These clusters contain many blue stars. They are among the youngest clusters (see Fig. 16.5). (Copyright by Akademia der Wissenschaften der DDR. Taken at the Karl Schwarzschild Observatorium.)*

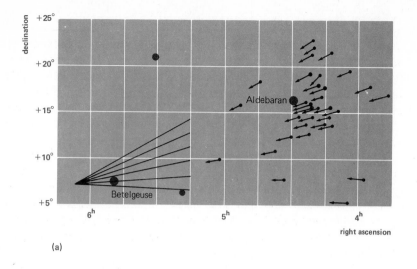

(a)

(b)

Figure 16.2 *Convergence of the Hyades cluster. (a) The stars of the cluster are converging toward a point in the sky east of the present position of Betelgeuse. The lengths of the arrows show the proper motions in 50,000 years. (b) A schematic representation of how the line connecting the star to the convergent point determines the angle.*

16.3 The Hyades cluster

The Hyades cluster is relatively so near us that the proper motions of its members offer a good example of the common motions of cluster stars. It comprises the stars of the V-shaped group itself, except the bright star Aldebaran which has an independent motion not shown in Fig. 16.2, and also stars in the vicinity within an area having a diameter of 20°. The cluster of at least 150 stars has its center 40 parsecs from the sun. It is moving toward the east and is also receding from us, so that the parallel paths of its stars appear to be converging toward a point in the sky east of the present position of Betelgeuse in Orion. This motion toward the convergent point allows us to compute the distance to the cluster independently of the method of parallax as was pointed out by L. Boss in 1908. Since there are many stars involved we have a valuable check on trigonometrically established calibrations. It is possible to show that this cluster passed nearest the sun 800,000 years ago at one-half its present distance from us.

The Hyades cluster, being less populous than the Pleiades but covering the same spatial volume, will probably suffer dissolution during the lifetimes of its stars. Some of its stars are grouped in **trapezium**-type systems (three to seven stars) and eventually only these systems going their separate ways will give evidence of the once great cluster. The Hyades differs from the Pleiades in that its brightest stars are reddish K-type giants of about zero absolute magnitude. Also, nebular material is not very much in evidence in the Hyades due to its age.

16.4 Convergent motion

It is often stated that if we know the tangential motion and the radial velocity of a star then we can find the star's space motion. This is not strictly true. What we should say is that if we know the radial velocity and the tangential velocity then we know the star's space motion. The difference is not trivial because to get the tangential velocity from the tangential motion, or proper motion, we must know the star's parallax, that is, its distance. In the case of a converging (or diverging) cluster we can convert each star's proper motion into a velocity and hence determine its distance directly.

Suppose a group of stars is moving through space with motions parallel to each other. When viewed from a distant vantage point these motions will appear to converge at some distant point, as other parallel lines in space do. Irrespective of the sun's own motion, an imaginary line drawn from the sun to the convergent point is parallel to the motions of the moving group, and the angle θ formed by this line and a line from the sun to an individual star in the group is the angle between the direction of the star's space motion v_s and its radial velocity v_r with respect to the sun. The ratio of v_r to v_s is the cosine of the angle θ.

Simple trigonometry tells us that the tangential velocity v_t is determined once we know the angle θ and the radial velocity since the ratio of v_t to v_r is the tangent of θ. Knowing the tangential velocity in kilometers/second and the tangential motion in seconds of arc per year we can determine the parallax *(p)* and hence the distance *(d)*. We must determine the angle subtended by one astronomical unit at the distance of the object. This is the same problem as the one we did earlier concerning the expansion of a nebula around a supernova. We leave it as an exercise for the student to show that

$$d = \frac{1}{p} = \frac{v_r \tan \theta}{4.72\mu},$$

where μ is the proper motion.

Thus the convergence method allows us to calculate the distances to moving clusters. This method is of great importance, because for the four or five nearest moving clusters we have three independent methods for determining their distances: the convergent method we have just discussed, trigonometric parallaxes, and spectroscopic parallaxes. The latter is often replaced with a curve-fitting technique that fits the main sequence of stars in the cluster to the main sequence as defined by the nearby stars, as we will see in the next section.

We cannot stress the point of independent methods of checking a measurement too strongly. In this case, it happens that the Hyades cluster is used to calibrate other more distant clusters containing both bright and variable stars that obey the period–luminosity relation. These stars are used in turn to calibrate other stars that then are used to calibrate other objects, and so on. Thus the distance calibrations in astronomy rest heavily (but not entirely) upon the Hyades cluster and any error compounds itself many times as we go farther into the cosmos.

16.5 The color–magnitude diagram

The color–magnitude diagram is an array of points representing the colors of stars in some standard system plotted with respect to their magnitudes. It is a variation of the H-R diagram. In the diagram of the Praesepe cluster (Fig. 16.3) the stars become redder from left to right—from blue stars such as Sirius at the left to yellow stars like the sun near the middle and to red stars at the right. Color has a known relation to spectral type, which also advances as the temperature diminishes; it is preferred to the spectral type in this and following diagrams, because color can be determined more easily and precisely. The stars become brighter from the bottom to the top of the diagram. In this case the apparent, or observed magnitude is plotted.

We prefer here to use color and magnitude as defined by the U, B, V photoelectric system.

The diagram of the Praesepe cluster can inform us of the distance and relative age of the cluster, as follows.

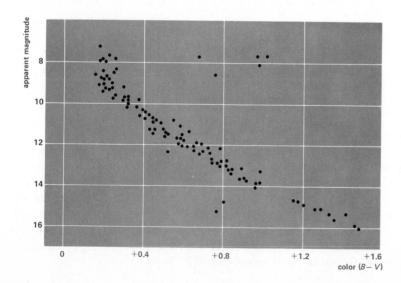

Figure 16.3 *A color–magnitude diagram of the Praesepe cluster. Note the evolved stars moving away from the main sequence at the upper left. (Adapted from a diagram by H. L. Johnson.)*

Distance of the Cluster

Compare the Praesepe diagram with the standard spectrum–absolute magnitude diagram (Fig. 12.18). From the latter diagram and Table 12.4 we note that the sun's color is about +0.6. The sun's absolute V magnitude is 4.8. From the cluster diagram, a main-sequence star in the cluster having color of 0.6 has an apparent magnitude $m = 10.8$. By the formula $\log d = (m - M + 5)/5$, we find that $\log d = 2.20$. Thus the distance d of the Praesepe cluster is accordingly about 158 parsecs.

It is important to remember that this main-sequence fitting technique depends upon the calibration of the main sequence. This calibration is done by determining the trigonometric parallaxes and colors of the nearby stars.

The distances of 18 galactic clusters are given in Table 16.1. The clusters were selected from a very extensive list by W. Becker in which he combined the best measurements of a number of observers including his own. After making allowance for the dimming and reddening of the cluster stars by intervening interstellar material, the main sequence of each cluster was then matched with that of a standard color–magnitude diagram. The difference, apparent minus absolute magnitude, at corresponding points in the two scales at the left of the diagram, gave all the

Table 16.1
Distances of Selected Galactic Clusters[a]

Cluster	Constellation	Distance (parsecs)
Hyades	Taurus	42
Coma	Coma Berenices	80
Pleiades	Taurus	127
Praesepe	Cancer	158
M 39	Cygnus	255
IC 4665	Ophiuchus	330
M 34	Perseus	440
NGC 1647	Taurus	550
NGC 2264	Monoceros	740
M 67	Cancer	830
NGC 4755	Crux	1100
M 36	Auriga	1260
NGC 2362	Canis Major	1500
NGC 6530	Sagittarius	1560
M 11	Scutum	1710
NGC 2244	Monoceros	2200
h Persei	Perseus	2250
χ Persei	Perseus	2400

[a] Abstracted primarily from an extensive list by W. Becker.

information needed for calculating the distance of the cluster by the formula we have already noted. The difference $m - M$ is called the **distance modulus** for obvious reasons.

$m - M$	Distance (pc)
0	10.00
1	15.85
2	25.12
3	39.81
4	63.10
5	100.0
6	158.5
7	251.2
8	398.1
—	—

In the margin we have given a convenient tabulation of distances in parsecs associated with various distance moduli through $m - M = 8$ magnitudes. Note that the distance for $m - M = 6$ is 10 times that of $m - M = 1$ and $m - M = 7$ is 10 times that of $m - M = 2$, etc. Therefore, $m - M = 9$ will be a distance of 631 parsecs, $m - M = 14$ is equivalent to 6310 parsecs, and so on. For really large distances it is often convenient to use the distance modulus rather than parsecs.

We have already noted that the open or galactic clusters receive their names from the fact that they lie in or close to the Milky Way or galactic plane. If we were to plot their directions in the galactic plane and their distances, we would find that these clusters are not distributed at random. The galactic clusters seem to be arranged along lines in space (Fig. 16.4). We will see later that these lines agree quite well with the arms of the Milky Way Galaxy as defined by other means.

16.6 Star clusters of different ages

The theory of stellar evolution described in Chapter 15 traces the stars from their birth in the nebulae to positions of temporary stability on the main sequence of the color–magnitude diagram. Here they are arrayed in order of color, from blue on the upper left to red at the lower right; the more massive stars being the blue stars and the less massive the red stars. We have also noted that the more massive stars evolve more rapidly than the less massive, therefore we should see these effects in the color–magnitude diagram.

This is revealed in Fig. 16.3. Note that the top of the main sequence of the Praesepe cluster bends to the right and that a few of the brightest stars have moved still farther to the right. Eventually they will become red giant stars. The bending point or break-off point is a measure of the cluster's age. A cluster's age will be revealed by this break-off point; the older a cluster is, the farther down the main sequence is the break-off point.

A composite color–magnitude diagram for ten galactic clusters and one globular cluster is shown in Fig. 16.5. The colors of the cluster stars were plotted against their absolute V magnitudes, and the trends of the points so plotted are represented here by broad lines. The ages of the clusters in years are read from the scale of ages at the right opposite the points where the curves begin to break from the main sequence. Above its point of departure a particular cluster has no stars remaining on the main sequence.

Figure 16.4 *The distribution of galactic clusters taken from material by W. Becker. The young clusters are not distributed at random (see also Fig. 17.8). The direction to the center of the Milky Way is toward the bottom of the page.*

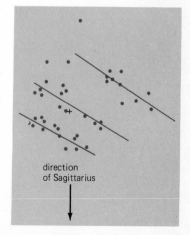

direction
of Sagittarius

Figure 16.5 *Color–magnitude diagrams for ten galactic clusters superposed plus the globular cluster M 3. Color is plotted against absolute visual magnitude. Age of stars evolving away from the main sequence can be read on the scale to the right. (Diagram by A. Sandage.)*

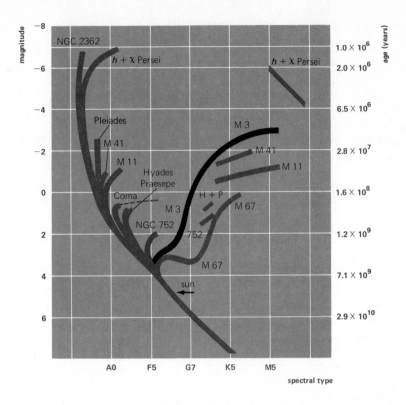

The cluster NGC 2362 is the youngest cluster represented in Fig. 16.5. The double cluster in Perseus is also in its youth. The Pleiades cluster is middle aged, and the Hyades and Praesepe clusters are approaching old age. M 67, age 4 billion years, has had a longer life than most galactic clusters; it departs from the main sequence where the stars are not much bluer than the sun. A still older galactic cluster, NGC 188 (not shown in the figure), leaves the sequence at about the sun's position. Its stars are represented by a curve similar in form to that of M 67.

Note in Fig. 16.5 the roughly triangular area avoided by the curves of the galactic cluster stars. This vacant area, known as the **Hertzsprung gap,** is not completely explained. Also awaiting explanation is the evidence that field stars of the Milky Way and of the Magellanic clouds do not avoid a similar area of their color–magnitude diagrams.

In using this technique of main-sequence fitting for comparative purposes we should be aware of a subtle difficulty often overlooked. The luminosities of stars depend upon their initial composition, that is, the amount of hydrogen, helium, and other elements. Thus, the main sequence of a young cluster may lie higher in the H-R diagram than that of an old cluster, because the young stars may be formed from enriched material.

16.7 Associations of young stars

Several years ago V. A. Ambartsumian called attention to the fact that the early-type stars tend to occur in groups. He noted that the groups of O and B stars were not held together firmly enough by mutual gravitation to prevent their eventual dispersal by separate motions in different directions. However, he pointed out that the groupings could not be accidental and therefore the stars are to be associated with a common origin. He called these groupings **associations.** Evidently the associations are so young that they have not had time to disperse. Associations are designated by the predominant star type. Thus an association made up mainly of O stars is called an O association. Groups of T Tauri stars meeting Ambartsumian's definition of associations are called T Tauri associations.

An example is the zeta Persei association, named after its brightest member. At the distance of 160 parsecs from us, the 17 stars of the group are moving at the rate of 14 kilometers/second away from a center where they were presumably born only 1.3 million years ago.

In one association of blue stars in Orion, at the distance of 400 parsecs, the stars are withdrawing from their center generally at the rate of about 8 kilometers/second, so that they must have begun to separate 2.8 million years ago. Three "runaway stars" of the group, however, have speeds of 130 kilometers/second; these are 53 Arietis, AE Aurigae, and μ Columbae, which have already moved out of Orion into neighboring constellations, as is indicated by their names.

The type O9 star AE Aurigae has reached the vicinity of the nebula IC 450, which it is now illuminating. An unusual feature is that part of the nebula gives an emission spectrum, as would be expected from the presence of the very hot star, whereas another part gives a reflection spectrum and has more nearly the color of the star itself. It would seem that there has not been time enough since the star came near the nebula for the radiations of the star to complete the dispersal of the nebular dust.

Globular Clusters

Globular clusters are spheroidal assemblages often of many tens of thousands of stars; they are much the larger and more dense of the two types of clusters.

16.8 The brighter globular clusters

About 120 globular clusters are recognized in the vicinity of our galaxy and many more are likely to be hiding behind its dust clouds.

Instead of crowding toward the Milky Way, as the galactic clusters do, they form a nearly spherical halo around the Galaxy. They are somewhat more numerous toward the center of the Galaxy. Relatively scarce in space, not one of them has been seen within the distance of 6100 parsecs from us, where all our galactic clusters are observed. Their luminosity makes them visible afar in our galaxy and even around nearby galaxies.

The brightest globular clusters for us are omega Centauri, near the northern edge of the Milky Way in the south polar region, and 47 Tucanae, also in that region. They appear to the unaided eye as slightly blurred stars of the fourth magnitude and were given designations as stars before their true character was recognized. These two are the nearest globular clusters, at the distance of about 4600 parsecs, and are among the richest in stars.

M 13 in Hercules (Fig. 16.6) is faintly visible to the unaided eye, as is M 22 in Sagittarius. M 5 in Serpens, M 55 in Sagittarius, and M 3 in Canes Venatici can be glimpsed under favorable conditions without a tele-

Figure 16.6 *The beautiful cluster M 13 in Hercules. The stars are so dense in the center that the negative is "burned" out. (Official U.S. Navy photograph.)*

scope. The Hercules cluster, M 13, at the distance of 7000 parsecs, is well known to observers in middle northern latitudes, where it passes nearly overhead in the early evenings of summer. This cluster covers an area of the sky having two-thirds the moon's apparent diameter. Its linear diameter is 50 parsecs, or the distance of Spica from the sun.

More than 50,000 stars of the Hercules cluster are bright enough to be observed with present telescopes, although the stars in the central region are too crowded to be counted separately. The total membership is estimated as one-half million stars, including the yellow and red stars of the main sequence, which are too faint to be observed. The slightly elliptical outline of the cluster, less than that of Jupiter's disk, suggests flattening at the poles by slow rotation; no other evidence of rotation has been detected.

The stars in the compact central region of the Hercules cluster have an average separation of 20,000 times the earth's distance from the sun, or about one-twentieth of the spaces between the stars in the sun's vicinity. For anyone observing from there the night sky would have a splendor quite unfamiliar to us. Probably a hundred times as many stars as we see in our skies would be visible to the unaided eye, and the brightest ones would shine as brightly as the moon does for us. This, of course, would have its drawbacks for astronomers located there.

In some globular clusters the proximity of astronomers would have partial advantages. Very dense cores of globular clusters are the source of x-ray radiation. It has been suggested that a centrally located black hole gives rise to this radiation. It would be very interesting to study such an object free from interstellar gas, dust, and other obscuring matter.

16.9   Variable stars in globular clusters

The presence in the globular clusters of many stars that are variable in brightness has been recognized for a long time. The majority of the variable stars are RR Lyrae variables, having periods of about one-half day. Originally called cluster variables, they are now known to be even more abundant outside the clusters. More than 1600 variable stars were reported in 1961 in the 80 clusters that had been searched for such objects; M 3 and omega Centauri are the richest in known variables. Variable stars of other kinds in the globular clusters include type-II cepheids and Mira-type variables.

RR Lyrae variables have about the same absolute magnitude (+0.7) and hence are excellent distance indicators to the distances to which they can be seen. The type-II cepheids are brighter and follow the lower curve in the period–luminosity diagram (Section 13.9); hence, they serve as distance indicators for even greater distances. Unfortunately many globular clusters are lacking or poor in these reliable types of variable stars and

Figure 16.7 *The color–magnitude diagram for the globular cluster M 3. Note the break in the horizontal branch at V = 15.5; it is the location of the RR Lyrae type variable stars. Below about V = 19 the main sequence is still evident. (Diagram by permission of the Astrophysical Journal, © American Astronomical Society.)*

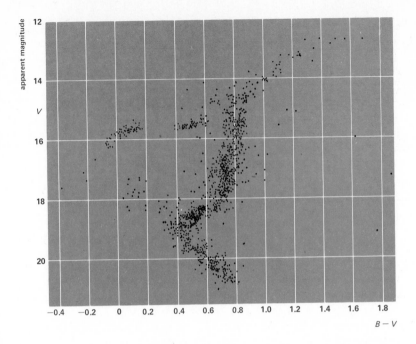

Figure 16.8 *A schematic diagram of how H. Shapley could find the direction and distance to the center of the Milky Way Galaxy. Let the eight globular clusters define the corners of a square box. Determining the distance and direction of each globular cluster then allows one to determine the distance and direction of the center of the box (light line) even though there is no cluster located there and even if the center cannot be seen.*

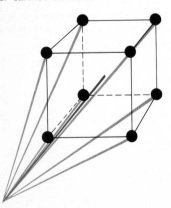

other methods must be used to find their distances. Usually the tenth brightest star in a globular cluster has the same brightness as the tenth brightest star in most other globular clusters so this is used for determining the distances of clusters lacking variable stars. The curve-fitting technique is not very useful for globular clusters. They are so distant that only a small portion of the main sequence can be observed. Also their H-R diagrams are quite different (Fig. 16.7), being dominated by red giant stars.

16.10 The system of globular clusters

In 1917 H. Shapley made the first important modern step toward the present understanding of our galaxy by determining the arrangement of its globular clusters in space. His purpose was to answer a question of long standing: Do the stars go on out into space indefinitely, or do they form a system of limited extent? His idea was that the globular cluster system should have about the same dimensions and center as the system of stars.

Shapley measured the distances of the clusters using RR Lyrae stars as distance indicators. Having found the distances and knowing their directions from us, he then made a model of the cluster system, showing the sun's position in it (Fig. 16.8). With later correction for the effect of intervening dust in magnifying the measured distances, it was determined that the cluster system occupies a spherical volume of space 30,000 parsecs in diameter surrounding the flat disk of the galaxy proper. The center of the system is about 10,000 parsecs, or 10 kiloparsecs, from the sun in the direction of Sagittarius and in the region of the sky where one-third of the globular clusters are found (Fig. 16.9). This classic survey established the dimensions of our galaxy and the eccentric location of the sun in it.

Stellar Populations

The H-R diagram, or color–magnitude diagram of stars in the sun's vicinity (Fig. 12.18) differs from the corresponding diagram for a normal globular cluster. Interest in this matter was increased by the discovery in 1943 that a similar difference is found in the galaxy M 31 in Andromeda between stars in the spiral arms and in the central region. Furthermore, the stars in the two elliptical companions of the Andromeda galaxy had H-R diagrams similar to the central region of M 31 and the globular clusters. The two arrays were called populations I and II, respectively.

16.11 Two stellar populations

The **type-I population** is represented by our region of the Galaxy, and was accordingly the first to be recognized. Population I frequents

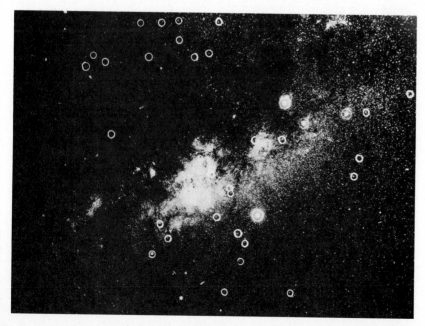

Figure 16.9 *Distribution of globular clusters in the direction of Sagittarius. The clusters are circled for easier recognition. This one photograph contains one fourth of all the Milky Way's globular clusters. Note their absence around the dark clouds due to obscuration. (Yerkes Observatory photograph.)*

Figure 16.10 *A schematic but accurate H-R diagram (color–magnitude diagram) for typical population I stars.*

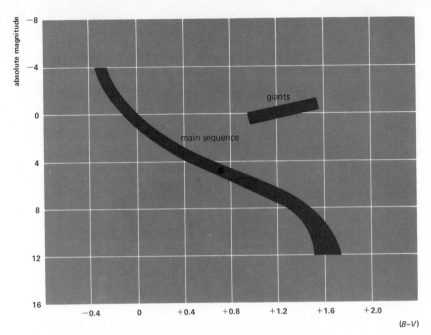

regions where gas and dust are abundant. Its brightest members are hot blue stars of the main sequence, which is intact (Fig. 16.10). Its giant members are mainly red stars around absolute magnitude zero. Open clusters are typical of a type-I population.

The **type-II population,** represented by a normal globular cluster, occurs in regions generally free from diffuse gas and dust. Its brightest stars are K-type giants of absolute visual magnitude −2.4, where the giant sequence begins. On the downward slope in the diagram this sequence divides into two branches, one of which runs horizontally to the left around magnitude zero. The second branch continues in about the original direction until it reaches the type-I main sequence. To the left of the junction there are no original main-sequence stars remaining (Fig. 16.11).

Almost all of the globular clusters in our galaxy are composed of type-II populations. In the Magellanic clouds (Section 18.4), however, there is about an equal division between populous red clusters and a formerly unfamiliar blue variety.

An extension of the original type-II diagram from photographs of the globular cluster M 3 is shown in Fig. 16.11. RR Lyrae stars were not included in this survey; a gap in the horizontal branch where they would appear suggests that this may be exclusively the domain of these variable stars. From the lower end of the vertical giant branch the remnant

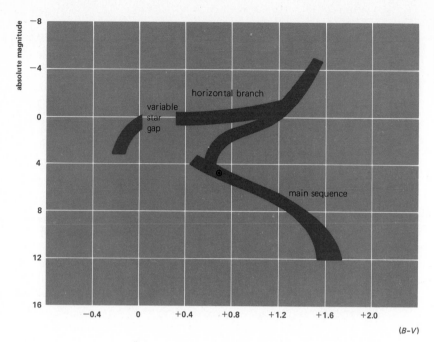

Figure 16.11 *A schematic but accurate H-R diagram (color–magnitude diagram) for typical population II stars to the same scales as Fig. 16.10.*

of the cluster's main sequence is shown extending down and to the right to yellow stars at the limit of faintness for the largest telescope.

Population types I and II represent extremes of young and old populations of stars. Intermediate types are needed to indicate the relative ages of stars in different parts of our galaxy and of other galaxies. It is now accepted that the two population types merely mark the extremes and that intermediate types exist as well. The youngest stars correspond to the original type-I population and are now referred to as **extreme population I.** Typical stars for this population are the classical cepheid variables, the T Tauri stars, and the blue supergiants. The next group contains the so-called **older population I** made up of the A stars, for example. In the middle of the two extremes is the **disk population** made up of planetary nebulae, ordinary novae, and the shorter-period RR Lyrae stars. Older yet is the **intermediate population II,** which is made up of the long-period variables and the so-called high-velocity stars. Finally, corresponding to the original population II is the **halo population II** composed of the globular clusters, subdwarfs, and longer-period RR Lyrae stars. A simplified table of populations is shown in Table 16.2.

The term "high-velocity stars" used above is somewhat misleading. The intermediate-population-II stars and the halo-population-II stars are traveling on highly elliptical orbits around the center of the Milky Way

Table 16.2 Population Types	POPULATION I NEW SYSTEMS		POPULATION II OLD SYSTEMS		
	Extreme	Older	Old Disk	Intermediate	Halo
	Gas, dust Open clusters Spiral arms Supergiants Classical cepheids T Tauri stars	Sun Nearby Stars	Planetary nebulae RR Lyrae of short period Novae White dwarfs	High-velocity stars Long-period variables W Virginis stars	Globular clusters Longer-period RR Lyrae Metal-poor stars

Galaxy. Their motion is generally perpendicular to the direction of the motion of the sun and so they have little or no radial velocity with respect to the sun. Hence, the radial velocity that we measure is actually the motion of the sun itself. We will see in the next chapter that the sun is moving on a nearly circular orbit and traveling at a velocity of about 250 kilometers/second.

16.12 The origin and evolution of clusters

Because the two types of clusters are associated with the extremes of the population types, we should expect their histories to be quite different. The globular clusters are quite old as evidenced by the composition of their stars. Therefore they must be quite stable and were probably formed about the time that the Milky Way Galaxy was formed, because we do not find any older stars. The globular clusters have highly elliptical orbits around the center of the Galaxy and some of the orbits of these clusters have very high inclinations to the plane of the Galaxy. Several of the globular clusters have orbits that actually take them through the central region of the Galaxy where the star density is quite high. These few clusters must suffer disruptions and loss of stars with each passage, but the number of such passages is rather small, perhaps ten since the Galaxy was formed. There is a question whether or not actual star collisions occur when a dense globular cluster passes through the dense central region. It is certainly possible, but the likelihood of its occurrence cannot yet be determined.

All evidence points to the fact that the extreme population type-II stars are the oldest stars in the Galaxy. Since the globular clusters are perfect examples, we can say that these clusters were formed from the same material from which the Galaxy was formed. The atmospheres of the globular cluster stars then tell us what this material was; we find that 78% was hydrogen and about 22% was helium, and there was just a trace of the other elements present. The halo-population-II stars are

Figure 16.12 *An open cluster and associated nebula M 8. (Steward Observatory photograph.)*

metal deficient, or perhaps we should say that the population-I stars have an overabundance of metals.

The galactic clusters, on the other hand, show a great spread in ages. Some are almost as old as the globular clusters and others, such as NGC 2362, are less than 1 million years old. The only way these young stars could form is from the great interstellar clouds. Most young open clusters are imbedded in interstellar clouds (Fig. 16.12), which supports this point of view. The interstellar clouds have been formed from the metal-enriched material released by previous generations of stars. If this is so, then the stars in the young clusters should show higher abundances of the heavier elements, which they do. Very young clusters, such as h and χ Persei, have about 70% hydrogen, 26% helium, and 4% heavier elements as compared to the sun with its 71% hydrogen, 26.5% helium, and 2.5% heavier elements or metals, as astronomers call all elements heavier than helium.

The open clusters, moving through fairly dense stellar regions, are subject to forces that tend to disrupt them. Their orbits of revolution around the center of the Galaxy will serve to string out and disrupt a loose open cluster in about two or three orbits or much less than one-billion years. The denser, tighter clusters can resist this shearing effect and only slowly dissolve to a minimum number of stars by ejection. This is the case with the Pleiades, which we discussed earlier.

Questions

1 What are the characteristics of open clusters? Give a second designation for these clusters.
2 Explain the convergent motion method for getting distances to moving clusters.
3 Which is the oldest cluster, the Pleiades or the Hyades?
4 What are some of the problems associated with distances determined by the main-sequence fitting method?
5 Define distance modulus.
6 Estimate the age of Praesepe from Fig. 16.5.
7 Describe the system of globular clusters.
8 Are globular clusters old or young?
9 Gas and dust are characteristics of which population?
10 Pulsars show a strong concentration in the Milky Way. What population do most of them belong to?

Other Readings

Burbidge G., and E. M. Burbidge, "Stellar Populations," *Scientific American* **199(2),** 44–65 (1958).

Clark, G. W., "X-Ray Stars in Globular Clusters," *Scientific American* **237(4),** 42–55 (1977).

Iben, Jr., I., "Globular-Cluster Stars," *Scientific American* **223(1),** 27–39 (1970).

Philip, A. G. Davis, "The Thin-Prism Spectra of Messier 4 in Scorpius," *Sky & Telescope,* **51,** 244–246 (1976).

chapter 17

The Galaxy

The galactic system, or system of the Milky Way, is so named because the luminous band of the Milky Way around the heavens is an impressive feature of the system in our view from inside it. This spiral stellar system, of which the sun is a member, is commonly known as the Galaxy as distinguished from the multitudes of other galaxies.

The Milky Way

The Milky Way, long a sight admired by humans, has only recently had its true nature divined. The luminous band is evidence of the disk of the Galaxy. The sun and solar system are located in this disk, and from this location we look out at the full expanse of the universe beyond.

17.1 The Milky Way in summer and winter

The full splendor of the Milky Way is reserved for one who observes it on a clear moonless night from a place removed from artificial lights. The view with the unaided eye or with a very-wide-angle camera is best for the general features. Photographs show the details more clearly than the eye alone can detect. The Milky Way is produced by the combined light of great numbers of stars that are not separately visible without the telescope. Its central line is nearly a great circle of the celestial sphere, so highly inclined to the celestial equator that it takes quite different positions in our skies in the early evenings of the different seasons.

At nightfall in the late summer in middle northern latitudes the Milky Way arches overhead from the northeast to the southwest horizon. It extends upward through Perseus, Cassiopeia, and Cepheus to the region of the Northern Cross. Here in Cygnus the Great Rift formed by cosmic dust clouds divides the Milky Way apparently into two parallel streams, which go on southward through Aquila into Sagittarius and Scorpius. This spectacular part of the Milky Way (Fig. 17.1) contains the bright star clouds of Scutum and Sagittarius; the latter is near the direction of the center of the Galaxy.

In the early evenings of late winter in middle northern latitudes the Milky Way again arches overhead, now from northwest to southeast. It passes through Cepheus, Cassiopeia, Perseus, and Auriga, which is near the zenith in the early evenings of February. Here it is narrowed by a succession of relatively nearby dust clouds, which angle from northern Cassiopeia through Auriga to southern Taurus. This is the direction away from the center of the Galaxy. The Milky Way then continues on past Gemini, Orion, and Canis Major, where it becomes broader and less noticeably obscured by dust.

The part of the Milky Way in the vicinity of the south celestial pole is either out of sight or else never rises high enough for favorable view in the United States. This region from Centaurus to Carina (Fig. 17.2) contains some fine star clouds and the dark Coalsack near the Southern Cross.

Except for the stars of Orion, the brightest stars of the winter sky are noticeably displaced to the west of the Milky Way. This band of stars, called Gould's belt, contains the Pleiades and Hyades clusters and is a local manifestation of the structure of the Galaxy.

Figure 17.1 *The Milky Way from Scutum to Scorpius. The great Sagittarius star clouds are near the center of the picture. (Hale Observatories photograph.)*

17.2 Galactic longitude and latitude

In descriptions of the Galaxy and objects composing it, it is convenient to denote positions in the heavens with reference to the Milky Way. For this purpose an additional system of circles of the celestial sphere is defined as follows.

The north and south **galactic poles** are the two opposite points that are farthest from the central line of the Milky Way. By international agreement for the year 1950 they are, respectively, at right ascension

Figure 17.2 *The Milky Way in the region of the Southern Cross. The obvious large dark cloud is the famous Coalsack. (Harvard College Observatory photograph.)*

12^{h}49^m, declination +27°4, in Coma Berenices, and 0^{h}49^m, −27°4, south of beta Ceti. Halfway between these poles is the **galactic equator,** a great circle inclined 62°36′ to the celestial equator; it crosses the equator northward in Aquila and southward at the opposite point east of Orion. The galactic equator passes nearest the north celestial pole in Cassiopeia and nearest the south celestial pole in the region of the Southern Cross. Thus the earth's equator is inclined almost 63° to the principal plane of the flattened Galaxy.

Galactic longitude was formerly measured in degrees along the galactic equator from its intersection with the celestial equator in Aquila. By decision of the International Astronomical Union in 1958, the zero of galactic longitude is now changed to the direction of the galactic center (Section 17.4) in Sagittarius. As before, the longitude is measured through 360° in the counterclockwise direction as viewed from the north galactic pole. **Galactic latitude** is measured from 0° at the galactic equator to 90° at its poles and is positive toward the north galactic pole.

To designate these two coordinates we use l^{II} for galactic longitude and b^{II} for galactic latitude. The roman II superscript indicates that the coordinates are the new revised system. In the near future the superscript will be dropped and all positions will be assumed to be on the new system unless otherwise stated. Most of the older radio maps shown here are on the old system.

Structure of the Galaxy

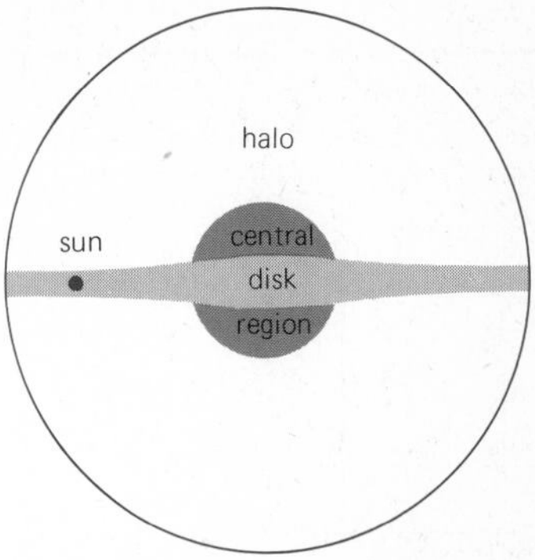

The Galaxy is an assemblage of the order of 100 billion stars. Its spheroidal central region is surrounded by a flat disk of stars, in which spiral arms of gas, dust, and stars are embedded. The center of the Galaxy is about 10,000 parsecs from the sun in the direction of Sagittarius. The disk rotates around an axis joining the galactic poles. Surrounding the disk is a more slowly rotating and more nearly spherical halo, containing globular clusters, scattered stars, and tenous gas.

Figure 17.3 *The halo, disk, and central regions of the Galaxy and the location of the sun all to scale. The disk is rather thin for its diameter.*

17.3 The disk of the Galaxy

The stars around us are concentrated toward the Milky Way. Stars visible to the unaided eye are 3 or 4 times as numerous around the galactic equator as they are in similar areas near its poles; the corresponding increase for large telescopes exceeds 40-fold, despite the greater obscuration by cosmic dust in the lower galactic latitudes. This shows that the majority of the stars of our galaxy are assembled in a relatively thin disk along the plane of the galactic equator. When we look in the direction of this equator, we are looking the longest way out through the disk and therefore at many more stars. The flat disk is estimated as 30,000 parsecs in diameter; its thickness is at least 3500 parsecs at the center and 1500 parsecs at the sun's distance from the center. Thus the disk is very thin, being only 5% of its diameter (Fig. 17.3) at the distance of the sun from the center.

17.4 The galactic center

The position of the center of the Galaxy was first determined by H. Shapley in 1917. He had determined the position of the center of the system of globular clusters that form a nearly spherical halo around the disk of the galaxy and had reasoned correctly that this must be the position of the galactic center itself. Shapley pointed out that the correctness of the position is indicated by the greater brightness and complexity of the Milky Way in this region.

As defined by international agreement in 1958, the direction of the galactic center is remarkably close to the original one. It is situated in Sagittarius at R.A. 17^{h}42.4^m, Decl. −28°55′ (1950), and is taken as the new zero of galactic longitude. The position of the galactic center in space is about 10,000 parsecs from the sun. Thus the sun is situated in the disk of the Galaxy about two-thirds of the way from the center to the edge the nearest part of which is in the direction of the constellation Auriga.

Figure 17.4 *Stars closer to the galactic center have larger orbital velocities than that of the sun. Stars farther away from the center than the sun have smaller orbital velocities.*

Figure 17.5 *Effect of rotation of the Milky Way Galaxy upon the observed proper motions and radial velocities. Stars nearer the galactic center seem to be passing the sun while stars farther away from the center seem to be lagging behind. This gives rise to the observed motions plotted in Figs. 17.6 and 17.7.*

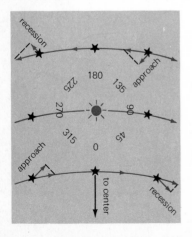

17.5 Differential galactic rotation

If the concept that the Milky Way is like other galaxies is correct and if Shapley's picture of the sun's position is correct, then a galaxy in rotation should show certain systematic effects. The sun is sufficiently far from the center of the Galaxy that it and the stars nearby are following circular orbits and Keplerian motion. Stars nearer the center are revolving about the center at greater speeds than those farther away. Following the reasoning of J. Oort we can say that if this is so, the proper motions will show one effect and the radial velocities another. Consider Fig. 17.4. Stars close to the galactic center are moving faster than the sun and those further away are moving slower than the sun. Since we observe from the sun it would appear that the stars nearer the galactic center are getting ahead of the sun and those further away from the center are falling behind the sun. The situation is as shown in Fig. 17.5. Proper motions in the direction of the center of the Galaxy should be positive in reference to the celestial sphere. In the direction of the revolution of the sun (the apex, galactic longitude $l^{II} = 90°$) the proper motions should be zero. In the direction of the anticenter ($l^{II} = 180°$) the proper motions should be positive and again, in the direction of the antapex ($l^{II} = 270°$) the motions should be zero. A curve for the expected proper motions as a function of galactic longitude is shown in the top of Fig. 17.6. The observed motions derived from the averages of many nearby stars are plotted in Fig. 17.7(a).

The radial velocities show a different pattern. Consider Fig. 17.5 again. Stars in the direction of the center of the galaxy should exhibit no radial velocities, those in a direction 45° from the center will show a positive or recessional velocity, those toward the apex zero velocity, those at 135° from the center show velocities of approach or negative velocities. In the direction of the anticenter the radial velocities will again be zero and so on around the plane of the Milky Way. The expected radial velocity curve is shown schematically in the bottom of Fig. 17.6. The observed velocities are shown in Fig. 17.7(b).

The observed curves of Figs. 17.7(a) and 17.7(b) are in such good agreement with the predictions that we must conclude the Milky Way is in rotation as hypothesized. Furthermore, the observed curves can be described by simple geometrical relations that have multiplying constants dependent upon the mass of the Galaxy and the distance to the center of the Galaxy. These constants appropriately are called the **Oort constants.**

The sun's circular velocity in the galactic rotation is about 250 kilometers/second toward galactic longitude 90° in Cygnus. The sun's distance from the center of the galaxy is taken here to be 10,000 parsecs for convenience. The period of revolution at this distance is 240 million years, sometimes referred to as a **cosmic year** or **galactic year.** The velocity of the galactic rotation diminishes with increasing distance from the center beyond the sun's distance following Kepler's third law; at the distance of

20,000 parsecs from the center the velocity is reduced to 194 kilometers/second.

The direction of the galactic rotation is clockwise as viewed from the north galactic pole. When we look toward the center in Sagittarius, the direction of the rotation between us and the center is toward the left.

17.6 The spiral structure

After the existence of galaxies as independent systems of stars was conclusively demonstrated by E. Hubble's detection of cepheid variables in the Andromeda galaxy in 1924, the frequently stated opinion that our own galaxy might be a spiral remained without firm observational support for more than a quarter of a century. Hubble pointed out that galaxies divided rather nicely into three general classes: elliptical, spiral, and irregular. In the concept of population types the spiral galaxies were characterized by lanes of gas and dust and bright stars often surrounded by a halo containing globular clusters. The irregular galaxies have plenty of gas and dust, but no organized distribution. The elliptical galaxies are almost without exception free from gas and dust. Even from our location inside the Milky Way we can deduce that our galaxy is a spiral because we see a flat Milky Way and plenty of gas and dust.

In 1949 W. Becker devised what he called a **synthetic galaxy;** this

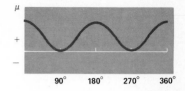
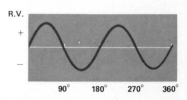

Figure 17.6 *The predicted pattern of proper motions (top) and radial velocities (bottom) as deduced from Fig. 17.5 for various galactic longitudes. The amplitudes of the two curves are arbitrary.*

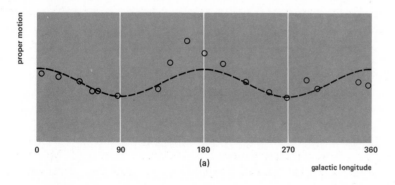

(a)

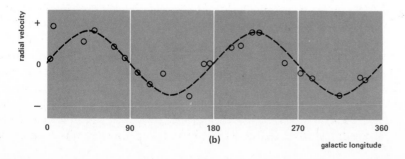

(b)

Figure 17.7 *The average of the proper motions (a) and the radial velocities (b) of many stars plotted against galactic longitude. Note how well these two curves agree with the predicted curves in Fig. 17.6.*

is as close a model as we could hypothesize at that time to prove that our galaxy is a spiral. Becker's procedure was to take a well-defined spiral galaxy (M 51), place the sun approximately two-thirds of the distance from the center to the edge, and sketch what the Milky Way would look like. One particular location reproduced the general features of the Milky Way as we actually see it quite well. This elegant proof left astronomers with an uneasy feeling, but it was the best that could be done. Fortunately this condition did not last long.

Studies of other galaxies suggest several features of spiral arms to search for. The bright blue stars of types O and B all lie in or close to spiral arms of other galaxies, the same should be true for our galaxy. The H II emission nebulae also define the arms of the nearby spiral galaxies. These objects should make good spiral tracers in the Milky Way. As work progressed, it was noted that the classical cepheids with periods greater than 10 days also define the spiral arms of other galaxies quite well. This latter discovery was most fortunate because such stars are supergiant stars visible from great distances through a considerable amount of dust and obscuring material. In addition their variability attracts our attention.

The most effective optical technique and the one used by researchers in the early 1950s is to photograph the Milky Way through a narrow filter that isolates the Hα line. The H II regions, by definition, will show strong Hα emission and hence stand out. Obscuration by dust limited observations to distances of about 3 kiloparsecs. Such distances are small compared to the scale of the Galaxy and a more powerful technique is required. Even so, some hint of structure might be expected. Astronomers plotted the distances and directions of the visible nebulae of the Milky Way on the galactic plane (Fig. 17.8) and found clear indication of two arms, one passing near the sun referred to as the **Orion arm** and another more distant one referred to as the **Perseus arm.** A third arm was suggested in the direction of the galactic center. Evidence for the third arm is now conclusive, and it is referred to as the **Sagittarius arm.** These arms coincide with the distribution of the bright O and B stars, young open clusters, dust, and hydrogen gas.

The Orion arm contains the Orion nebula, the North American nebula in Cygnus, the Great Rift, and the Coalsack. The Perseus arm, more distant from the galactic center, passes some 2.2 kiloparsecs from the sun and contains the clusters h and χ Persei. The Sagittarius arm passes about 1.5 kiloparsecs from the sun and nearer the center of the Galaxy. The sun is located on the inner side of the Orion arm so the spacing between arms in this region is about 2 kiloparsecs.

In the very year (1951) that the H II region results were published, H. Ewen and E. Purcell detected the λ 21-centimerer line of hydrogen suggested by H. van de Hulst in 1944. As we have noted in Chapter 14, the λ 21-centimeter radiation passes through the dust with little difficulty

Figure 17.8 *Parts of the spiral arms of the Galaxy traced by the directions and distances of emission nebulae. The position of the sun is indicated. The galactic center is at C. (After a diagram by S. Sharpless.)*

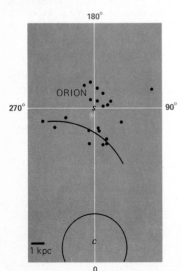

and, furthermore, it will pass through intervening hydrogen provided there is a slight difference in the velocities along the line of sight. We have already noted that the spiral arms of galaxies are defined by gas and dust; much of the gas must be hydrogen since it makes up more than 70% of the mass of the universe. Differential galactic rotation provides the velocity differences we require. It is almost as if the radio technique was made to order.

To explain the technique let us assume a simple model of the Galaxy, one where all material—stars, gas, and dust—follow circular orbits around the center of the Galaxy. Now let us look through the Orion arm at the Perseus arm in a direction about 135° from the galactic center. First, as in Fig. 17.9, we will see the hydrogen in the Orion arm; it will fill the field of view of our antenna because it is so close and it will appear to be approaching us at a very slight velocity, perhaps 8 kilometers/second. This is because it is so close so that its radius from the galactic center differs only slightly from that of the sun. Then we will see the hydrogen from the Perseus arm which does not fill our antenna beam but we will see more atoms and they will appear to be approaching us at a higher velocity, perhaps 45 kilometers/second, because it is located considerably farther from the galactic center than the sun. Our model, then, predicts two peaks in an intensity–velocity diagram. What do we see? The predicted curve is shown schematically in the sketch at the top of Fig. 17.10. Since the near arm is relativly close we see few atoms so the intensity is not large. For the somewhat more distant arm we see all of the atoms and hence its signal is more intense. This is so even though the intensity falls off with the square of the distance. The emitting volume increases with the cube of the distance. Surprisingly enough, we see exactly what we predict and even more as seen in the bottom of Fig. 17.10. We see an additional tail that may indicate an arm beyond the Perseus arm. Now if we change our antenna direction slowly we can follow the peaks and hence trace out the spiral structure.

There is a guidepost that prevents the picture from becoming chaotic. As we follow an arm around the plane of the Galaxy toward the center it will at one point have its greatest radial velocity. At this point the motion of the arm is perpendicular to the center of the Galaxy (Fig. 17.11). If we know the distance from the sun to the center of the Galaxy we can then derive the distance from the center to the Galaxy to this tangent point. The various tangent points must give a reasonable picture or the scale (the distance to the center of the Galaxy) is wrong. A schematic radio picture of the Galaxy is shown in Fig. 17.12.

Since the H II regions are excellent optical tracers of spiral arms, it is a pity that they are visible only to distances on the order of 3000 parsecs. Because of the low densities in H II regions, it has been suggested that very-high-level electron capture and a resulting cascade to the ground level may occur. This is called recombination and such lines are observed.

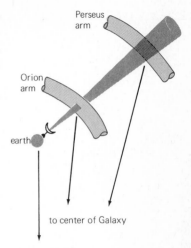

Figure 17.9 *A radio telescope looking through two arms in the plane of the Galaxy. The beam "sees" only part of the near arm so its signal is less intense than that received from the far arm. Both arms are observed because of their different radial velocities.*

Figure 17.10 *Predicted hydrogen line profile (top) and observed hydrogen line profile (bottom) in the direction of longitude 135°. The two arms are apparent and the agreement between the prediction and observation is quite good. (Observed curve after B. Burke et al., Carnegie Institution of Washington.)*

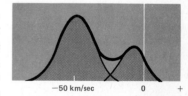

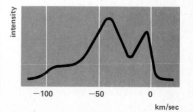

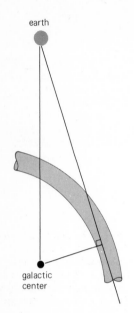

Figure 17.11   *The maximum radial velocity in a given direction occurs at the point where the line of sight is perpendicular to a line passing through the center of the Galaxy.*

One such line is the transition of an electron from level 110 to 109 (recall that the Hα line arises when the electron goes from level 3 to level 2). This line presents the opportunity to observe H II regions by radio techniques at distances far removed from the sun.

Fortunately optical techniques have now been extended to 6000 parsecs and the scale of the Galaxy is being determined with increasing confidence. The radio and optical values must agree and it appears that the distance from the sun to the center of the Galaxy is about 8.9 kiloparsecs. For ease of computation and simple calculations we will continue to use 10 kiloparsecs.

17.7 The nature and origin of the spiral arms

What are the spiral arms and where do they come from? Following B. J. and P. Bok we define the spiral features in the Milky Way and other spiral galaxies as long connected streamers of hydrogen gas and young stars. We have seen that the bright hot young stars, characteristic of population I, are found primarily in the spiral arms and that the hydrogen gas can be followed to great distances in a spiral pattern. Other objects help reveal the spiral features, but as we move toward older and older stars, that is, more intermediate population types, the correlation is less strong. Thus, if we study the distribution of the ordinary type-G giants (luminosity class III of the MK system, Section 12.20), we find no evidence of a spiral pattern at all.

We do not understand the origin of the arms, but there is one theory that has made some progress toward explaining them. Its principal proponents are C. C. Lin, F. H. Shu, and W. W. Roberts. They assume that the gravity field in the plane of the galaxy has a spiral field impressed upon it. The gas, dust and stars are on circular orbits. When they reach the spiral they slow down, so there is a piling up of gas and dust material along the pattern. Here the density is high, and hence the term **density wave** is attached to the theory. A large dense cloud can move into the spiral, suffer compression and then collapse to form stars.

In this theory the spiral pattern is a semipermanent feature of a spiral galaxy while the gas and stars that form it are constantly changing. An analogous phenomenon is familiar to traffic engineers. If an obstruction is placed on a high-density, high-speed freeway, the traffic quickly backs up and forms a compressed region of cars. Long after the original obstruction is removed the compression wave is evident although the population forming the compression wave has changed many times.

The beauty of the density-wave theory is that it makes predictions that can be tested by observation and so far several remarkable confirmations have been found. For one, the spiral structure in the Milky Way should lie between 5000 parsecs and 15,000 parsecs from the center. Another confirmed prediction is that there will be only two arms in a sprial galaxy. A third is that there exists a relation between the spacing of the arms and their pitch angle where the pitch angle is defined as the angle between the tangent to the spiral arm at that point with a line perpendicular to the direction to the center of the galaxy at the same point. A fourth is that the dark clouds will be located on the inside, i.e., the trailing side, of the spiral arms.

17.8 Gould's Belt

When we observe the winter sky with the unaided eye our first attention is drawn to the constellation of Orion in the south and the galactic clusters of the Hyades and Pleiades above and to the right of Orion. Of course, for observers in the southern hemisphere the situation is reversed. After letting the eye adapt to the dark, the Milky Way becomes apparent just to the east of Orion. In this region of the sky the Milky Way is rather faint. The bright stars seem to run down through the Pleiades and Orion, joining the Milky Way to the south of Orion. This band of stars is referred to as **Gould's Belt,** but was first pointed out by W. Herschel.

Gould's Belt is a spur of a spiral arm that is tilted with respect to the plane of the Milky Way. The distortion is considered to have been caused by a gravitational interaction with a passing galaxy, perhaps one of the Magellanic clouds. Or perhaps it is a distortion in the density wave in this region of the Galaxy.

17.9 The radio view of the Galaxy

Many interesting features of the galaxy are found with radio techniques, although occasional fleeting glimpses by other techniques may be obtained. By analogy with other spiral galaxies our galaxy has a spherical central region surrounded by a great halo and a very thin disk (Fig. 18.2). The central region is about 5000 parsecs in radius. There is a great ring of H II regions where the disk meets the central region. The spiral features begin there and spiral out to perhaps 15,000 parsecs where another great ring of neutral hydrogen is to be found. The disk itself is only about 1500 parsecs thick at the sun.

This vast region has been surveyed at various radio wavelengths; one of these surveys is shown in Fig. 17.13. Several features are always present. The plane of the Galaxy is well defined and there is a general continuum centered on the Galaxy extending far out of the plane. At certain wavelengths, in particular those of neutral hydrogen, the continuum out of the plane of the Galaxy shows some interesting structures. A large feature in the general direction of the center of the Galaxy is associated with hydrogen moving up out of the Galaxy. This is referred to as a

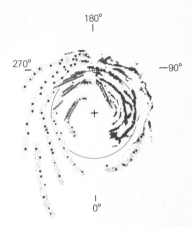

Figure 17.12 *Spiral structure of the Milky Way galaxy as deduced from observations made in Holland and Australia. The location of the sun is shown by the solar symbol. The galactic center is at the plus mark. The structure suggests a rather loose open system of arms with many streamers.*

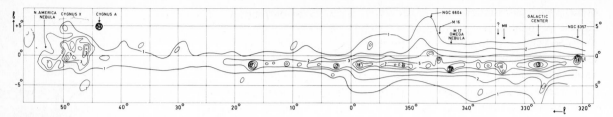

Figure 17.13 *An early hydrogen survey at a wavelength of 21 centimeters by G. Westerhout from Cygnus to beyond the center of the Galaxy. Some interesting features are marked. Longitude is given in the old system hence the center of the Galaxy is at 327°.7 rather than 0°.*

spur and appears at wavelengths as long as 2 meters. Looking toward the galactic poles we often see weak concentrations of hydrogen moving down toward the galactic plane with velocities greater than 100 km/sec. These are interpreted as clouds falling into the Galaxy, perhaps from intergalactic space or perhaps ejected material from the center of the Milky Way and are referred to as **high-velocity clouds.**

Every radio survey shows discrete sources or concentrations. Some are individual clouds, others are objects of an unusual nature, some are even stars. These sources can be studied at many wavelengths, analogous to photographing stars in different colors, and the intensity I plotted against wavelength λ. If a source is thermal in origin, i.e., if it approximates a black body it will be a straight line sloping upward as the wavelength decreases. If the source is thermal but has a large amount of ionized gas present, the intensity line will rise but then flatten out. If the source is nonthermal the intensity–wavelength line will slope upward toward longer wavelengths. The nonthermal-source radiation is generally attributed to synchrotron radiation. The three simple possibilities are shown schematically in Fig. 17.14. Some sources, for example, the active sun, show an intensity plot that is a combination of a thermal and nonthermal source.

As the resolution of radio telescopes improved, many very large concentrations of radio radiation, such as Sagittarius A (the first radio source found in the region of Sagittarius), began to show structure and eventually proved to be numerous small sources. These particular sources are concentrated in the direction of the center of the Galaxy and are presumably located there. There are at least five discrete sources comprising Sagittarius A. This name is now attached to the very strong continuum source located in the very center of the Galaxy. It is referred to as a continuum source because it is radiating at all wavelengths and in this case the origin of the continuous radiation is nonthermal. This source, because of its location, is occasionally occulted by the moon so we can study it in detail. It is about 3 minutes of arc in diameter, corresponding to about 10 parsecs at the distance of the galactic center. In the very center of this source is an exceptionally strong point source with a diameter less than 1 parsec. This point source is also a very strong source in infrared wavelengths. Sagittarius A, that is the continuous, nonthermal source, shows no structure as large as 1 parsec, the limiting resolution of the lunar occultation technique and equipment used to date.

The sources surrounding Sagittarius A are thermal sources emitting the hydrogen line at λ 21 centimeters and OH lines at λ 18 centimeters. The neutral-hydrogen emission at λ 21 centimeters and the recombination lines (particularly hydrogen level 110 to 109) look exactly like that from H II regions. Our general description of the Galaxy describes the central region as pure population II, but here is clear evidence for the presence of gas even in population II regions.

Absorption due to hydrogen and some molecules can be seen against the Sagittarius A sources. The hydrogen is seen to be rotating very rapidly even within a few hundred parsecs of the center of the Galaxy and expanding outward in all directions as well. At about 5 kiloparsecs from the center, the rotating, expanding ring ceases and the spiral arms begin.

Unfortunately we cannot see the center of the Galaxy visually. Scans in the infrared show the great bulge characteristically seen in the radio maps, but even the infrared photons do not penetrate the great clouds of gas and dust located between the sun and the galactic center. We are certain, however, from our radio observations cited above, that the galactic nucleus is a very active and energetic region similar perhaps to those found in many other galaxies.

17.10 Discrete radio sources

All radio surveys show discrete emission (and absorption) areas other than those in the galactic center and the galactic plane. Some of these regions are quite large in angular extent and even show detailed structure. Others are essentially point sources. These discrete sources are distributed all over the sky.

Many of the extended sources can be associated with emission nebulae (H II regions) and show the characteristic ionized thermal spectrum. Objects such as the Orion nebula, the Rosette nebula (see color plate), and many others are so identified. Other extended sources can be associated with emission nebulae but show nonthermal spectra. The most intense source of this type is Cassiopeia A, followed by Puppis A, and Taurus A. There are some traces of optical nebulosity in Cassiopeia A, and Taurus A is centered directly on the Crab nebula. There is no optical counterpart to Puppis A.

Most of the nonthermal galatic radio sources can be associated with supernova remnants. The Crab nebula is the most striking example, but there are others including Cassiopeia A and the Cygnus loop. Another ring-type nonthermal radio source is found in the direction of Tycho's supernova, and there can be little doubt that it is the remnant of that supernova.

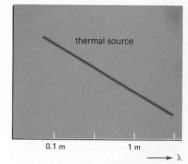

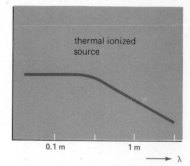

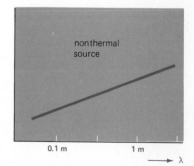

Figure 17.14 *The three forms of radio spectra shown schematically.*

Most of the discrete radio sources are very small excepting, of course, the sun. The sun dominates the radio sky in the daytime. Other stars have been observed by radio techniques. UV Ceti, the prototype of flare stars, shows bursts of radio noise coincident with flares that are observed optically. Some M-type supergiants have been observed in the water-vapor lines and the eclipsing binary beta Lyra has been observed to have a nonthermal radio spectrum, which can be associated with the ionized gas streams between the two stars.

The great majority of the small discrete radio sources cannot be associated with stars. They are below the resolution of our instruments and at first they were called radio stars. However, it soon became clear that many of these point sources could be identified with other galaxies, especially peculiar galaxies. Almost all of these sources have nonthermal radio spectra and show high-excitation emission-line spectra in the optical region.

17.11 General distribution of stars

The overall picture of the Galaxy is that of a large somewhat squashed ball having a diameter of 30 kiloparsecs called the halo. The halo increases in stellar density toward the center of the Galaxy where the star density becomes more than one star per cubic parsec. Scattered symmetrically throughout the halo are the globular clusters, themselves spherical concentrations of stars. There is almost no gas or dust in the halo. All of the stars of the halo are old stars. Embedded in the halo are distributions of stars showing more and more flattening as the stars become younger and younger until finally we get to the very youngest stars which are found only in the plane of the Galaxy. The youngest stars are not spread over the entire plane but are concentrated in the spiral arms along with gas and dust, Fig. 17.15.

To illustrate the picture more completely we shall select the K-type giants of luminosity class III. Stars of this type in the halo have very weak metal lines in their spectra whereas similar stars in the young clusters of the spiral arms show stronger metal lines. To a first approximation, the strength of the metal lines is an indication of the age of the star; weak lines are old stars, strong lines are young stars.

Looking perpendicular to the plane of the Galaxy we find the density of all of the ordinary K giants falls gradually to the limits of our ability to detect and identify them. If now we divide these stars according to the strength of their metal lines, into three classes—strong, moderate, and weak—we find all three classes in about equal numbers in the plane of the Galaxy. At a distance of 1000 parsecs above the plane the strong-line K giants make up only 2% of this type of star while the moderate- and weak-line stars are about equal in number. Strong-line stars do not exist 5000 parsecs above the plane of the Galaxy. The moderate-line stars

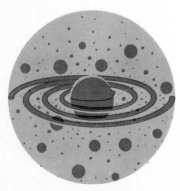

Figure 17.15 *A schematic drawing of a spiral galaxy. Basically it is a large spherical system with smaller spherical concentrations (globular clusters) embedded in it. The disk is composed of spiral arms, shown as tubes here, containing the gas, dust, and young stars.*

are very few in number and the weak-line stars make up the overwhelming proportion of the K giants this far out of the plane. Thus the distribution of stars hints of an evolutionary process within the Galaxy or in the formation and evolution of the Galaxy.

The description of the halo is characteristic of a general class of galaxies called elliptical galaxies. The description of the Galaxy is characteristic of another general class of galaxies called spiral galaxies. We now turn our attention to other galaxies.

Questions

1 What region of the Milky Way is prominent in the northern hemisphere in the summer?

2 Approximately how far is the sun from the center of the galaxy?

3 In what constellation is the galactic center?

4 What is meant by differential galactic rotation?

5 List four good spiral arm tracers.

6 What are the names of the three optically detectible arms?

7 What is Gould's Belt?

8 Distinguish between thermal and nonthermal sources in the radio region of the spectrum.

9 What is the halo of the Galaxy?

10 Where do we find the young stars of the Galaxy?

Other Readings

Fitzgerald, M. P., and A. F. J. Moffat, "Penetrating Puppis," *Sky & Telescope* **52,** 104–107 (1976).

Geballe, T. R., "The Central Parsec of the Galaxy," *Scientific American* **241(1),** 60–70 (1979).

Gordon, M. A., and W. B. Burton, "CO in the Galaxy," *Scientific American* **240(5),** 54–67 (1979).

Sullivan, III, W. J., "A New Look at K. Jansky's Original Data," *Sky & Telescope* **56,** 101 (1978).

chapter 18

Galaxies

Cosmology is the study of the universe. We begin this study by looking at the building blocks of the universe—the galaxies—in some detail. The galaxies are structurally of three main types: spiral, elliptical, and irregular galaxies. Galaxies are assembled in large clusters and also in smaller groups such as the Local Group in which the Milky Way galaxy is located. The redshifts of their spectral lines are increasingly positive as the galaxies are more remote. Certain quasistellar objects (quasars) have the largest redshifts yet observed and are difficult to interpret.

Structural Features of Galaxies

Galaxies may be classifed conveniently into three classes that correlate with stellar population types. The classes are called elliptical, spiral, and irregular, and they are composed principally of population type II stars, a mixture of types I and II stars, and population I stars, respectively.

18.1 The great debate

Modern study of galaxies began with V. M. Slipher at Lowell Observatory in late 1912 when, in a letter to his former teacher J. Miller, he presented a table of radial velocities of nine extragalactic nebulae (at the time the term extragalactic meant avoiding the galactic plane) obtained using the Lowell 61-centimeter refractor. Slipher pointed out that seven objects showed increasing recessional velocities that correlated with their apparent magnitudes. Furthermore, the velocities were astounding, the largest one being 1800 kilometers/second, far larger than that of any star. Miller counseled Slipher to be cautious and to add a few more observations before releasing this material. Finally in 1914 Slipher presented his observations on 13 extragalactic nebulae to the American Astronomical Society and received a standing ovation. The ovation was as much for the perseverance shown, a characteristic to this day of observational cosmologists, as for the very large radial velocities observed. Some of Slipher's spectrograms covered 30 hours and required four different nights to complete the exposure. Little noticed at the time were Slipher's statements that the spectra were all composite stellar-type spectra and that one of the objects, the Sombrero nebula, showed rotation. He concluded that the extragalactic nebulae of the type showing composite stellar spectra were separate island universes of stars, and all but the nearest ones were receding from us.

At about the same time A. van Maanen, using the greatest telescope in the world of his day, measured the proper motion of the arms of the spiral nebula called M 101. Clearly any object showing proper motion cannot be very far away, and so a contrast of views prevailed. The differing viewpoints became so sharp that a so-called great debate was arranged by the U.S. National Academy of Sciences where H. Shapley and H. Curtis presented the opposing points of view. The great debate took two different aspects, the first discussed on the first day concerned the nature of the Milky Way, and the second discussed on the second day concerned the nature of the extragalactic nebulae. In the first part of the discussion Shapley was right, clearly the Milky Way was centered at a point far removed from the sun. In the second part, Curtis, leaning heavily upon Slipher's observations, claimed the extragalactic nebulae were island universes of stars. Shapley, leaning upon van Maanen's proper-motion observations, claimed the contrary. In the second part of the discussion Curtis eventually was right. The stage was set for modern cosmology.

In 1885 a supernova was observed in the Andromeda galaxy that aroused the curiosity of some astronomers. However, it was not until 1924 when Hubble observed cepheid-type variable stars in M 31 (the Messier Catalogue designation for the Andromeda galaxy) that astronomers were convinced that the island-universe concept was correct. Hubble, using Shapley's calibration of the period–luminosity relation, assigned a distance to M 31 of 1.5 million light years (460,000 parsecs), which was later revised to 2.2 million light years or 675,000 parsecs using the revised calibration of the period–luminosity relation. Hubble and M. Humason fully confirmed Slipher's work and Hubble was able to classify galaxies into three principle types: spiral, elliptical, and irregular. Hubble refined this classification. The elliptical galaxies were classified as E0 for spherical systems through E1, E2, and so on to E7 depending on the degree of ellipticity. The spiral galaxies were classifed as SO, Sa, Sb, or Sc depending upon how large the nucleus was and how tightly wound the arms were, with Sc galaxies having the smallest nuclei and the least wound arms. Irregular galaxies are not classified by definition. Hubble added a parallel classification to that of the spiral galaxies because a significant number of spiral galaxies showed a bar across their nuclei. These barred spiral galaxies he designated by SBO, SBa, SBb, and SBc. Photographs of some of the Hubble-type spirals are shown in Fig. 18.7.

Many astronomers have tried to attach evolutionary significance to the Hubble classification scheme (Fig. 18.1). For example, some astrono-

Figure 18.1 *The sequence of types of galaxies following the original classification scheme of Edwin Hubble.*

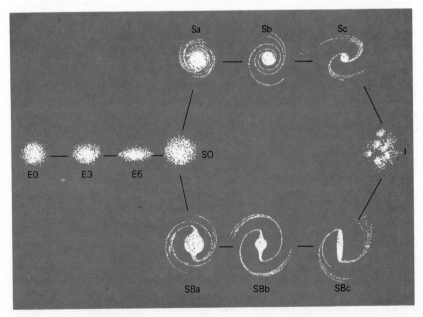

mers have held that the galaxies evolve from irregular galaxies to elliptical galaxies. Since elliptical galaxies have the oldest stars, this has some merit. However, others argue that evolution goes the other way. Still others feel that, based upon our knowledge of the lifetimes of stars and clusters of stars and the age of the universe (Chapter 19), one type of galaxy could not evolve into another type even if we are wrong in our estimate of the age of the universe by a factor of 2 or more. Except for cataclysmic events in the nuclei of certain galaxies, the galaxies are not now very much different from what they were 10^{10} years ago nor will they be much different 10^{10} years in the future except for the gradual evolution of the stars making up the galaxies.

As we noted earlier, new stars are formed from the debris of other stars. The debris from which stars are made is enriched in heavy elements. This follows from the course of stellar energy production. Thus, stars formed recently are formed in the plane of a galaxy where the debris of the other stars is to be found. Newly formed stars have a high metal content to start with and hence they will evolve more quickly. Indeed, newly formed stars may well evolve faster than stars which were formed from the original cosmic mixture.

The net effect of stellar evolution upon galaxies is that the interstellar material is gradually locked up in stars and the stars and the galaxies become redder and redder in color.

18.2 Spiral galaxies

Spiral galaxies are structurally of two types: normal and barred spirals. **Normal spirals** have lens-shaped central regions. From opposite sides of this central region two arms emerge and at once begin to coil around the centers in the same sense and the same plane. They are divided into three classes. **Class Sa** spirals have large central regions and thin, closely coiled arms; examples are NGC 4594 in Virgo (the Sombrero galaxy; Fig. 18.2) and M 81 (Fig. 18.3). In class Sb the centers are smaller and the arms are larger and wider open; an example is M 31 (Fig. 18.6). In class Sc the centers are smallest, and the arms are largest and most loosely coiled; an example is M 33 in Triangulum (Fig. 18.4) and probably the Milky Way.

Barred spirals have their two coils starting abruptly from the ends of a bright bar, which projects from opposite sides of the central region. They are classifed in the series SBa, SBb, and SBc, paralleling the series of normal spirals. Through this series the central region diminishes while the arms build up and unwind. A third and less familiar type of spiral galaxy has its arms beginning tangentially from opposite sides of a bright ring around the center.

Spiral galaxies are presented to us in a variety of orientations. In the flat view they appear nearly circular. M 51 (Fig. 18.5) is an example;

Figure 18.2 *The Sombrero galaxy (NGC 4594), an Sa galaxy, is seen almost edge-on. The disk is extremely thin, the apparent thickening at the outer reaches is due to perspective and the hat brim effect. (Kitt Peak National Observatory photograph.)*

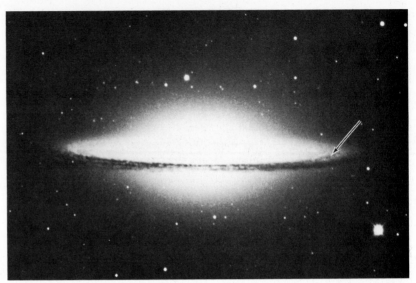

Figure 18.3 *M 81, an Sb spiral galaxy in Ursa Major. The peculiar galaxy M 82 is nearby. (Copyright by Akademia der Wissenshaften der DDR. Taken at the Karl Schwarzschild Observatorium.)*

it has an irregular companion projected near the end of one arm. When they are moderately inclined to the plane of the sky, as in the cases of M 31 and M 81, another giant galaxy, these galaxies appear to have an oval shape. In the side view they are reduced to bright streaks around thicker central regions. Characteristic of the spirals seen on edge is a dark band that sometimes seems to cut them in two; an example is the Sombrero galaxy, shown in Fig. 18.2. The dust in their arms obscures their equatorial regions, just as the dust clouds of the Milky Way obstruct our view of what lies beyond.

One notable feature of spiral galaxies is the extreme thinness of the absorbing disk in comparison to the diameter of the disk. The Sombrero shows this clearly, the apparent thickening of this layer toward the edges can be seen as due to a turning down of the left edge and a turning up of the right edge. This is known as the **hat-brim effect** often seen in spiral galaxies. We have noted already the thinness of the disk of the Galaxy and should point out that the hat brim effect in the Galaxy's layer of neutral hydrogen has been observed. The Andromeda galaxy shows the same effect.

The nearest spiral galaxies have been studied at λ 21 centimeters and show two distinctive rings, one at about 20% of the diameter and the other just beyond the normal visible limits of the arms. The arms are located between these two rings and can be well represented as logarithmic spirals.

Any study of deep-sky photographs shows many spiral galaxies. At first they seem to be the preponderent type of galaxy, but this is a selection effect. First, the obvious arms attract our attention, and second, the spirals as a group are generally large galaxies having diameters ranging from

Figure 18.4   *M 33, an Sc-type galaxy in Triangulum. (Hale Observatories photograph.)*

25 to 60 kiloparsecs. Even so, present counts indicate that spiral galaxies make up about 50% of the galaxies. A typical, though large, spiral galaxy is M 31, which is near enough so that we can study it in detail.

The spiral galaxy M 31 in Andromeda is the brightest spiral system and is one of the two spirals that are nearest us. Marked in Map 4 in Andromeda, it is visible to the unaided eye as a hazy spot about as long and half as wide as the moon's apparent diameter. Only the central region appears to the eye alone and, for the most part, to the eye at the telescope. Fainter surrounding features come out in the photographs, where the object is shown in its true character as a flat, double-armed spiral inclined 13°

Figure 18.5   *The almost face-on Sc galaxy M 51. In long exposures, the small galaxy at the end of one arm and M 51 show interacting streamers. (Kitt Peak National Observatory photograph.)*

from the edgewise presentation. The flat disk is slightly turned up on one end of the major axis and down on the other as can be seen in Fig. 18.6. This is the hat-brim effect noted earlier.

In the photograph the tilted, nearly circular, spiral appears as an oval 3° long. Separate stars in the arms were first observed in photographs in 1924. The cepheid variable stars among them served to determine the distance of the spiral and to show it as another galaxy far beyond our own Milky Way.

The lens-shaped central region of the Andromeda spiral is surrounded by a flat disk of stars. Embedded in the disk are the spiral arms, which contain all the features observed in the Milky Way around us. Young stars, gas, and dust are localized in the arms. Patches of the gas appear as emission nebulae, made luminous by the radiations of blue stars in the vicinity. Based on these nebulae the tracing of the arms was extended beyond their more conspicuous parts, increasing the diameter of the disk to 4°5. At a distance of 675,000 parsecs from us, the diameter of the Andromeda spiral is about 55,000 parsecs or larger than the diameter

Figure 18.6 *The Great Andromeda galaxy (M 31) an Sb galaxy, and two of its elliptical galaxy companions. (Hale Observatories photograph.)*

assigned to the disk of our own galaxy. Globular clusters of stars surround the disk, as they do in the halo around the center of our galaxy although M 31 seems to have 3 to 4 times as many.

Before leaving spiral galaxies we should consider some interesting observations of M 51 (Fig. 18.5). M 51 is a relatively nearby spiral with a rather large companion galaxy partially hidden by the end of one of the arms. M 51 is only slightly tilted to us and presents the classic appearance of a spiral galaxy. It is almost the same size as our galaxy and shows all of the features that we see in the Milky Way—H I regions, H II regions, novae, supernovae, gas, dust, and an inner and an outer ring of hydrogen.

Recently M 51 has been studied in the radio continuum using the supersynthesis telescope at Westerbork. The results show a long continuous ridge following the spiral pattern. This leaves us with the strong suggestion that the spiral arms are long connected streamers as defined by the Boks and that the density-wave theory has credence. A pictorial summary of the types of spiral galaxies is given in Fig. 18.7.

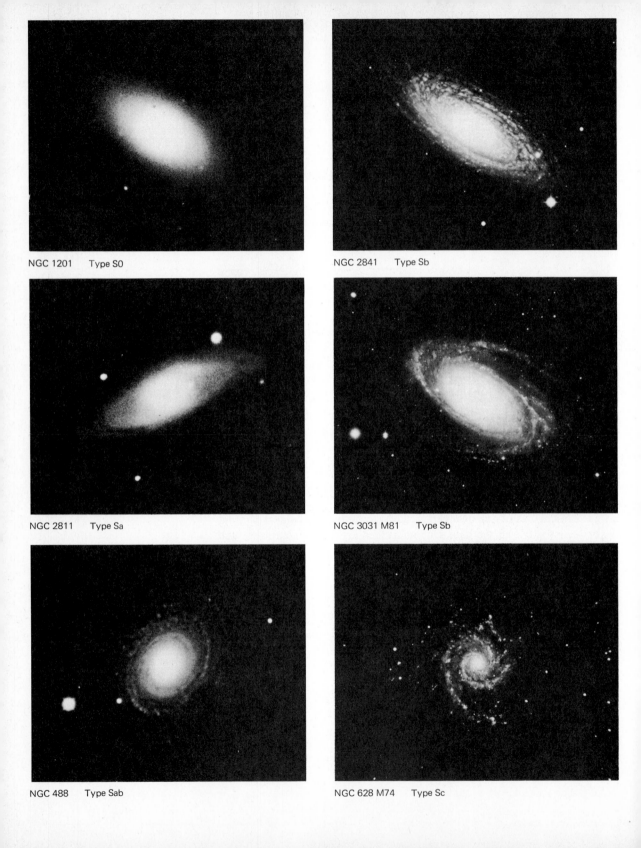

NGC 1201   Type S0

NGC 2841   Type Sb

NGC 2811   Type Sa

NGC 3031 M81   Type Sb

NGC 488   Type Sab

NGC 628 M74   Type Sc

18.3 Elliptical galaxies

Elliptical galaxies are so named because they appear with the telescope as elliptical disks having various degrees of flattening. They are designated by the letter E followed by a number that is 10 times the value of the ellipticity of the disk. The series runs from the circular class E0 to the most flattened E7, where the object resembles a convex lens viewed on edge. Examples are M 32, the nearly circular class E2 companion of the Andromeda spiral, and the second companion, NGC 205, of class E5. Another is NGC 3115, Fig. 18.8. Despite the observed flattening of elliptical galaxies there is no evidence of rotation such as we find for the disk of spiral galaxies.

These galaxies are systems of stars that are generally dust free. Almost all the gas and dust available for star building seems to have been exhausted. Individual red giants, the brightest stars of this older population, are visible in photographs of the nearer elliptical galaxies.

This is the general picture of elliptical galaxies, but there is a great variety of elliptical galaxies and a few do show gas and dust. These form a small fraction of all elliptical galaxies—less than 1%. The elliptical galaxies range from the largest and brightest galaxies, designated by gE (g for giant), in the Universe with diameters of 85 kiloparsecs to the smallest galaxies, designated by dE (d for dwarf), in the universe with diameters as small as 200 parsecs. The giant elliptical galaxies are generally very

Figure 18.7 *(Opposite page) The principal types of spiral galaxies. (Hale Observatories photograph.)*

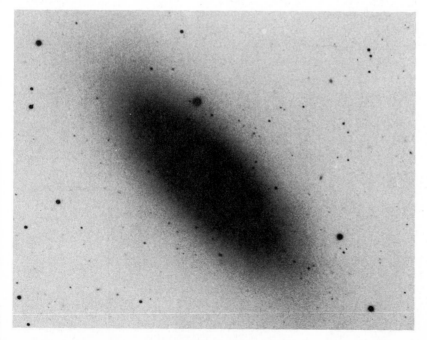

Figure 18.8 *The highly elliptical galaxy NGC 3115. There is no evidence of gas or dust in this galaxy. (Hale Observatories photograph.)*

strong radio sources. The giant elliptical galaxies are visible to the limits of our optical techniques, although a type of object, the quasistellar objects, may be visible at even greater distances and some unidentifiable radio sources may be at the limit of the observable Universe. Elliptical galaxies make up about 40% of the observable galaxies.

18.4 Irregular galaxies

Hubble's third class of galaxies, the irregular galaxies, is just as descriptive as the other two. These galaxies present a picture of an unorganized accumulation of stars, gas, and dust. They show all of the typical population I objects with gas and dust accounting for 50% of their mass. Despite this, many irregular galaxies contain globular clusters and red giants typical of population II. Irregular galaxies have a considerable size range. Some are as large as the large spiral galaxies, whereas others are comparable to the smallest elliptical galaxies. Five percent of all galaxies are irregular galaxies, although many galaxies may be classified as irregular when in fact they are peculiar. Typical irregular galaxies are the Milky Way's nearest neighbors, the Magellanic clouds.

The Magellanic clouds, named in honor of Ferdinand Magellan, are two satellites of our galaxy. Plainly visible to the unaided eye, they are too close to the south celestial pole (Map 6) to be viewed north of the tropical zone. As they appear generally in the photographs, the clouds have apparent diameters of $7°7$ and $2°5$. At distances of 52 and 63 kiloparsecs, their linear diameters would be 7 and 3 kiloparsecs, respectively. The neutral hydrogen of the clouds extends farther than the stars and forms a link between the two objects. The clouds are classified usually as irregular galaxies, abbreviated I and are ragged assemblages that lack rotational symmetry. Yet the rotations of both clouds have been suspected. Arguments for a bar in the center of the Large Megallanic Cloud (Fig. 18.9) have been advanced.

Both clouds have variable stars, novae, planetary nebulae, etc. The small cloud even has an x-ray source that is a binary star. In general, the southern hemisphere of the celestial sphere has been lying fallow despite the fact that the center of the Galaxy is located there as are the Galaxy's two nearest companions. B. J. Bok reminds us that the Large Magellanic Cloud (LMC) and the Small Magellanic Cloud (SMC) are galaxies and may be studied using 50-centimeter telescopes with the same precision with which the Andromeda galaxy is studied using the 5-meter Hale reflector.

18.5 Other types of galaxies

Many galaxies, like stars, do not fit into the Hubble classes nor, for that matter, into more refined systems of classification. Some galaxies show

Figure 18.9 *The Large Magellanic Cloud, an irregular galaxy which covers 8° of the sky. (Hale Observatories photograph.)*

strange but symmetrical forms suggesting that a superviolent event, usually in the nucleus, has occurred that is disrupting the galaxy. These galaxies are referred to as **peculiar galaxies** and often show the characteristic indicators of highly energetic events such as we have found in supernovae and their remnants.

M 82 is such a peculiar galaxy (Fig. 18.10). It shows filaments running out of its center perpendicular to its plane at velocities as great as 10,000 kilometers/second and even larger. These filaments emit polarized light, typical of a synchrotron source. This galaxy is a strong nonthermal radio source, an x-ray source, and an infrared source as well. Its spectrum shows numerous high-excitation forbidden lines. All indications are that M 82 was a normal galaxy, probably a spiral, that has now been torn asunder by an event in its nucleus of such violent proportions that it is disrupting the entire galaxy.

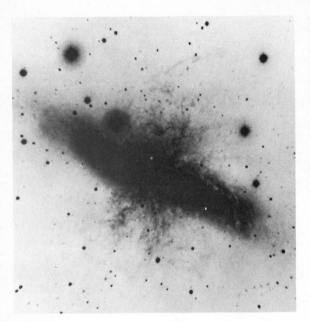

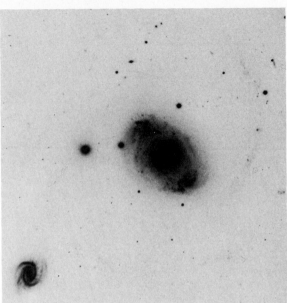

Figure 18.10   *M 82, a peculiar galaxy. The central region of this galaxy has undergone a catastrophic event that is tearing it apart. (Hale Observatories photograph.)*

Figure 18.11   *A strong source of radio emission is this Seyfert galaxy known as NGC 4151. This exposure is too long to show the intense central source. (Hale Observatories photograph.)*

Another type of galaxy is the **radio galaxy** (Fig. 18.11). Actually all galaxies emit energy in the radio region so we limit the name to the very strong radio emitters. Most of these galaxies are the giant elliptical galaxies, peculiar galaxies, N-type galaxies, and unidentifiable sources that have spectra typical of radio galaxies and that we presume are galaxies, perhaps at the edge of the Universe. Occasionally radio galaxies are variable in their energy output, not just in the radio region but in the infrared, visual, and x-ray regions. The time for these variations is quite short. This implies that the emitting region is very small.

N galaxies are galaxies having bright star like nuclei and strong emission in the radio spectrum. They often show faint, wispy halos. **Seyfert galaxies,** first discussed by C. Seyfert and named for him, are similar to N galaxies. They have very bright starlike nuclei that show high-excitation emission lines, which are very broad. Some, but not all, Seyfert galaxies are radio emitters. The Seyfert galaxies show traces of spiral structure and may be an early stage of the N galaxies. To complete this picture we mention the **quasistellar objects** and **BL Lacertids** (both of which we discuss later) that show many of the features of Seyfert and N galaxies, but which are essentially starlike. Most of them are not strong radio sources, but those that are are also infrared sources and x-ray sources.

Dwarf galaxies are divided into two classes: dwarf elliptical galaxies and dwarf irregular galaxies. They are really the faint, low-mass galaxies extending the two major classes already discussed. The dwarf elliptical galaxies resemble globular clusters except that the largest globular cluster has a diameter of about 100 parsecs, whereas the smallest dwarf elliptical galaxy has a diameter of 200 parsecs. The majority of the known members of the Local Group of galaxies are dwarf elliptical galaxies (Section 18.7). These galaxies are essentially free of gas and dust. Exhaustive studies have failed to detect even a trace of interstellar material in these galaxies. They appear to be examples of pure population II objects. It is possible that they are only found as companions to the large or giant galaxies.

Dwarf irregular galaxies are probably very numerous but even at 1 megaparsec they are extremely difficult to detect. Two of the Local Group (Table 18.1) are dwarf irregular galaxies, NGC 6822 and IC 1613. These

**Table 18.1
The Local Group**

Member	Type	Distance (kpc)
Milky Way	Sc	—
LMC	I	52
SMC	I	63
Draco	dE	67
Ursa Minor	dE	67
Sculptor	dE	85
Ursa Major	dE	115
Sextans C	dE	138
Fornax	dE	170
Pegasus	dE	174
Leo I	E4	230
Leo II	dE	230
Carina I	dE	240
NGC 6822	I	470
NGC 205	E5	640
NGC 147	E5	660
NGC 185	E3	660
M 32	E2	660
M 31	Sb	670
M 33	Sc	730
IC 1613	I	740
WLM	I	870
Sextans A	I	1000
Leo A	I	1100
Maffei I	gE	1100
IC 10	S	1260
Maffei II	Sa	2630[a]
Capricornus	dE	[a]
Serpens	dE	[a]
Sextans B	dE	[a]
Simonson I	I	[a]
Phoenix I	I	[a]

[a] Unknown or uncertain values.

two galaxies are much smaller than the Magellanic clouds but contain typical population I type objects such as cepheids, gas, dust, and H II regions. Dwarf irregular galaxies make up a significant fraction of all galaxies and are at least as numerous as the spirals. Dwarf galaxies of both types are, in general, always associated with large galaxies or clusters of galaxies.

Distribution of Galaxies

Galaxies appear to be evenly distributed over the entire sky, but they tend to group together in clusters. The Milky Way galaxy is a member of a small cluster of galaxies.

18.6 The arrangement of galaxies in the sky

The apparent distribution of the nearer galaxies over the face of the sky is shown in Fig. 18.12. This diagram is an equal area projection of the sky with the galactic equator across the center or major axis and the north and south galactic poles at the top and bottom. Few galaxies are visible within 10° of the galactic equator, which is represented by the heavy curve in the figure; here they are concealed generally behind the dust of the Milky Way. Their numbers increase toward the galactic poles, where the least amount of dust intervenes. The increase would be fairly symmetrical in the two celestial hemispheres except for the conspicuous Virgo cluster at the left. Because of their apparent avoidance of the Milky Way the galaxies were called "extragalactic nebulae" before their true significance was recognized.

The more remote galaxies are similarly arranged in the sky, as has been shown by more penetrating surveys. If the obscuring dust were not

Figure 18.12 *The distribution of galaxies in old galactic coordinates. The blank areas are regions of the sky not observed from Mt. Wilson. Eliminating equal areas above the galactic equator leaves a distribution showing that galaxies are evenly distributed everywhere. The filled dots are areas of heavy concentration, circles are areas of normal concentration, and dashes are areas where no galaxies are seen. The symbols correlate with galactic latitude showing the effect is due to our galaxy. (Diagram by E. P. Hubble, Mt. Wilson Observatory.)*

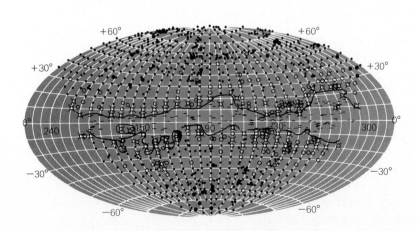

present, the galaxies would be about equally numerous in all regions of the celestial sphere. The fact that galaxies are distributed uniformly over the sky is important. If one removes the large concentration of galaxies in Virgo, the distribution in the two halves of the sky is about the same.

The spatial distribution of galaxies can be considered when their distances as well as their directions from us are known. Determining distances establishes what is known as the **cosmic distance scale.** Originally the distances were derived in three successive steps: (1) by use of cepheid variable stars that are found in M 31, M 33, and other relatively nearby galaxies; (2) by observing the apparent magnitudes of what seemed to be the most luminous stars in more remote galaxies, where the cepheids were too faint to be seen; (3) by measuring the apparent brightness of still more distant galaxies, assuming that galaxies of the same class have the same absolute brightness. Later investigations showed the need for revising the early scale of distances.

One reason for the revision was the discovery (1952), that the classical cepheids are, on the average, 1.5 magnitudes more luminous than they were previously supposed to be. On this account and with the availability of the more precise photoelectric standards for the apparent magnitudes, the former values of the distances of most galaxies required multiplication by a factor of 2.

More recently it has been pointed out that what seemed to be the brightest stars in intermediate galaxies with the smaller telescopes are in fact small patches of emission nebulae. It is estimated that these may appear 1.8 magnitudes brighter than the brightest stars in a particular galaxy and that the distances originally assigned to most galaxies might need to be multiplied by a factor as great as 10.

Present methods follow the original method rather closely. Cepheid variables are observed in nearby galaxies and the distances obtained on the assumption that the period–luminosity relation holds. This is checked by observing individual giant and supergiant stars as well as RR Lyrae stars. Whenever novae occur they are calibrated and cross checked. Globular clusters are visible around many galaxies and H II regions are visible in spiral galaxies and serve as secondary standards. As we move to increasingly distant galaxies we depend more and more on the less reliable of these indicators until finally we assume that the tenth brightest galaxy in any given cluster of galaxies has very nearly the same brightness as the tenth brightest galaxy in any other cluster of galaxies. Fortunately a spectral feature appears that allows us to go to more distant galaxies; it is referred to as the redshift and we will discuss this feature later in Section 18.15.

Novae occur in the galaxies, just as they do in our own. They are of two general types, normal novae and supernovae, with respect to the order of luminosity they attain.

Normal novae resemble those found in our galaxy in their greater abundance and their lower luminosities at maxima. Surveys of many of these novae in the spiral M 31 reveal their characteristics more clearly with the elimination of the distance factor. About 26 normal novae flare out annually in the Andromeda spiral; one-fourth of these are likely to be concealed by the dust clouds of this galaxy or occur in its brighter regions. The absolute magnitudes at the maxima range from -6.2 to -8.5.

Supernovae attain much higher luminosities and are less frequent. These spectacular outbursts of the more massive stars occur in a single galaxy only two or three times a century. They are most likely to be detected in repeated photographs of clusters of galaxies. Supernovae are divided into two groups. In **group I** they rise to absolute magnitude -19, or more than 100 million times as luminous as the sun. The extreme width of the lines in their spectra indicate the violence of the explosions. These supernovae are deficient in hydrogen. An example is the supernova of 1959 in the spiral galaxy NGC 4725 (Fig. 18.13). Supernovae of **group II** attain about absolute magnitude -17 at maximum brightness. The supernova that formed the Crab nebula is believed to be of this group. They probably constitute the more numerous group, but being fainter are less readily detected. Supernovae of both groups are good secondary distance indicators.

18.7 The Local Group

Our galaxy is a member of a loose cluster of some 32 galaxies referred to as the **Local Group.** It occupies a roughly ellipsoidal volume of space some million parsecs in its longest dimension. The Milky Way and the

Figure 18.13 *A supernova event in NGC 4725. The photograph on the left was taken on 10 May 1940 and the one on the right was taken on 2 January 1941. (Hale Observatories photograph.)*

Andromeda galaxies are on opposite ends of the long axis. These, M 33 in Triangulum, IC 10 and Maffei II are the spiral members. There are nine irregular galaxies of which the Magellanic clouds are the largest and are satellites of the Milky Way. The remaining 18 galaxies are elliptical galaxies.

Maffei I is a giant elliptical galaxy and the largest galaxy of the Local Group. There are six elliptical galaxies and 12 dwarf elliptical galaxies. The first dwarf elliptical galaxies discovered were the Sculptor and Fornax systems. Recent photographs reveal two dwarf ellipticals that are satellites of the Milky Way. The Andromeda galaxy has four dwarf elliptical companions and four regular elliptical companions. The members of the Local Group of galaxies are listed in Table 18.1.

The Local Group, considered a small irregular cluster of galaxies, shows the preponderance of small low-luminosity galaxies; hardly one of the smallest members would be visible at a distance of 1 megaparsec. It would appear that more members are associated with M 31 than the Milky Way, but this could be an observational effect. Five members of the Local Group do not have distances determined for them.

18.8 Clusters of galaxies

Clusters of galaxies are believed to be the rule rather than the exception. They range in population from a very few to several thousand galaxies. As many as 10,000 rich clusters of galaxies are listed in F. Zwicky's catalog of clusters north of declination −30°. A **rich cluster** is here defined as containing more than 50 members having a photographic brightness within three magnitudes of the brightest galaxy in the cluster. Some of the more prominent clusters, such as the Virgo cluster and Perseus cluster (Fig. 18.14), are often designated by the names of the constellations in which they appear.

A **compact cluster** in the catalog has a single concentration of galaxies that appear close together in the photographs. An example is the Coma Berenices cluster; it has a membership of 9000 galaxies and a preponderance of SO galaxies, especially in its densest central region. A **medium compact cluster** has either a single concentration where the galaxies are separated by several of their diameters, or else a number of pronounced concentrations. The Virgo cluster is an example; it has many spiral and giant elliptical galaxies.

An **open cluster** has no outstanding concentration. An example is the Ursa Major cluster, which includes the great spirals M 81 and M 101.

G. Abell, in contrast to Zwicky, prefers to classify clusters of galaxies as **regular** and **irregular.** Regular clusters show spherical symmetry, they have a strong tendency toward a central condensation, their brightest mem-

Figure 18.14   *NGC 1275, a cluster of galaxies in Perseus. SO spirals dominate this cluster. (Hale Observatories photograph.)*

ber is a giant elliptical galaxy, and the appearance of large spiral galaxies is rather rare. Regular clusters range in membership from the great Coma cluster with 9000 members to clusters having 1000 member galaxies.

Irregular clusters are amorphous in appearance and have a large spiral galaxy as their brightest member. Their composition ranges from the Virgo cluster and its 1000 members through the Local Group with its 32 members to groupings as small as 10 members. It is significant that the dense or compact clusters are usually dominated by SO or giant elliptical galaxies. The normal spiral galaxies are found in the less dense outer regions of these clusters. On the other hand open clusters of galaxies seem to be dominated by spiral galaxies. This suggests that the star-forming gas and dust in spiral galaxies in dense clusters is stripped out, halting the star-formation process. The spiral galaxies in the outer regions have fewer interactions hence retain their normal appearance. Since the dense clusters are often dominated by giant elliptical galaxies, it has been suggested that these galaxies form by cannibalism, that is, they sweep up the numerous dwarf galaxies.

Recently astronomers have come to recognize clustering among clusters of galaxies. This is referred to as superclustering and the objects are referred to as **superclusters.** However, evidence is growing that super-clustering is just a chance arrangement and superclusters may be a projection effect.

18.9 Galaxies as radio sources

Galaxies differ in the character and strength of their radio emissions. The radiations of the majority are mainly of the thermal type; they are weaker in the longer than in the shorter radio wavelengths. Examples are our galaxy and M 31. The radiations of some other galaxies are nonthermal; they are generally of the synchrotron type, which is stronger in the longer wavelengths and attains remarkable strength in some of these sources. A number of these sources have characteristics that are so much alike they may represent a class that occurs rather frequently. An example is Cygnus A, the first radio source found in Cygnus.

The optical counterpart of the radio source Cygnus A (Fig. 18.15) is a pair of overlapping bright spots centered 2″ apart and surrounded by a larger dim halo. The appearance was interpreted at first as two colliding galaxies that had interpenetrated until their centers were only a few parsecs apart. The collision hypothesis for this and similar galaxies was favorably received by most astronomers until 1960, when it began to seem improbable that such strong emission could be produced by collisions. Moreover, the radio emission of Cygnus A comes mainly from two areas centered about 80″ on either side of the optical pair.

In 1960, it was concluded that Cygnus A is most likely a generically related double galaxy. A possible interpretation could be inferred from a hypothesis of the instability of some clusters of galaxies, proposed in 1954 by Ambartsumian. Radio galaxies such as this one might be single originally, possessing enormous and unexplained energy that is tearing them apart. They might be splitting into pairs of galaxies with the production of intense radiation.

Very strong radio radiation is often associated with giant elliptical galaxies. The radio radiation does not originate in the galaxy, but in two gigantic clouds on either side of the galaxy. The radiation often is most concentrated on the sides of the clouds facing away from the centrally located galaxy (Fig. 18.16). Radiographic pictures give the impression that the central galaxy is somehow beaming energy into these clouds. The main problems with this explanation are the enormous energies involved and the tremendous distances. From the edge of one cloud to the outer edge of the other is sometimes as large as one million parsecs. Considering the clouds and galaxy as a single object, these are the largest objects in the universe.

Figure 18.15 *Optical object identified with the strong radio source Cygnus A. The radio emission comes from two lobes far removed from the optical object as in Fig. 18.16. (Hale Observatories photograph.)*

Figure 18.16 *A radiograph of the giant double radio source DA 240. The centrally located object is a rather normal looking elliptical galaxy. (Courtesy of J. Oort, G. K. Miley, and R. G. Strom, Sterrewacht Leiden.)*

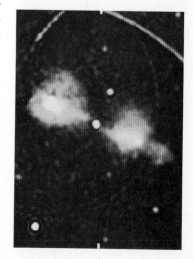

Figure 18.17 *The bright radio source Centaurus A. This object, known as NGC 5128, is an unusual galaxy as can be seen by studying it on the color plate. (Hale Observatories photograph.)*

Figure 18.18 *Two interacting galaxies, NGC 4038 and 4039. On the original negative the arms stretch more than 10 times the diameter of the galaxies. (Hale Observatories photograph.)*

Another example is Centarus A (Fig. 18.17 and color plate). This galaxy has enormous regions emitting radio radiation above and below the disk which is defined by the dark absorbing material. There is clear evidence of a disturbance in the central regions of this galaxy.

18.10 Interactions between galaxies

We have noted several early conjectures that SO galaxies might be spirals which lost the gas and dust from their arms in collisions and that radio sources such as Cygnus A might be powered by interpenetrating galaxies. Although collisions between galaxies are perhaps rare enough to be disregarded, effects of interactions between close pairs of galaxies are frequently observed. In some cases the structural features of the galaxies are distorted. In other cases bright filaments join the components of the pairs or appear in their vicinities, reminiscent of the gas streams around close binary stars. The original negatives of the pair NGC 4038 and 4039 (Fig. 18.18) show long filaments above and below the galaxies.

F. Zwicky has observed many pairs of interacting galaxies. In 1959 B. A. Vorontsov-Velyaminov published an illustrated catalog of 355 such pairs. The interacting pairs are believed to be revolving binary galaxies (Fig. 18.19). Some astronomers believe that interacting pairs of galaxies and even groups of interacting galaxies are the rule rather than the exception. Here we use interacting to mean having a direct disturbing effect rather than a gravitationally stable system such as a cluster of galaxies. These interactions can often be interpreted as ejected members of clusters or parts of galaxies in agreement with Ambartsumian's proposal.

18.11 The intergalactic medium

It is still a matter of considerable argument whether or not there is an intergalactic medium, that is, whether or not there is a significant amount of material between the galaxies. If there is an intergalactic medium it must be composed primarily of hydrogen since more than 70% of the universe is hydrogen by mass.

If the medium is cold, a condition that we might expect intuitively, then we should detect the neutral-hydrogen radiation at λ 21 centimeters. There is no evidence for any neutral-hydrogen radiation outside of galaxies. If the medium is not cold then perhaps it is so hot that it is completely ionized? While there is no direct evidence that there is a hot intergalactic medium there is some indirect evidence.

Some galaxies show definite tails streaming behind them, Fig. 18.20. The interpretation is that the galaxy is moving through an intergalactic medium that is resisting its passage and is stripping the gas from the galaxy. Also there is a general, almost isotropic, soft x-ray background radiation. This could be due to the interaction of electrons with protons in a hot intergalactic medium.

18.12 Galaxy formation

Our knowledge of galaxy formation is in a very primitive state. A few points seem important; rich, dense clusters are dominated by giant elliptical galaxies while low-density clusters are dominated by spiral galaxies. It is thought that in dense clusters star formation was carried out efficiently and so all of the initial material was used up forming elliptical galaxies. In less dense regions star formation was less efficient, so elliptical galaxies were formed with gaseous material remaining. This material rapidly settled into the central plane perpendicular to the axis of rotation and spiral galaxies resulted.

The irregular galaxies and other galaxies are the result of events having nothing to do with the original process of galaxy and star formation. We have already noted that most of the peculiar galaxies are involved in some great internal explosion. Ring galaxies (Fig. 18.21) are the result of a direct collision between a small (not dwarf) elliptical galaxy and another galaxy. Irregular galaxies are formed from material either pulled out of a spiral galaxy by the chance encounter with another galaxy or from the gas left over when two spiral galaxies collide.

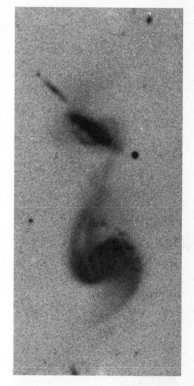

Figure 18.19 *A pair of interacting galaxies referred to as Arp 87. Note how the arms wrap around and the jet from the smaller member. (Hale Observatories photograph, courtesy of H. Arp.)*

Spectra of Galaxies

The spectra of galaxies are composites, as would be expected for assemblages of stars of various spectral types. The lines are also widened and weakened by the different radial velocities of the individual stars. Doppler effects in the spectrum lines show the rotations of galaxies and redshifts of the lines; the latter increases as the distances of the galaxies from us are greater.

18.13 Rotations shown by spectra

The flattened forms of regular spiral galaxies suggest that these galaxies are rotating. The character of the rotations may be determined from the spectra of spiral galaxies having their equators considerably inclined to the plane of the sky. When the slit of the spectroscope is placed along the major axis of the projected oval image of the spiral, the spectrum lines show Doppler displacements that depend on the direction of the rotation and its speed at different distances from the center of the spiral. V. M. Slipher was first to appreciate this technique and to apply it to the galaxies when he measured the rotation of the Sombrero galaxy.

As an example of such spectroscopic studies we use studies of M 33. The rotation period of the spiral M 33 is the same within 18′ from its center. In the outer parts, between 18′ and 30′ from the center, the period increases with increasing distance, as it does in the outer regions of our own galaxy. Thus the inner regions of spirals rotate as a rigid wheel whereas the outer regions are in Keplerian motion.

If we can correct our calculations for the projection effects of the galaxy we are measuring, we then know the velocity of the star's orbit at a certain distance from the center of the galaxy. If this distance is sufficiently removed from the center we can assume that the mass between it and the center is completely concentrated at the center. The above assumption means that we are dealing with Keplerian motion. Now if we assume a circular orbit we can obtain the mass of the galaxy. Galaxies range in mass from 10^9 to 10^{13} solar masses. The Milky Way galaxy has a mass of 2×10^{11} solar masses. Dwarf galaxies are considerably less massive.

A final question involves the direction of rotation of the spiral arms: Are the arms leading or trailing the rotation? Or, more simply, are the arms unwinding or are they winding up? Slipher's conclusion was that the arms are trailing and this conclusion was a point of controversy for several decades. His procedure was to determine the inclination of the spiral and then assign the closer arm by the arm having dust lanes close to and projected upon the nucleus of the galaxy under study. More refined techniques applied over the past 20 years have, in every case, verified

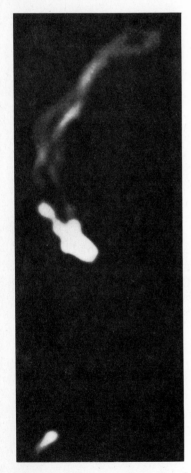

Figure 18.20 *A radiograph of the galaxy 3 C 129. This galaxy is located in a cluster of galaxies about 125 Mpc away, and the trailing gas is evidence of an intracluster medium. (Courtesy of J. Oort. G. K. Miley, and R. G. Strom, Sterrewacht Leiden.)*

Figure 18.21 *A ring galaxy designated ESO-034 IG 11. From the locations of the galaxies in the field and the shape of the ring, this galaxy was penetrated by the small elliptical galaxy to the upper left. (European Southern Observatory photograph.)*

Slipher's position and there is no doubt that the arms in spiral galaxies wind up. Thus the spiral M 33 (Fig. 18.4) would be expected to rotate in the clockwise direction.

18.14 Motions of galaxies

Besides the rotation of the spiral galaxies, each galaxy moves through space. Galaxies in clusters move about in the cluster and the cluster as a whole is moving with respect to other clusters. The random motion of any given galaxy or cluster of galaxies may be as high as 1000 kilometers/ second with respect to the Milky Way. That is, a galaxy may be approaching, or going sidewise, or going away from us with velocities as large as 1000 km/sec.

In addition, if we study the motions of galaxies at some reasonably large distance from us, say twenty million parsecs or so away, we find a systematic motion among the galaxies. The technique applied and the interpretation of it is the same as establishing the solar apex with respect to the distant stars. The distant galaxies seem to be drifting out of one region on the celestial sphere and towards the opposite point in the sky. The explanation for this is that the whole local group is moving in a particular direction. It happens that this direction is in the direction of the constellation Gemini and is 90° to the direction of the great Virgo cluster of galaxies. It appears that the local group is orbiting the Virgo cluster just as the earth orbits the sun.

Beyond these distances a general trend in the motions of the galaxies sets in. Everywhere we look the galaxies are receding from us.

18.15 Redshifts

As galaxies of similar types become fainter and smaller and presumably more remote, a spectral feature becomes more and more obvious— their spectral lines become more and more shifted to longer wavelengths, that is, to the red side of the spectrum. The redshift of the galaxies was discovered by V. M. Slipher in 1912 and is generally interpreted as a Doppler shift. We have already seen that this relation is given by the equation

$$\frac{\Delta\lambda}{\lambda} = \frac{v}{c},$$

where $\Delta\lambda$ is the shift in wavelength units from the normal wavelength λ, v is the velocity of the object, and c is the velocity of light. For velocities where v is much smaller than c, this simple relation holds to a high degree of accuracy. Thus a shift of 1 Å in the titanium line at λ 4300 Å corresponds to a velocity of 70 kilometers/second.

For large shifts in the spectrum this simple formula no longer holds. If the shifts are due to large velocities, we must appeal to the theory of relativity. A little manipulation places the special relativity equation in the form

$$\frac{\Delta\lambda}{\lambda} = \left(\frac{1 + v/c}{1 - v/c}\right)^{1/2} - 1.$$

We can immediately see the difference between the two relations by calculating a few values as has been done in Table 18.2. Astronomers often use the abbreviation $Z = \Delta\lambda/\lambda$.

For velocities of 30,000 km/sec or less, the differences between the results from the two relations are insignificant (Table 18.2). However for objects like the quasistellar object 3 C 9 with a measured Z of 2, the differences are quite large. The first formula would require a velocity of twice the speed of light, which means that we would never be able to see the source we are seeing.

An empirical law given by Hubble relates the redshift or, more correctly, the velocity of recession to the distance of the object,

$$v = H \times r.$$

Here v is the radial velocity of the source, H is the Hubble constant, and r is the distance. An approximate value of H is 100 km/sec/Mpc, and r is given in units of millions of parsecs. For our example of 3 C 9 we obtain a value for its velocity of 240,000 kilometers/second from Table 18.2 and from Hubble's relation we find the distance to be 2.4 billion parsecs, or about 8 billion light years. According to the Hubble relation a first approximation to the radius of the observable universe is 3 billion parsecs or 10 billion light years.

It is important to note the direct linkage between distance and time. When we observe a redshift that indicates an object is 10 billion light years away we are observing it as it was 10 billion years ago. Galaxies having velocities that are large with respect to the velocity of light are said to be at cosmological distances. We generally consider velocities of one-tenth the velocity of light and larger to be large with respect to the velocity of light. Studies of galaxies with large redshifts require great care in order to treat them in the same way that we do nearby galaxies. Their energy curves must be corrected for the redshift, for example.

Redshifts (Fig. 18.22) have been observed for many galaxies since Slipher's first observations. In addition, A. Sandage has completed a uniform set of observations in order to obtain the corrected visual magnitudes of galaxies. His diagram relating these magnitudes to redshifts (i.e., distances) for galaxies in many clusters is shown in Fig. 18.23. The points fall on a straight line, which agrees with the Hubble relation. Cluster 1410, the most remote of the points, corresponds to the quasistellar object 3 C 295 and has a velocity of recession of about 132,000 kilometers/second. Possibly other points could be placed on this diagram by including other quasistellar objects. However, the redshifts may not be due to velocity, especially those of certain quasistellar objects. In such cases we will have to find a new explanation of their redshifts.

Table 18.2

Velocity (km/sec)	$\Delta\lambda/\lambda$ First Order	$\Delta\lambda/\lambda$ Relativistic
30,000	0.1	0.105
60,000	0.2	0.225
90,000	0.3	0.363
120,000	0.4	0.528
150,000	0.5	0.732
180,000	0.6	1.000
210,000	0.7	1.381
240,000	0.8	2.000
270,000	0.9	3.359

Figure 18.22 *Spectra of galaxies arranged to show the linear relationship between redshift and distance. Each of the galaxies is a giant elliptical galaxy located in the cluster of galaxies named in the figure. The diagonal line shows how the amount of redshift increases proportionally with distance. (Hale Observatories photograph.)*

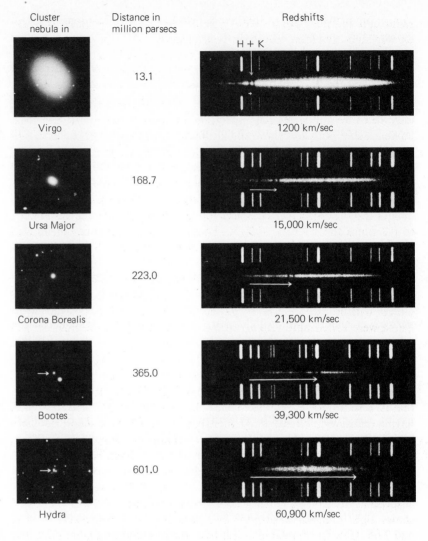

Cluster nebula in	Distance in million parsecs	Redshifts
Virgo	13.1	1200 km/sec
Ursa Major	168.7	15,000 km/sec
Corona Borealis	223.0	21,500 km/sec
Bootes	365.0	39,300 km/sec
Hydra	601.0	60,900 km/sec

18.16 Quasistellar objects

Most galaxies, like stars, become very interesting individually if studied carefully enough. Those that first attract our attention are apt to be abnormal. For example, the strong nonthermal radio source 3 C 8A was finally identified with NGC 1275, which when studied carefully showed high-excitation forbidden lines of oxygen, neon, carbon, etc. This was considered strange and peculiar until it was discovered that all Seyfert and N galaxies showed the same lines in the visual regions. They differed only in the radio and infrared regions. Many apparently normal galaxies can also exhibit high-excitation spectra in their centers.

In 1960 astronomers called attention to several radio sources with unusual characteristics. Despite increasing resolution the sources could not be resolved and their energy spectra seemed to have nonthermal components in the long-wavelength spectrum. Five of the objects were soon identified with optical sources. Spectrograms were obtained that displayed unidentifiable emission lines and a weak continuum with an excess of ultraviolet light. Finally, in 1963, M. Schmidt fit the emission lines of 3 C 273 uniquely to hydrogen if one assumes a redshift of the spectral lines corresponding to 0.16 times the velocity of light. Another such object 3 C 48 was found to have a redshift corresponding to 0.3 times the velocity of light and the object 3 C 9 has been found to have a redshift corresponding to 0.8 times the velocity of light. These objects show emission lines of ionized carbon, silicon, etc., and occasionally absorption lines of the same elements as can be seen in Fig. 18.24.

Soon many objects very similiar to the quasistellar objects were found. Many had no optical counterpart and were called quasistellar radio sources. Others had no radio emission, but showed large redshift emission lines. These were called blue stellar objects. All such objects are now lumped into a category called **quasar** or **QSO.**

The measured values of the redshifts of many quasistellar objects soon exceeded 1 and then even 2. The largest redshift observed up to late 1979 was for a quasistellar object called OQ 172 having $Z = 3.53$, that is, a redshift that is 91% of the velocity of light. Quasistellar objects having emission-line redshifts greater than about 0.7 show absorption lines and, in general, the absorption lines are very sharp. The absorption lines also yield redshifts, some differing quite markedly from the emission-line redshifts and always in the sense of smaller values. For example, the emission-line redshift of one object, PHL 938 is 1.955, whereas one set of absorption lines has a redshift of 1.906 and another of 0.613. PHL 957 shows eight different absorption line redshifts lying between $Z = 1.26$ and 2.66. How can the same object have three or eight different redshifts? How can the absorption lines differing by 1.3 be so sharp? The physics of this problem is too detailed for discussion in this text and always leads to results open to interpretations that remain contradictory.

Here are unfamiliar objects, all of which are far away if we interpret the redshift in the customary cosmological way. Furthermore, they have very small dimensions but great radio and optical power output, as Table 18.3 shows. Our knowledge of spectroscopy requires that the forbidden transitions originate in a high-temperature, low-density condition. The spectra are quite like those of planetary nebulae. The problem with this interpretation is that the QSO's are radiating enormous amounts of energy.

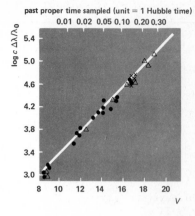

Figure 18.23 *Corrected magnitude* V, *which is a measure of distance, plotted against velocity for the brightest galaxy in various clusters of galaxies. (Diagram as presented by A. Sandage in* Galaxies and the Universe.*)*

Table 18.3
Energy Output of Sources

Type of Source	Radio Power (ergs/sec)	Optical Power (ergs/sec)
Normal galaxies	10^{38}	10^{44}
Radio galaxies	10^{43}	10^{44}
Quasars	10^{44}	10^{46}

Take 3 C 273 as an example. Its redshift of 0.16 implies that its distance is about one-billion parsecs. Therefore its distance modulus is 40. The apparent magnitude of 3 C 273 is about +13 so the absolute magnitude of 3 C 273 is about −27. The absolute magnitude of large galaxies is only about −21. Thus the problem: how can such a small, low-density object be radiating 250 times more energy than that of a whole galaxy?

There might be other explanations of these objects. H. Arp has pointed out that quasistellar objects and peculiar galaxies occur together (Fig. 18.25) with a frequency greater than that of a chance coincidence; therefore certain quasistellar objects are associated with peculiar galaxies that we know are not at extremely great distances. Therefore these redshifts are not cosmological. If this is so we must then find an explanation for the large redshifts of these objects.

Several hypotheses have been advanced. It has been argued that the redshifts are due to gravitational collapse of a large mass. However, this

Figure 18.24 *The large red-shifted (Z = 1.95) quasistellar object 3 C 191 showing emission and absorption lines of highly ionized elements. The absorption lines are extremely sharp. (Kitt Peak National Observatory photograph.)*

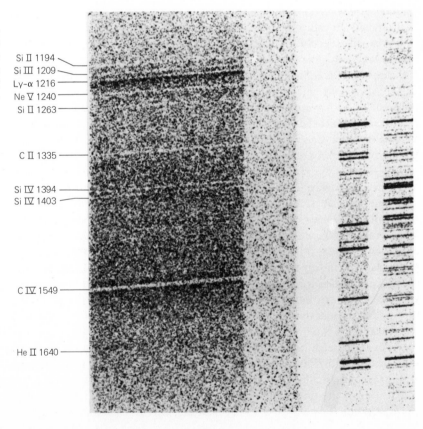

cannot be the case from energy considerations. In addition, this mechanism does not reproduce the sharp emission lines, so for the moment the gravitational argument does not appear promising.

H. Arp, who has gone further than anyone else in trying to find an empirical explanation, has suggested that a catastrophic event caused two highly condensed objects to be expelled from a galaxy in opposite directions. Such a catastrophic event would explain the peculiar galaxy well enough as well as the paired symmetry that Arp finds in his studies. However, it would appear that some of these condensed objects should be expelled more or less toward us and that we should see almost as many blueshifts as redshifts among the quasistellar objects. So far we have not observed any blueshifts. Also, energy considerations show that any event powerful enough to eject such objects would undoubtedly rend the entire galaxy asunder. Along the same lines, J. Terrell has proposed that the source of the objects was the nucleus of the Milky Way. We see all the objects redshifted because the event took place some 10^7 years ago and hence they have long since passed the position of the sun in the Galaxy. If this were the case some of the objects might exhibit proper motions, which they do not.

Also suggested by Arp is the possibility that the quasistellar objects are plasma balls formed well outside a galaxy and are now falling into the nearest galaxy. Such a cause could hardly exhibit the symmetry Arp finds. We should also see such objects blueshifted as well and in about equal numbers which, as pointed out above, we do not. All of which indicates that we really do not have an explanation for the quasistellar objects that stands up on all counts. In this discussion we have presumed that redshifts are caused by Doppler shifts.

A very recent piece of evidence supports the idea that the QSOs are at distances given by the Hubble relation. Two QSOs, located very close together, have the same colors and identical redshifts. It is thought that these are actually images of a single QSO formed by a gravitational lens. It appears that there is a giant elliptical galaxy located on the sky on a line between the QSOs, but the galaxy is very faint. The elliptical galaxy is very far away and in order for it to act as a gravitational lens the QSO must be even more distant. If this interpretation is true we have yet another proof of the general theory of relativity as well as proof that the QSOs are at cosmological distances.

Some QSO's appear to have wisps of material around them. The suggestion is that these are the central region of galaxies at a very early stage just after galaxy formation. If this is so then the energy problem is still to be explained as it is for some apparently related objects.

18.17 BL Lac objects

As the designation indicates (Section 13.7), BL Lacertae is the 70th variable star found in the constellation of Lacerta and it was so designated in 1929. Forty years later when it was noted to have strong emission at radio wavelengths interest in BL Lacertae renewed. The problem with BL Lac and other objects like it was that it did not have obvious emission lines like the QSO's. However there is a strong resemblence to the QSO radiation in the radio region.

BL Lacertae has a nonthermal radio spectrum and, like some QSO's, is quite variable in its energy output. In fact, the BL Lac objects are in general much more variable than QSO's. Recently, weak emission lines have been identified in the BL Lac spectra and it is clear that these objects are extragalactic and have redshifts intermediate between the normal and Seyfert galaxies and the QSO's. Detailed studies show that these objects are a small central condensation in an otherwise normal giant elliptical galaxy.

Our present knowledge of the BL Lac objects is that they form an intermediate object between the QSO's and other galaxies. This seems to say that the QSO's are galaxies at a rather early stage of formation. As they settle down they become highly variable and are BL Lac objects. Then they become Seyfert galaxies and then they become normal galaxies.

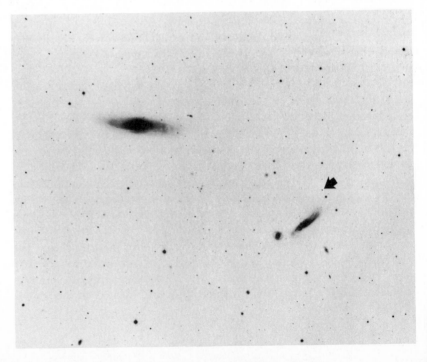

Figure 18.25 *A QSO is marked by the arrow and shows a true starlike appearance. Its Z value is 1.94. The nearby distorted galaxy is NGC 5682 with Z = 0.01. The strange ringed galaxy is Markarian 474 with Z = 0.04. (Hale Observatories photograph, courtesy of H. Arp.)*

18.18 Active galaxies

We have noted that radio galaxies, Seyferts, BL Lacertids, and QSOs show one form of activity or another. Only recently have we begun to notice that many galaxies that appear to be normal actually give evidence of activity on a time scale that is short when compared to the lifetimes of galaxies. The effect that is found most often is one that appears to be beams of material (Fig. 18.26) emanating from the center of a galaxy. Sometimes condensations are seen in the beams. In a few cases the condensations mimic the properties of QSOs.

The detailed study of galaxies, begun by V. M. Slipher, is actually still in its infancy. With the addition of numerous large telescopes covering all regions of the electromagnetic spectrum and powerful new astronomical spacecraft we can expect a rapid increase in this area of astronomy.

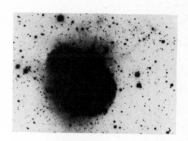

Figure 18.26 *The galaxy NGC 1097 showing beams to great distances from its center. (Hale Observatories photograph, courtesy of H. Arp.)*

Questions

1 What observation did Shapley rely on during the great debate to conclude that spiral nebulae were nearby? What observation caused Curtis to take a contrary position?

2 What are the three basic Hubble classes of galaxies?

3 List the observed differences between elliptical and spiral galaxies.

4 This chapter presents two fundamental observations about the universe; can you figure out what they are?

5 Why were galaxies originally called extragalactic nebulae?

6 List the steps used to establish distances to galaxies.

7 What is the largest galaxy in the Local Group?

8 What information about galaxies can be obtained from observing their rotation?

9 How can a redshift be greater than 1?

10 What do we think the BL Lac objects are?

Other Readings

Disney M. J., and P. Véron, "BL Lacertae Objects," *Scientific American* **237(2)**, 32–39 (1977).

Groth, E. J., P. J. E. Peebles, M. Seldner, and R. M. Soneira, "The Clustering of Galaxies," *Scientific American* **237(5)**, 76–96 (1977).

Larson, R. B., "The Origin of Galaxies," *American Scientist,* **65** 188–196 (1977).

Strom, R. G., G. K. Miley, and J. Oort, "Giant Radio Galaxies," *Scientific American,* **233(2)**, 26–35 (1975).

van den Bergh, S., "Golden Anniversary of Hubble's Classification System, *Sky & Telescope,* **52,** 410–414 (1976).

chapter 19

Cosmogony

Cosmogony is the study of the origin and evolution of the universe. Redshifts of galaxies increase with distance, which is interpreted as an expanding universe. This, plus a knowledge of the distribution of various types of galaxies and objects in the universe, allows an interpretation of the type of universe we live in and its age. To accomplish the necessary observations we must make use of instruments in space and intensify our ground-based efforts.

Theories of the Universe

Anyone wishing to discuss cosmogony in a rational way must begin from some certain well-defined rule. This is similar to baseball where we decide in advance that a fair ball hit beyond a certain distance on the fly is declared a home run, counting as one run (although it may result in as many as four runs). Imagine the chaotic results if you could run around the bases until the original ball was retrieved? So it is in cosmogony; we begin with the rule that the cosmological principle must be satisfied. The **cosmological principle** states that the universe must look the same at a given time to an observer located anywhere within the universe. Some propose a much more stringent condition. This condition, called the **perfect cosmological principle,** requires that the universe look the same to any observer regardless of time.

It goes almost without saying that we reject any theory that does not begin from some reasonable rule and make predictions that can be tested. We note that accepting either principle as our starting point has an immediate consequence—the universe must be infinite. It may be infinite and closed (finite expansion) or infinite and open (continuous expansion forever), but it must be infinite. If the universe were not infinite observers near the "edge" or boundary of the universe would see a different universe than those located elsewhere and this is not allowed by either of our ground rules.

19.1 Observations and Olbers' paradox

Prior to 1950 there were a few basic observations of the universe from which we had to deduce its origin and evolution: the universe is expanding; the universe on a large scale is isotropic and homogenous; the universe contains approximately 75% hydrogen, 25% helium, and a trace of deuterium and these abundances are remarkably uniform. These three observations apply to the universe only on a very large scale. We have already noted that individual galaxies may have random motions as large as 1000 kilometers/second. Thus the **cosmological expansion,** or **Hubble flow** as it is called, only becomes evident and reliable at distances greater than 20 megaparsecs. By the same token the universe is homogeneous only if volumes larger than 100 megaparsecs in radius are considered; then the universe is remarkably uniform and may be treated as a uniform-density fluid.

Cosmogony has developed slowly from the time of Copernicus although Thomas Digges made a major advance only 33 years after Copernicus published his heliocentric theory. Copernicus correctly ordered the known members of the solar system, but placed the stars on a distant celestial sphere. Digges proposed that the universe was infinite and that

Figure 19.1 *An old print of H. W. M. Olbers. (The Bettmann archive.)*

the stars were uniformly distributed throughout the infinite universe. This gave rise to a paradox first recognized and discussed by Kepler, but it is now referred to as Olber's paradox after H. W. M. Olbers (Fig. 19.1) rediscussed it in 1823.

The basic problem is that it is dark at night. However, in an infinite universe any line drawn outward to the infinitely distant celestial sphere must intercept a star somewhere and therefore the entire sky must glow as bright as the average star. The entire infinite universe must be bathed in an infinite amount of light. The fact that the sky is dark at night is often cited as a fundamental cosmological observation.

There are ways around this problem. Kepler's solution was to assume the universe is finite and not very large. Cheseaux's solution was to assume there is a pervasive medium through which light travels, causing its brightness to decrease faster than the inverse square law predicts. A fairly recent solution is to assume the universe is finite but unbounded (like the surface of a balloon). The most recent solution of a similar nature claims the answer lies in the expansion of the universe.

All of these explanations fail. A detailed discussion of the failing of each suggestion cannot be undertaken in this text but an explanation of the paradox can be given if we assume that the universe is young.

Let the observer be standing in a large forest where each tree has an average diameter of d and occupies an average area of A. Then in any direction that you look you will intercept a tree at an average distance of A/d. Suppose each tree is 40 centimeters (0.4 meter) across and occupies a square 10 meters on a side or 100 m². Thus the average look-out distance is 250 meters. The number of trees that this observer sees is $\pi A/d^2$ or 1963 trees.

The case for stars is analogous, only we must consider it in three dimensions. Let V be the volume occupied by a star and let a be the area of an average star's disk. The look-out distance is just V/a and the total number of stars is $4\pi V^2/3a^3$. In an infinite universe the number of stars is infinite but the observer sees only a finite number of them. If all stars on the average are like the sun the look-out distance is 10^{23} light years and the total number of stars seen is 10^{60}. Herein lies the explanation.

In order to fill the sky we must look out to a distance of 10^{23} light years. However, stars only live on the order of 10^{10} years so any stars beyond 10^{10} light years cannot yet be seen. The stars out to 10^{10} light years are only contributing $10^{10}/10^{23}$ or 10^{-13} the required amount of light to make a bright sky.

19.2 Model universes based on general relativity

Early models of the universe were produced by Einstein and others from Einstein's theory of general relativity. Einstein's early model (1915) was the first really successful attempt to produce a model of the universe.

stein believed that the uni-
expanding (or contracting)
just the required universe.
odel of the universe from
ing (there was no mass to
eresting because it allowed
and independently G. La-
ivity allowed for a matter
wed that Einstein's static
ncluded that the universe
om a dense origin or from
n a dense origin are called

he cosmological principle
s expanding. However, at
ubble constant was large;
so small that there was a
n the universe. To under-
is in (kilometers/second/
s are measures of length
e ($H = 1/T$). As a first
write $T = 1/H$. Suppose
ere are about 3×10^{13}
kilometers in a megapar-
e as 6×10^{16} seconds or
n to be much older than

anized his universe model
ent state, and then slow
expand to infinity. This
was so contrived that H. Bondi and T. Gold, later joined by F. Hoyle, developed a model that would take them around the age problem.

19.3 The steady-state universe

Bondi, Gold, and Hoyle proposed starting from the perfect cosmological principle. Since the universe is expanding, but must look the same to all observers at all times (Fig. 19.2), it must therefore have a constant density. As galaxies move apart new galaxies must form; that is, matter must be created from which new galaxies form.

This theory makes certain predictions: every large portion of the universe looks like every other large portion of the universe and matter is being created all of the time all over the universe. There is no progressive

Figure 19.2 *A comparison of the universe as depicted by the big-bang and steady-state theories. In the steady state the universe looks the same all of the time.*

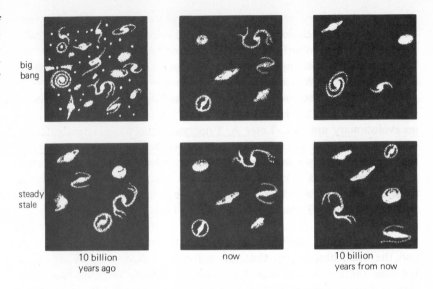

evolution of one large part of the universe compared to any other large part.

Some astronomers objected to this theory on the grounds that matter (hydrogen) is being created in little bits all of the time. In fact so little is created at a given time that the detection of the matter being created is impossible. Such an objection is not valid; there is no philosophical difference from matter being created continuously, or having the matter created at some time in the distant past or even all at once in the distant past.

We will see that this steady-state model fails because of relatively recent observations. All of the models discussed so far run into a fundamental difficulty with the abundances of the elements: there is too much helium and deuterium in the universe. The steady-state theory has less trouble with the amount of helium, but cannot explain the observed free deuterium since in the steady-state model deuterium would exist mainly as a transient element deep in the interiors of stars.

19.4 The hot big-bang cosmology

In 1948, R. Alpher, H. Bethe, and G. Gamow (note the spoof since papers are often referred to by the last names of the authors) proposed a general relativity model that began with the universe in a very hot dense state; the whole universe was energy. Something disturbed this energy or **primeval atom** and a violent event (explosion) took place. The event is called the **primeval fireball.**

As soon as the primeval fireball occurred the universe began to expand and cool. As any gas expands, it cools—much as the high pressure CO_2

at room temperature in a fire extinguisher forms a frost as it expands out of the cylinder. As the energy gas cooled, it first formed exotic particles. These particles tried to form stable particles but because of the intense radiation any stable particles immediately returned to the exotic state and radiation. After a few seconds the energy gas cooled to the point where neutrons, protons, and electrons formed. Protons and electrons are stable, but neutrons are not. Before the next 12 minutes passed the neutrons combined with some of the protons to form deuterium. The deuterium combined with protons and deuterium to form isotopic helium and finally normal helium. Some deuterium remained, but after 12 minutes the remaining neutrons decayed into protons and electrons. The initial atom-building stage, **nucleosynthesis,** was complete. Further nucleosynthesis would have to take place mainly deep in the interiors of future stars.

The universe continued to expand and cool uniformly for some two million years, the radiation and elementary matter all being at the same lowering temperature. After this time the expansion was sufficient such that the matter retained some heat while the radiation continued to cool. This is called the **decoupling** of matter and radiation; from this point on the radiation would continue to cool as an expanding gas while the matter was free to interact and form stars and galaxies (Fig. 19.3). Detailed calculations show that at the decoupling time there should have been about 78% hydrogen, 22% helium, and a trace of deuterium. Stars and galaxies must form much later, but this theory accounts for the large amount of helium and the trace of free deuterium.

Gamow and his collaborators pointed out a significant observational test of the model. The model predicts (in addition to the expansion of the universe and the relative abundances of hydrogen and helium) that since the radiation decoupled from matter and has been cooling ever since, there should be a general radiation field pervading the universe and observable now at a temperature characteristic of the expansion of the universe.

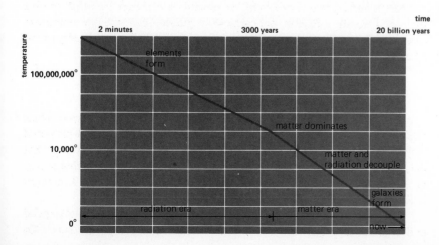

Figure 19.3 *How the universe cooled with time according to the big-bang theory. The various epochs are noted.*

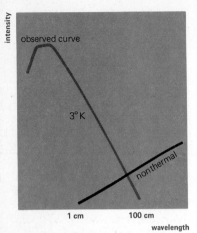

Figure 19.4 *A schematic 3 K radiation curve bounded on the long-wavelength side by nonthermal galactic noise.*

Based on the then deduced age of the universe, Gamow predicted a general isotropic radiation characteristic of a 5 K black-body radiator.

19.5 The 3-K background radiation

Until the early 1960s there was not much to recommend either the big-bang or the steady-state theory. However, the change in the zero point for the period–luminosity relation had removed the age problem associated with the big-bang theory. At this time P. J. E. Peebles reworked the big-bang model for the temperature of the isotropic radiation in terms of a newer calculation of the age of the universe based upon a Hubble constant of 100. The new value was 3 K. At that time the Bell Telephone Research Laboratories were researching sources of noise (shades of Jansky's discovery of cosmic radio radiation) at their Echo satellite receiving station. With this work complete, the equipment was made available to the staff's scientists.

In 1965 A. Pensias and R. Wilson observed an isotropic background radiation from space. Tedious, detailed studies revealed a radiation curve characteristic of a 2.7 K black-body radiator (Fig. 19.4). Further studies have confirmed this discovery which validates the hot big-bang model of Gamow and his associates. It was a remarkable confirmation of a prediction made 15 years earlier. The general relativity models and, in particular, the hot big-bang model must be on the right track.

19.6 Radio source counts

The number of radio sources, assuming uniform density, should be proportional to the volume investigated, which is proportional to the cube of the radius. In a sphere two times larger the number of sources is 2^3 more than in the first volume. The intensity *(S)* of the radiation from sources decreases as the square of the radius, that is, it is inversely proportional to the square of the radius. Hence in simple three-dimensional space, the number *(N)* of sources should be inversely proportional to the three-halves power of the intensity, or

$$N \propto S^{-3/2}.$$

Thus if we plot the logarithm of *N* against the logarithm of *S*, a straight line of slope $-3/2$ results, if we are in a simple constant universe. When distances become large in an expanding universe our relation becomes

$$N \propto S^{-3/2+\alpha},$$

where α is always positive, but depends upon the model of the universe selected. Hence the slope can never be larger than $-3/2$, or -1.5. At very faint (weak) intensity levels, the slope of the line appears to be −1.8 (Fig. 19.5), which substantiates the argument against a constant density of radio sources. There are too many sources in the distant past. The universe is evolutionary.

19.7 Other considerations

We have not made any use of the ubiquitous quasistellar sources. If they are cosmological, they clearly indicate an evolutionary universe and rule out steady-state models. Many astronomers will not commit themselves, but most recent research concludes that the redshifts of the quasistellar objects are cosmological in origin. The upper limit on their redshifts is 3.5, which represents about 91% of the age of the universe. Thus the quasistellar objects were "turned on" well after the universe was formed. They apparently represent a critical stage in the history of the universe. Why this stage is not closer to a redshift of 1000, which is the redshift when the matter decoupled from the radiation, raises new questions for us to answer.

The hot big-bang model leaves us with another problem. When particles are formed, by whatever method one chooses, antiparticles are formed. Regular particles are often referred to as matter and antiparticles as antimatter. Thus when an electron is formed, its antiparticle—a positron— is also formed. When the universe began, why was it formed primarily of matter? Where is the antimatter? These are unanswered questions.

Also why should a universe appear suddenly, constantly evolve and change, and yet obey fixed unchanging laws of physics? It is just as valid to expect an evolving, changing universe to obey changing laws of physics.

19.8 The expanding universe

The pressing question now is how to devise tests to see if the universe is open or closed; i.e., will the universe expand forever or will it expand to a limiting size and then fall back? The tests are very difficult but at present two lines of evidence seem to favor an open universe. There is no evidence that the universe is decelerating and the density of the universe seems to be too low to close the universe.

Work is continuing on the determination of the deceleration parameter and the density of the universe. The problems associated with determining the deceleration factor are good examples of this field of astronomy. First we plot the distances of galaxies against their redshifts (Fig. 19.6). If the most distant galaxies lie above a straight line, meaning that the expansion was more rapid in the past, we can say that the universe is slowing in its expansion. If the opposite is true then the universe is accelerating and is open.

The difficulty lies in determining the distances of galaxies, especially

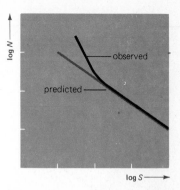

Figure 19.5   *A schematic source count diagram. The number of sources at weak flux levels, hence very distant, is too large for a steady-state universe.*

Figure 19.6 *Magnitude–velocity diagram similar to the one in Fig. 18.24 with various acceleration curves q$_0$ drawn in. If q$_0$ is +1 or larger the universe is slowing down and is closed. If this is not the case the universe is open. (Diagram by A. Sandage, in* Galaxies and the Universe.*)*

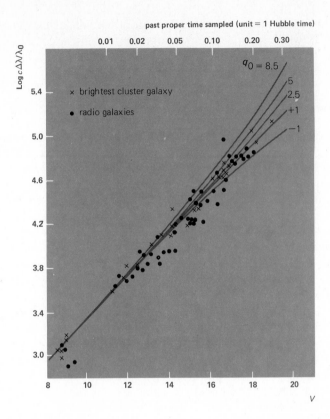

those very far away. Galaxies do not have the same magnitudes and their brightnesses and colors change with time—hence distance determinations change. The magnitudes must be corrected for absorption and other factors. The problem is so complex that there is no clear-cut answer, but there is a tendency to interpret the current information in favor of an open universe.

A second line of reasoning involves the density of the universe. Let the critical density of the universe be 1. This is the dividing line; on one side of this line there is enough density to close the universe, while on the other side the universe is open. At exactly 1 the universe expands to infinity but at an ever slower rate. In Fig. 19.7 we have plotted the density against the Hubble constant. Constraints on this diagram are possible; the youngest possible age for the universe is 8 billion years and the oldest perhaps 20 billion years. For each density we can calculate what value of *H* is required to give the observed universe if the youngest and oldest ages are used. In this way we get an upper and lower envelope (Fig. 19.8). The universe must lie somewhere between these two curves. While the region is very constrained, the universe may still be open or closed.

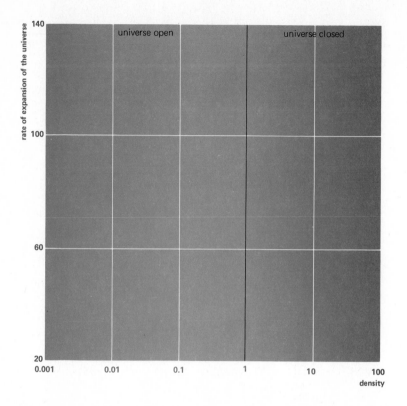

It is possible to constrain the diagram even further by considering the amount of deuterium present. The deuterium line is a gradually curving line (Fig. 19.9). To the right of this line there is not enough observed deuterium. This approach leaves a more restricted region where the universe may exist and the entire region lies in the open part of the diagram. Thus two independent lines of study seem to indicate the universe is open.

A natural question that arises from our observation that the universe is expanding is—where is the center? In fact, it appears that we are at the center. We must be careful here, for throughout history we have considered ourselves unique and at the center; we should not make this mistake again. In a universe obeying general relativity, every observer seems to be at the center; all are at the center and none are at the center. A very simple example of this is shown in Fig. 19.10. Selecting any dot as our galaxy we will find that the expansion is such that it yields a Hubble law. Every dot observes itself to be at the center.

In any case, at no point except the existence of the primeval atom have we invoked any arbitrary conditions for the evolution of the universe. This is also true for the origin and evolution of the stars and planets. Is this also true for the origin and evolution of life?

Figure 19.8 *A plot of the expansion rates for given densities for the youngest and oldest possible universe. This leaves only a narrow curving band as the region where the universe can lie. It still can be open or closed.*

Figure 19.9 *(Opposite page) A plot in which the amount of deuterium agrees with observation. To the right of the line the density of deuterium is far too low. This leaves a very restricted region where the universe must lay and it is entirely in the open portion.*

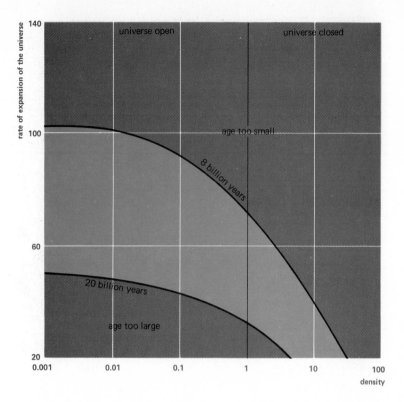

Origin and Evolution

The cosmogonical theories, in general, assume a uniform, isotropic, expanding universe. They do not really allow for the irregularities that become galaxies and clusters of galaxies, let alone stars and life.

19.9 The origin and evolution of galaxies

Galaxies exist, therefore, they originated somehow. We do not know how. Some astronomers assume an early instability, perhaps during the formation of the elements a few minutes after the universe began. These instabilities remained and eventually became detached clouds that condensed into galaxies, or, more likely, clusters of galaxies. Once galaxies exist we can then follow with reasonable certainty their evolution—the formation and evolution of stars, etc. Galaxies evolve very slowly.

The three basic types of galaxies must be related to their early history (just as with stars), and the only conceivable difference would be their initial rotation. If the rotation is small, an elliptical galaxy results; if the rotation is intermediate, a spiral galaxy results; and if the rotation of the cloud is high, an irregular galaxy forms.

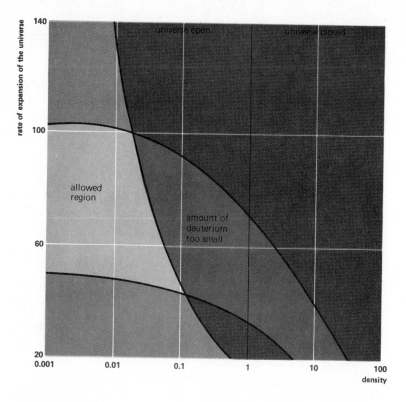

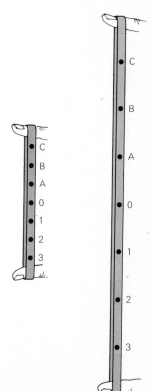

Figure 19.10 *A schematic explanation of the Doppler redshift in galaxies. The points on a rubber band (left) move away from each other uniformly as the rubber band is stretched. Point 1 has moved 3 times farther away from the arbitrary origin (O) during the time of stretching. The same is true for point 3; thus point 3 has had to move 3 times faster than point 1. This is true for any and all points with respect to all others. Thus, any point on the rubber band will see itself as the origin and the points farthest from it will be moving the fastest.*

In a galaxy stars will be formed, evolve, and return material to the galaxy; new stars will form, evolve, and so on. Near any one of these myriads of stars, planets may form, and life may exist.

19.10 The conditions of life

We do not really propose discussing in great detail all of the aspects of the origin of life. In this context we refer to life as a branch of carbon chemistry and assume that we are interested in "intelligent" life after establishing the chemistry.

We have seen how stars are formed. A cloud that forms stars always forms many stars, most in multiple systems but some as single stars. These single stars have a large revolving cloud in their plane of rotation with a cosmic composition that includes the interstellar molecules. Some planets can form from this cloud; just how and how many is a matter of opinion at the moment. The number can be anything equal to or greater than zero.

In the case of the solar system we are fairly certain the terrestrial planets formed by accretion; the gaseous planets may be original condensations or may also have formed by accretion. We know that the earth is about 5×10^9 years old and that 3.6×10^9 years ago a great event occurred, perhaps the capture of the moon, which may or may not be important to life. Life on earth began after this event, perhaps 3×10^9 years ago, thus the evolution of life requires something of the order of 3×10^9 years.

This sequence of events can occur anywhere in the universe. There is no reason to believe that the sun and its history are unique in any way. Of course the planet must be reasonably hospitable wherever life begins. If the initial step occurs by chance, then the conditions should be such that many chances take place. Temperature and pressure must not change too greatly nor too rapidly on a short time scale. For a given set of conditions those life forms which adapt most readily survive. The evolution of life forms will naturally converge toward the most efficient forms possible.

One of the conditions only recently appreciated is the role of the fundamental constants of physics. If these constants are varied only slightly our carbon chemistry will not work, the region of variation is highly restrictive. In a Dirac-type universe, which supposes that the constants are changing, life would exist only during the phase when the physical constants were balanced just right.

We cannot pursue this point since it is beyond the scope of the book. A hint must suffice. The fundamental constants are the coupling constants for electromagnetic interactions, gravitational interactions, and the strong and weak nuclear interactions. Very small changes in these constants cause significant changes in the charge on the electron that greatly affect the chemistry of the amino acids (subunits of proteins), for example. If the charge becomes too small the amino acid bonds break; if the charge becomes too large the amino acids crystallize (Fig. 19.11). Thus the constants must be constant over a period of time that is long compared to 3 billion years.

Another argument involves the fine structure constant. This constant determines the splitting of atomic energy levels. Spectral lines originating in the very distant past, 8 billion years if the quasistellar objects are cosmological, show the same splitting as they do in our laboratories today. We conclude, therefore, that the constants of physics are indeed constant.

19.11 Does life exist elsewhere?

This intriguing question has no answer at the present time. If you ask this question of any person the answer will merely reflect the personal bias of the person. All astronomers can do is exhaustively determine whether the conditions for life exist elsewhere. A sufficient proof for life elsewhere in the universe would be the reception of an understandable communication. The communication may be as simple as radar pulses.

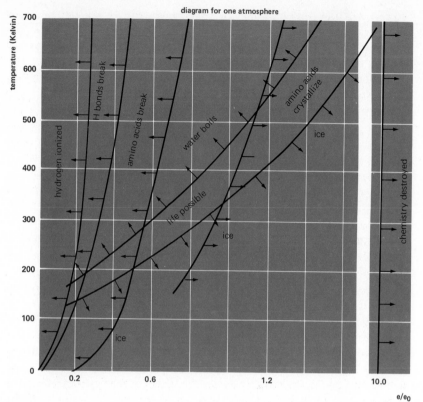

diagram for one atmosphere

Figure 19.11 *A schematic diagram plotting the ratio of the charge of the electron, e, to its present value, e₀, against the absolute temperature. The permitted region for our type of chemistry is restricted. (Diagram courtesy of W. Saslaw.)*

This is not an easy proof. Communication techniques change quite rapidly; thus the development of a civilization must reach the right stage at the right time and right distance or communication will not occur.

On the other hand, failure to receive a communication does not prove that life does not exist.

What are the conditions for life? Briefly summarized they are as follows:

1. The proper central star, nonvariable, neither too energetic nor too feeble—an *F* through *K* main-sequence star will do.
2. A planet with water located at the proper distance from its star, the size and mass being within 50% of that of the earth.
3. Time—at least 3×10^9 years since the solidification of the planet.
4. A universe where the fundamental constants are constant.

Then it is the task of the chemist and biologist to tell us if life will begin.

One of the exciting challenges for astronomy is to seek out those places where life may exist. The existence of stars with planet-sized companions, the existence of billions of stars to choose from, the existence of organic molecules in interstellar space all argue that life somehow resembling life on earth is highly probable most anywhere in the universe. If this is so can we learn about it or can it learn about us?

It seems likely that our only hope of detecting extraterrestrial life is through communication by means of electromagnetic radiation. We can conceive of an earthlike planet essentially covered with water and whose life forms are fish. We can even conceive of a plant covered planet without evolved animals. We cannot conceive of communicating with the life on such planets so our discussion immediately narrows down to extraterrestrial intelligent life in the sense that it has the technology to communicate with us. This is not to say that intelligent life on other planets will be like us; not at all. All we are saying is that they can build and operate things like radios, computers, and the like.

19.12 Where to search

Outside the earth's atmosphere the universe has a broad noise minimum centered roughly on the microwave region at a wavelength of 3 centimeters (Fig. 19.12). Intelligent beings will know this and concentrate

Figure 19.12 *A plot of wavelength versus noise temperature outside the earth's atmosphere. Transmitting where the noise is least is the most efficient use of power.*

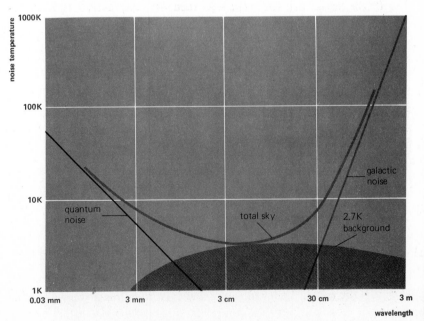

their efforts where the chances of success are most likely. It is argued that we should search near the hydrogen 21 centimeter line because intelligent beings will be interested in the galaxy and its spiral structure. Others have argued that searching anywhere in the "noise window" will yield success just as well.

We can try to actively contact extraterrestrial life or we can do it passively by listening. We are actually doing both; the former inadvertently, the latter by design.

The earth is leaking microwave radiation at very high levels from several major sources. There are some 2000 high-energy television transmitters located mainly in North America, Europe, Japan, and Australia. There are perhaps two dozen very high powered early-warning radars. The signals from all of these sources are concentrated towards the horizon (Fig. 19.13), so the earth's microwave signature is not simple and is dependent upon the declination of a distant receiver. Someone receiving our television signals would "see" the U.S. rise in power as it comes over the horizon, and drop in power, and then rise again as it sets on the other horizon (Fig. 19.14).

We cannot go into details, but an intelligent people could learn a lot about the earth without being able to decode our television signals. They could learn the periods of rotation and revolution, the size of the earth, the fact that the transmitters are concentrated in certain areas, the surface temperature, and many other things. To do this they would need a very large receiving telescope.

If the earth can be detected, then we can detect civilizations

Figure 19.13 *TV, FM, and high-powered radar signals are beamed toward the horizon. The transmission cone may or may not intercept a distant observer as the earth turns.*

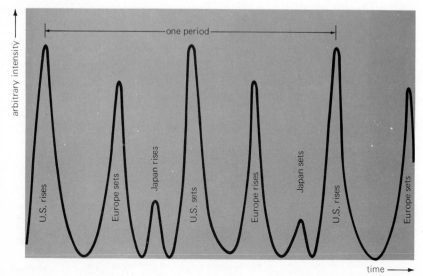

Figure 19.14 *A hypothetical and simplified reception graph for a distant observer receiving earth microwave radiation in a direction of the earth's equatorial plane. [Adapted from a more complete diagram by W. T. Sullivan III, Science 199, 377 (1978).]*

on other worlds. It is proposed that we build a huge listening system composed of 1000 radio telescopes each 100 meters in diameter (Fig. 19.15). This is being called **project Cyclops** and would be a tremendous undertaking. Why listen if you do not communicate?

Communicating has problems that involve how a technological society evolves and we do not know the exact details. Within just 20 years the earth has become very noisy in the microwave region, far outshining the sun. It is likely that with the development of more communications satellites and fiber optic cables in another 20 or 30 years the earth will become radio quiet. Thus, except for the very powerful military radars, there is only a brief period when it is possible to hear a technological society by passive listening techniques, if all such societies follow our example.

If a civilization is 200 light years away it will have had to reach its noisy microwave period 200 years ago in order for us to detect it now. If we wished to answer, and we do not have the capability to do so now, another problem arises. It takes our answer 200 years to reach that other civilization and by then that civilization may have discarded radio communications for a better technique. After all, 400 years will have passed.

Figure 19.15 *An artist's rendition of the multiple telescope array proposed for project Cyclops.*

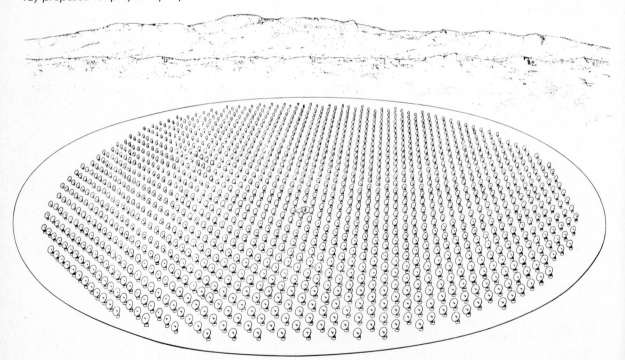

19.13 Active communication efforts

There is a very good sign of optimism here on earth in so far as contacting an extraterrestrial civilization goes. While the possibility of contact and communication is small, active efforts are being made and that augurs well for mankind. While active radio efforts have been mostly accidental except for one or two "this is what we can do" efforts, there have been several attempts at sending evidence of our presence here on earth.

Every major spacecraft that is destined to leave the solar system carries a plaque designed to inform any interceptor of where and when the spacecraft was launched. The plaque on the Pioneer spacecraft is shown in Fig. 19.16. A schematic of the spacecraft behind the human figures gives an idea of the size of the people who launched it. The scale is given also in terms of the hydrogen 21-centimeter radiation which is shown on the far right and is tied to the hydrogen atom depicted in the upper left. The solar system and its coded dimensions in terms of the hydrogen 21-centimeter radiation as a standard length are shown at the bottom. The schematic spacecraft is depicted as leaving the third planet, earth, and passing the fifth, Jupiter.

The star-burst-type rendition in the middle left shows the directions and periods of certain pulsars. This tells the interceptor the location of the solar system and the epoch of launch. The latter is obtained from

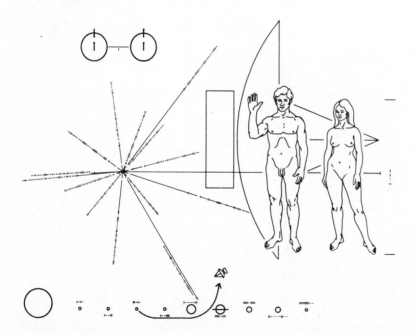

Figure 19.16 *The plaque carried aboard Pioneers 10 and 11. If discovered by other civilizations, it is hoped that they will decipher it as explained in the text.*

the fact that the periods of pulsars gradually lengthen. The interceptor can take the periods for the time of intercept and work back to where the periods would be as short as those shown.

A similar plaque (Fig. 19.17) is on the Voyager spacecraft. These spacecraft also carry simple tape recorders with various "sounds of the earth."

With every adventure there is a serious argument against any such efforts. This argument is forceful and must be heard. Our existence on earth is in a delicate balance. We should not attract attention because any visitors (an unlikely event in view of the distances, i.e., time, involved) may have germs that could wipe us out (of course the opposite may occur as was so forcefully written by H. G. Wells in his "War of the Worlds"). Finally, we should assume that we are relatively ignorant on a cosmic scale and any revelations of this ignorance to other beings only invites them to come and subjugate the earth to their needs.

Figure 19.17 *The plaque carried aboard Voyagers 1 and 2. It conveys the same information as the Pioneer spacecraft plaques. The Voyagers also carry tape recordings. (National Aeronautics and Space Administration photograph.)*

Astronomy from Space

The continuing advances in our capability to place large stabilized payloads in orbit around the earth has led astronomers to employ this useful tool to press forward the frontiers of astronomy. Space astronomy can be classified into three areas: (1) solar research, (2) planetary research, and (3) stellar research, where the term stellar is used in the very broadest sense. Also, the three areas overlap in many ways as would be expected from a frontier science.

19.14 Solar research

The sun is known to be quite variable in the ultraviolet and x-ray regions of the spectrum. This can be deduced from the effects of these wavelengths upon our atmosphere. The origin of this radiation is from high-temperature processes that are little understood. Hence, a detailed study of the sun in the high-energy region can be most fruitful. Also, a detailed study of solar flares and their triggering mechanism is necessary.

A start in this study has been made with the various Russian, American, and European satellites, such as OSO (Orbiting Solar Observatory). Additional information can be obtained from radiation probes scattered around the sun in positions far distant from the earth. After all, the earth samples only a small portion of the radiation emanating from the sun. Interplanetary probes, such as the Voyager (Fig. 19.18) and Venera series

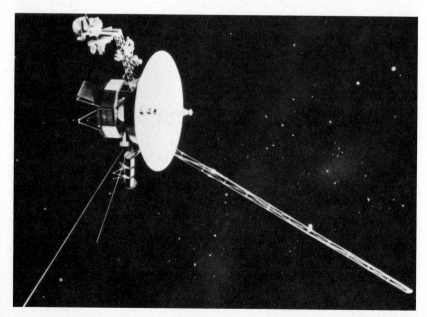

Figure 19.18 *An artist's rendition of the fully deployed Voyager spacecraft. It carries equipment to monitor the interplanetary medium. (National Aeronautics and Space Administration photograph.)*

are starting such studies. Soon there will be probes sent far out of the plane of the solar system.

19.15 Planetary research

In addition to studying radiation from the sun and the solar magnetic field, interplanetary probes allow astronomers a chance to sample interplanetary space itself. The size and mass of particles can be directly ascertained. Dramatic advances can be expected in our knowledge of the material making up the interplanetary medium.

Even more spectacular is the opportunity for obtaining close-up information of the planets. The recent photographs of Mars and Jupiter are examples of the ability of space probes to assist our studies by eliminating certain hypotheses from further consideration. We can expect similar results, possibly more exciting results from probes to other planets.

The continued torrents of information from Mars by Vikings 1 and 2 is providing work for years to come. Pioneer 10, after obtaining its marvelous pictures of Jupiter, is on its way out of the solar system, probing for the point where the interstellar and interplanetary mediums interact. Pioneer 11, after its encounter with Jupiter, has now encountered Saturn. Voyagers 1 and 2 have completed their missions to Jupiter to follow up on the studies begun by Pioneers 10 and 11. Hopefully, after encountering Saturn the Voyager spacecraft will be redirected to study Uranus, Neptune, Pluto, and maybe the comets before leaving the solar system completely.

Perhaps with these new tools, we can find the answer to the origin of comets. Is there a tie between the remnant solar nebula where the comets originate and the interstellar medium nearby? Perhaps such information will bring us closer to discovering the origin of the solar system and even the origin of the Galaxy.

19.16 Stellar research

Study of the interstellar medium is already in progress with the Orbiting Astronomical Observatory (OAO) series (Fig. 19.19). This study and its bearing upon cosmology has already been mentioned. From measurements of the absorption of ultraviolet radiation we will be able to deduce the total mass of interstellar hydrogen. Similar observations of high-temperature stars will have a bearing upon the interaction of such stars on the interstellar medium surrounding them (Fig. 19.20).

The search for x-ray point sources is very important because of their bearing on black holes. Fortunately, recent x-ray satellite experiments have turned up some 130 galactic and extragalactic x-ray sources, a number of which are point sources. A search for γ-ray sources in similar experiments is underway with very promising early results.

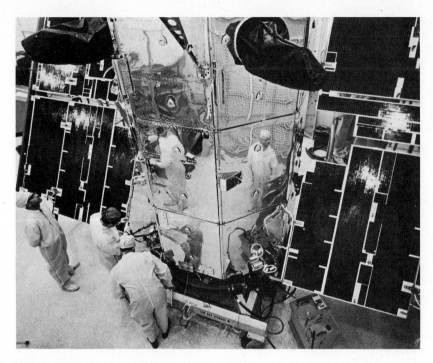

Figure 19.19 *The highly successful spacecraft Copernicus (OAO-C) during prelaunch checkout. This spacecraft is dedicated to the study of the interstellar medium. (National Aeronautics and Space Administration photograph.)*

At the other end of the electromagnetic spectrum, fruitful areas of research in space are open. The atmosphere cuts off radiation at a wavelength longer than 30 meters. Strong sources of low-frequency radio radiation would be nonthermal in origin. We can already predict that the Crab nebula will be such a source. A detailed study of this nebula coupled with ground-based observations will shed a great deal of light on the energy sources operating in it.

There is no purpose served in rating the results from the various space probes; any such list would be subjective. The effect of space technology upon mankind is undeniable. One example is the revolution in communications between people. It is also revolutionizing astronomy.

Removing the spectacular aspects, the lunar sampling missions have provided answers to long-standing questions and, of course, raised new ones. The various missions to Mars, particularly Vikings 1 and 2, have completely changed many of our ideas concerning Mars. The highly successful x-ray satellites UHURU, HEAO-A, HEAO-B, and HEAO-C have opened a whole new area of study, answered a host of questions, and raised many more. And so the story goes on—Venus, the sun, and the interstellar medium have all been studied. Answers have been found and questions have been raised.

During the past few years, some 30 gamma ray sources have been

discovered. A few are identified with known objects. We expect significant advances in the understanding of sources emitting these very high energy photons.

Further in the future we can anticipate the use of high-resolution, perhaps even diffraction-limited telescopes, some of which will be manned (Fig. 19.21). Such telescopes would add information to every area that we have studied so far. For example, a large space telescope would resolve many galaxies into stars and allow a detailed study of their contents. The variable stars in these galaxies could be studied and more reliable distances obtained. A telescope with such capabilities would open up vistas undreamed of, even now. In fact, all space research can be expected to yield unexpected results and new problems for study.

Astronomy from the Earth

In order for the full potential of the space astronomy effort to be realized, there must be a similar effort on the ground. Certain essential observations, especially those pertaining to tests of cosmological theories, require large telescopes and a considerable amount of time. More large- and intermediate-sized telescopes are needed to make further progress.

19.17 Optical research

As exciting as space astronomy is and will prove to be, ground-based astronomy is at least its equal. Space astronomy is relatively expensive,

Figure 19.20 *(Opposite page) Nebulae are interstellar gas illuminated by hot stars as in M 16. (Kitt Peak National Observatory photograph.)*

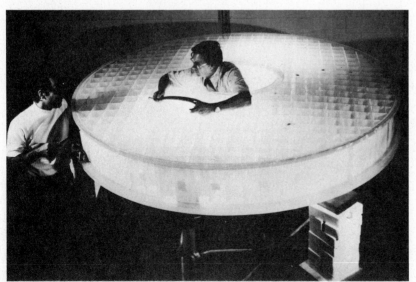

Figure 19.21 *The Space Telescope mirror being inspected by Perkin–Elmer engineers. The core, a lightweight structure made of small glass plates cemented in a vertical arrangement, is sandwiched between two glass plates, each about 3 centimeters thick. This results in a total weight of only 750 kilograms which is less than 10% the weight of a normal 2.5-meter mirror. (Perkin–Elmer photograph.)*

so we must use it wisely and apply it where other approaches are unavailable. We have already discussed one example where space astronomy can be applied to the specific task of looking at variables in galaxies unresolved from the ground. The integrated magnitudes for these space-resolved galaxies can be obtained cheaply from great telescopes on the earth and the statistical studies of fainter unresolved galaxies carried out. Thus, one should make maximum use of the two capabilities.

Another example is in the study of the masses of stars. A thorough study of the motion of an unresolved astrometric binary from the ground yields only the mass function. A few observations of the resolved system, coupled with the ground-based study, would quickly yield the separate masses.

In those areas where light-gathering power is the criteria, such as spectroscopy and photometry, ground-based work will press forward with newer and many more large telescopes over the next few decades. A 25-meter multiple-mirror telescope has been proposed and is under serious consideration. The success of the 7-meter multiple-mirror telescope (MMT) (Fig. 19.22) points the way. The detailed study of stellar spectra, populations, and variability will go on at an ever-increasing pace. Even the study

Figure 19.22 *The Multiple Mirror Telescope on Mount Hopkins. This telescope is composed of six 1.8-meter mirrors with an equivalent collecting aperture of a 4.5-meter telescope. The astronomer leaning on the telescope shows the scale. Much larger MMTs are being planned. (Smithsonian Institution Mt. Hopkins Observatory photograph.)*

of astrometry (parallaxes, motions of stars, time and fundamental reference systems) promises new and interesting advances. The success of the U.S. Naval Observatory's 1.5-meter (Fig. 19.23) reflector in obtaining the parallaxes of stars down to the 16th magnitude argues well for this branch of study. The establishment of an absolute reference frame by using external galaxies is another example of work best carried out from the surface of the earth.

The life cycles of stars can be used to predict the evolution of a galaxy. The low-mass red stars may live 2×10^{10} years and in the beginning of the life of a galaxy contribute very little to its brightness and color. The galaxy would be dominated by the light of its massive, bright, blue stars. As evolution proceeds fewer massive, bright, blue stars are born. As these stars rapidly evolve and burn out, the brightness of a galaxy must diminish until, finally, it is dominated by the light of the faint red stars. Such ideas as this can be tested within the next 100 years. However, before the above ideas can be fully tested, we must understand how a star changes in the H-R diagram as it evolves. To do this there must be thorough observational studies of all types of stars in minute detail. This means large telescopes where coudé plates can be taken of faint stars (Fig. 19.24). Only now are we starting to make a beginning toward understanding the peculiar A stars and we have not even scratched the surface of the variable-star problem. The observations must be obtained so that the theoreticians will have a reasonable footing for their work.

Figure 19.23 *A modern astrometric reflector dedicated to finding the distances of the nearby stars. (Official U.S. Navy photograph.)*

Figure 19.24 *The great Mayall 4-meter reflector showing the prime-focus cage (top left) and the Cassegrain cage (bottom right). The observer rides in either cage with the equipment being used. (Kitt Peak National Observatory photograph.)*

Another most interesting problem of high priority is a thorough explanation of the quasistellar sources already discussed. The key to the explanation may well lie in the determination of their true angular size, a problem that may tax the imagination of astronomers for decades to come. Most observations needed for testing cosmological consequences can be carried out from the ground. Intermediate- and large-sized telescopes are needed to carry this work forward. Just as important, of course, is the need for trained astronomers to use these instruments and interpret the results.

19.18 Radio research

This relatively young branch of astronomy is enjoying a healthy growth that will go on over several decades, and the search for weak sources will go on with more and more refined equipment. The study of known sources over a greater spectral range should intensify as techniques develop.

The desire for a more detailed study of the structure of sources is leading to the rapid refinement of radio interferometry. Presently coming on line is the Very Large Array (VLA) composed of nine 25-meter telescopes spaced along each 21-kilometer arm of a huge Y for a total of 27 telescopes in all (Fig. 19.25). This telescope will yield high-resolution radiographs (Fig. 19.26) of many curious high-energy radio sources. Increasing use of very-long-baseline interferometry is also certain. The trick here is to tape record signals received by two distant telescopes and to bring the tapes to a common laboratory for analysis. The tapes can be synchronized by inserting time marks from previously synchronized atomic clocks (Section 2.16) as the recordings are being made. Such experiments by

Figure 19.25 *An aerial photograph of the VLA nearing completion. Compare this with the artist's rendition in Fig. 4.17. (National Radio Astronomy Observatory photograph.)*

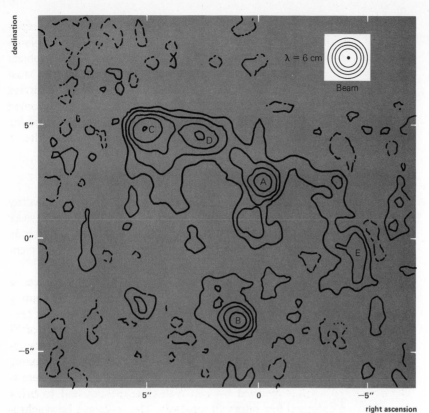

declination

5″

0″

−5″

5″   0   −5″

right ascension

Figure 19.26 *A very high resolution radio intensity map made using the VLA of the double quasar 0957+561. The strong intensity peaks are labeled A,B,C,D, and E. C-C-A-E appear to be connected and B is a separate source. Originally the double quasar was thought to be due to a gravitational lens such as a black hole. (Courtesy of D. H. Roberts, P. E. Greenfield, and B. F. Burke.)*

Canadian and National Radio Astronomy Observatory astronomers have been successful. These recent experiments have shown that many quasistellar objects have diameters of less than 0.01 second of arc. Similar experiments using intercontinental baselines are becoming a matter of routine.

19.19 Gravity research

According to the general theory of relativity certain objects release gravitational radiation that distorts the local geometry as it moves away from its source. The most likely sources of gravitational radiation are very-short-period binary stars, the collapse phase of the supernova event, and rapidly rotating neutron stars.

Efforts to detect the distortion due to the passage of gravity waves have been under way since 1969. In the very near future highly sensitive new detectors are to be in operation. The successful detection of gravitational radiation will herald the onset of a whole new era of astronomical discovery.

19.20 Theoretical research

There are two factors that have shaped and accelerated theoretical
investigations in astronomy, namely, the rapidly increasing number and
quality of observations and the development of modern computers and
techniques (Fig. 19.27). In the first factor we include the results of physicists
in determining the lifetimes of energy states and the like. In the latter
we include the calculation of important nuclear cross sections and other
long, laborious calculations.

The carrying through of the calculations of the declining era of a
star was impossible prior to the development of modern computer technology. We can look forward to even more exact computations of model
stellar atmospheres, interiors, and evolutionary tracks.

19.21 Summation

The oldest of the sciences is presently experiencing a growth that is
unparalleled in all its long history. No one denies that this growth was
accelerated with the coming of the space age, but the results have merited
the growth. The discovery, study, and interpretation of the quasistellar
radio sources, not to mention pulsars, are good examples.

The excitement and promise of modern astronomy is breathtaking to behold. Perhaps the most recent comparable period in this science was during the early part of this century when V. M. Slipher discovered the receding galaxies and H. Shapley discovered the true extent of the Milky Way. These great steps had been preceded by the H-R diagram, the discovery of the period–luminosity relationship, and fundamental work in atomic physics, both theoretical and experimental.

We are now witnessing major advances in nuclear physics and have before us the discovery of the quasistellar sources. We can look forward to the new surprises and expanding knowledge that always accompany major developments.

Questions

1. What is meant by an open or closed universe?
2. What two lines of reasoning lead us at present to think that the universe is open?
3. Which two observations cause us to favor hot big-bang models of the universe?
4. Outline the requirements for the development of life and some of the supporting observations.
5. What is meant by the term "the earth's microwave signature"?
6. What principle would allow an extra-solar-system civilization to determine the rotation and revolution periods of the earth from the earth's microwave radiation?
7. What feature of a large telescope can be taken advantage of in space that is sacrificed on the ground?
8. Why is the discovery of x-ray point sources interesting?
9. How does earth-based astronomy support space astronomy?
10. What was Thomas Digges' contribution to cosmology?

Other Readings

Callahan, J. J., "The Curvature of Space in a Finite Universe," *Scientific American* **235(2)**, 90–100 (1976).

Geller, M. J., "Large Scale Structure of the Universe," *American Scientist* **66(2)**, 176–184 (1978).

Linsley, J., "The Highest Energy Cosmic Rays," *Scientific American* **239(1)**, 60–70 (1978).

Malkin, M. S., "The Space Shuttle," *American Scientist* **66**, 718–723 (1978).

Overbye, D., "The X-ray Eyes of Einstein," *Sky & Telescope* **57**, 527–534 (1979).

Penzias, A. A., "The Riddle of Cosmic Deuterium," *American Scientist* **66(3)**, 291–297 (1978).

Smarr, L. L., and W. H. Press, "Our Elastic Spacetime: Black Holes and Gravitational Waves," *American Scientist* **66(1)**, 72–79 (1978).

Spitzer, Jr., L., "The Space Telescope," *American Scientist* **66(4)**, 426–431 (1978).

Glossary

A

Acceleration   a change in the velocity or direction of a moving body.

Achromat   a compound lens of two or more simple lenses corrected to remove chromatic aberration over a broad region of the visual spectrum.

Albedo   the reflectivity of a body compared to the reflectivity of a perfect reflector of the same size, shape, orientation, and distance. Usually given as a percentage.

Almanac   a yearbook containing astronomical tables of positions of celestial bodies, sunrise, sunset, moonrise and moonset, etc. The American Ephemeris and Nautical Almanac are the almanacs most used by astronomers in the United States.

Apastron   that point in the true relative orbit farthest from the primary star.

Aperture   technically the clear diameter of a telescope, but often the diameter of the principal element or of the entrance pupil.

Apex   usually the solar apex, the direction toward which the sun is moving. The apex of a cluster's motion is either the convergent point or the divergent point.

Aphelion   the farthest point from the center of the sun in a solar orbit.

Apogee   the farthest point from the center of the earth in an earth orbit.

Apparant   as observed or measured.

Apsides   the two points where the major axis of an orbit intersects the orbit.

Asteroid   a minor planet.

Astration   the processing of material through stars.

Astrometry   the measurement of stellar and planetary positions.

Astronomical unit   the mean radius of the earth's orbit.

Astrophysics   the study of the physical properties of the sun, planets, and stars. It includes solar physics and planetary physics.

Atmosphere   the low-density, gaseous sphere surrounding the earth and various planets. Also the gaseous layers lying above a star's photosphere.

B

Balmer limit   the atomic transition from the second electron orbit of hydrogen to the continuum. The confluence of the Balmer lines near this point in a stellar spectrum.

Binary (binary star)   a pair of stars orbiting each other by their mutual gravitational attraction.

Bipolar   having two poles. A magnetic field having both a plus pole and a minus pole is said to be a bipolar magnetic field.

Bolometer   a device to measure infrared radiation.

Bolometric magnitude   the magnitude assigned to a body when the total energy from far-ultraviolet to long radio waves is used.

Bright points   small regions of intense x-ray radiation on the sun.

C

Carbonaceous chondrites   the oldest solar system material.

Catalyst   an agent that assists or speeds a reaction with no change in itself in the end product. Carbon is a catalyst in the carbon–nitrogen cycle.

Celestial mechanics   that branch of astronomy that deals with the orbits and motions of planets and stars.

Centrifugal force   the tendency of a body in motion to move away from a centrally restraining force.

Cepheids   intrinsic variable stars with periods between 1 and 60 days.

Chromosphere   that part of a stellar atmosphere lying just above the photosphere. It is about 15,000 kilometers thick.

Circumference   usually the measured boundary of a plane figure.

Cislunar   the region between the earth and the moon.

Coelostat   a stationary telescope looking at a flat mirror that is in a polar mount. The drive is arranged to keep the sun or a star field fixed in the telescope.

Collimator   optical lens (lenses) and/or mirror (mirrors) designed to render light in parallel rays. A telescope is a reverse collimator.

Color excess   the difference between the measured color of a star and the normal color for a star of its type.

Color index   the difference in magnitudes between intensities measured in two different colors. More precisely, the difference between visual and photographic magnitudes.

Conjunction   two celestial bodies on the same longitude. Inferior conjunction occurs when a planet or body passes between the earth and sun. Superior conjunction occurs when a planet is on the opposite side of the sun from the earth.

Constellation   a historical grouping of the bright stars. Constellations are useful for indicating specific portions of the celestial sphere.

Convection   in astronomy, the transfer of energy by the bodily transport of a hot plasma to a cooler region where it can cool by radiation.

Coplanar   lying in one plane.

Coronagraph   a telescope so arranged as to block out the bright disk of the sun in order to photograph the corona.

Cosmogony   theories and hypotheses dealing with the origin of the Universe.

Cosmology   the study and measurement of the Universe.

Coudé focus   a fixed focus in a room in line with and south of the polar axis.

D

Declination   the astronomical coordinate measuring the position of a body above (+ or N) and below (− or S) the celestial equator. It is always given in circular measure (e.g., degrees).

Density   (1) the mass of a body divided by its volume (units are usually grams per cubic centimeter); (2) the number of objects per unit volume, such as the number of G stars per cubic parsec.

Deuterium   an isotope of hydrogen having twice the mass of hydrogen.

Distance modulus   the apparent magnitude of a star minus its absolute magnitude, $m - M$. The term is derived from the distance relation: $m - M = 5 \log r - 5$.

Diurnal   an event taking place each day.

E

Eccentricity   the measure of the degree of flattening of an ellipse. The letter e is generally used in astronomy to denote this parameter.

Eclipse   a celestial body passing between the observer and another body.

Ecliptic   the path the sun seems to follow and, therefore, where solar eclipses tend to occur. Actually the plane of the earth's orbit projected to the celestial sphere.

Elongation   the angular distance from the center of motion, generally a planet's angular distance from the sun.

Ephemeris   a table of predicted positions. Generally a table of predicted positions as contained in the American Ephemeris and Nautical Almanac.

Ephemeris time   the time interval of one second as based upon the earth's annual orbit in 1900. Clocks using this second are keeping ephemeris time.

Epoch   a precise date used for referencing an astronomical observation or event.

Equinox   a point where the sun's path (ecliptic) crosses the celestial equator. When the sun is moving north it occurs around 22 March and is called the vernal equinox; when the sun is going south it occurs around 22 September and is called the autumnal equinox.

Erg   the amount of work done by a mass of 1 gram moving 1 centimeter.

Escape velocity   the velocity needed to escape the gravitational attraction of a body. It is equal to the parabolic orbital velocity.

Evection   slight irregularities in the moon's motion caused by the sun and planets.

F

Fluorescence   the emission of light at a longer wavelength than that of the absorbed radiation.

Focal length   the distance between the objective and its focus when looking at parallel incoming light.

Focus   the point where incoming radiation is imaged by an optical system.

Forbidden transitions   changes in energy levels not normally allowed of an atom or molecule.

Fundamental stars   a set of carefully chosen stars distributed over the whole sky whose positions are carefully measured and which serve as positional standards for other stars and objects.

Fusion   (1) the nuclear process joining atoms together to form heavier elements and releasing energy; (2) the point of change from solid to liquid.

G

Galaxy   (1) a great assemblage of billions of stars, gas, and dust gravitationally restrained; (2) when capitalized, the Milky Way galaxy is implied.

Galactic wind   the outflow of material in a galaxy.

Gamma radiation   that region of the spectrum shortward from a wavelength of 8 Å.

Geiger counter   an electron tube that discharges each time a charged particle passes through it. It is used by astronomers to discover cosmic rays, X rays, and gamma rays.

Geodesy   the science dealing with measuring the size and shape of the earth.

Geomagnetic storm   Disturbance of the earth's ionosphere and magnetic field by an unusual influx of charged particles from the sun.

Grating   (1) bars across the objective causing diffraction of the light. This would be called an objective grating. (2) A ruled grating, usually blazed, used to diffract the light for spectroscopic purposes.

Gravity   the term given to the attractive force between two masses.

H

Harmonic law   Kepler's third law.

Heliostat   a coelostat used to observe the sun.

High-velocity stars   stars with high radial velocities with respect to the sun. In reality their large radial velocities reflect the high velocity of the sun with respect to them.

Hubble diagram   a plot of the distance of galaxies against their redshift.

Hydrocarbon   compounds of carbon and hydrogen; methane, CH_4, is important to the gaseous planets.

I

Igneous rocks   rocks formed from molten materials.

Inertia   the property of matter to remain in its current dynamical state unless acted upon by an outside force.

Infrared radiation   that region of the spectrum extending from a wavelength of 7500 Å to about 1 millimeter.

Insolation   solar energy received by the earth.

Interferometer   any instrument that recombines electromagnetic radiation with a known change in phase.

International Astronomical Union (IAU)   the international organization of astronomers.

Ionosphere   the shell of atmosphere lying between 70 kilometers and 320 kilometers above the surface of the earth. It contains a high density of ionized atoms and particles.

J

Julian Day (Julian date)   day numbers beginning at noon and starting at an arbitrary date B.C. Used especially by variable-star observers.

K

Kiloparsec   1000 parsecs, equivalent to 3200 light years.

Kinetic energy   energy due to mass and motion ($\frac{1}{2}mv^2$).

Kirkwood gaps   zones of avoidance in the asteroid belt.

L

Leap year   that year every four years when an extra day is added to the month of February to keep the civil calendar in step with the sun. Century years not divisible by 400 are not leap years.

Luminescence   visible glow of a material induced by invisible radiation.

M

Magnetometer   an instrument for measuring a magnetic force.

Main sequence   the concentrated band of stars in a luminosity—temperature (H-R) diagram.

Mass   the property of a body that resists a change in motion. It is a unique property and, in the body's own reference frame, remains constant regardless of where the body is located.

Megaparsec   1 million parsecs, equivalent to 3,200,000 light years.

Metal   in astronomy, any element heavier than helium.

Microwave   short radio waves; their wavelengths lie between 1 centimeter and 1 meter.

Minor planet   any of thousands of small bodies in orbit around the sun with mean orbital radii between the distances of Mars and Jupiter.

Molecule   two or more atoms chemically united forming the smallest unit possessing the properties of a compound.

Monochromatic   literally, one color. Usually light of a sufficiently narrow wavelength that can be treated as having a single wavelength.

N

Nadir   the direction opposite the zenith.

Nebula   (1) a cloud of gas surrounding a star; (2) an interstellar gas cloud.

Nightglow   the same as airglow.

Node   the intersection points of an orbital and an arbitrary plane, such as the plane of the sky.

O

Objective   in astronomy, the main lens or mirror of a telescope.

Oblate   a circle or sphere flattened at the poles.

Observatory   the astronomer's laboratory where the telescopes are mounted.

Occultation   a larger astronomical body passing between the observer and a smaller body.

Opacity   a measure of the blocking of radiation.

Opposition   a planet is in opposition when it, the earth, and the sun are in a line.

P

Parallax   the apparent change in location of a body due to a change of position of the observer.

Parameter a measurable quantity characteristic of a system to be described.

Parsec a parallactic second, equivalent to 3.26 light years.

Penumbra (1) the partial shadow in a solar and lunar eclipse; (2) the gray outer regions of a sunspot.

Periastron the closest point to the center of the primary star in a stellar orbit.

Perigee the closest point to the center of the earth in an earth orbit.

Perihelion the closest point to the center of the sun in a solar orbit.

Photosphere that level where the atmosphere of the sun becomes opaque. The visible surface of the sun.

Planet a satellite of a star, in nearly circular orbit, which was never capable of shining by a self-sustained energy reaction.

Plasma an ionized gas.

Platonic year the time required for the earth to complete its precessional cycle—approximately 25,800 years.

Polarizer any substance that polarizes light or radiation passing through it.

Precession (1) the slow drift of the poles of a spinning body, due to a force or set of forces acting to try to tip its poles. (2) In astronomy, when unqualified, it usually means the change in coordinates due to the precession of the earth's poles.

Primary mirror the main mirror of a reflecting telescope.

Prism a wedge-shaped piece of glass used to disperse light into its familiar rainbow colors in spectroscopy.

Proper motion the cross motion of a star as seen projected on the sky and always given in seconds of arc (usually per year but occasionally per century).

Protostar the central region of a condensed cloud that is about to become a self-luminous star.

Q

Quadrature a planetary elongation 90° east or west of the sun.

R

Radial velocity the line of sight velocity of an object given in kilometers per second.

Radiant usually refers to the point from which a meteor shower appears to be coming.

Radiation (1) the processes of emitting energy; (2) often used in terms of the energy emitted.

Radio astronomy　that branch of astronomy dealing with the study of the cosmos at radio wavelengths.

Relativity　generally one of two theories by A. Einstein. The special theory treats time and distance as depending upon the motion of the object and the observer. The general theory relates the structure of space to gravitation.

Reticle　cross-hairs or finely etched reference lines at the focus of an eyepiece.

Retrograde　the apparent westward motion of a planet through a star field.

Right ascension　the astronomical coordinate measured along the celestial equator eastward from the vernal equinox. It is usually given in units of time but occasionally in angular measure.

S

Satellite　(1) any body that orbits a planet; (2) also, any inferior body that orbits a larger body.

Secondary mirror　usually the mirror at the front end of a reflector with the purpose of changing the direction of the light reflected to it by the primary mirror.

Sedimentary rocks　rocks formed by deposition, either by settling out of water or by precipitating out of a solution.

Seeing　the atmospheric effects on an image as seen through a telescope.

Solstice　times when the sun appears to stand still in the annual north and south motion.

Spectrogram　a photograph of a spectrum produced by a spectrograph.

Spectrograph　an instrument that renders light parallel upon a dispersing agent (prism or grating) and then photographs the resulting spectrum.

Spectroheliograph　a special instrument that photographs the sun in monochromatic (one-color) light.

Specular reflection　reflection, as from a mirror.

Spicule　small grass-like jets at the surface of the sun.

Superluminal　apparently faster than the speed of light.

T

Telescope　a somewhat misnamed instrument used by astronomers to collect radiation from celestial objects.

Telluric lines　lines in a spectrum arising from atoms and molecules in the earth's atmosphere.

Thermocouple　a junction of two wires of different metals that changes resistance when the amount of heat falling upon it changes.

Totality   commonly used during total solar and lunar eclipses to denote the total phase of the eclipse.

Transit   (1) a smaller astronomical body passing between the observer and a larger body; (2) a celestial body crossing the observer's meridian; (3) a telescope mounted in such a way that it can observe an object only when that object crosses the local meridian.

Transverse   lying across. Transverse vibrations are vibrations across the direction of motion or perpendicular to the line of sight.

Triple alpha process   a nuclear process, first discussed by E. Salpeter, where three helium nuclei combine directly to form a carbon nucleus.

Turbulence   irregular, random motions in a fluid or gas.

U

Ultraviolet radiation   that region of the spectrum extending from a wavelength of 100 Å to that of about 4000 Å.

Umbra   (1) the black shadow in a solar or lunar eclipse; (2) the dark center of a sunspot.

Universal time   the local time of the $0^h0^m0^s$ meridian (Greenwich, England).

Universe   the totality of physical reality.

X

X-radiation   that region of the spectrum extending from a wavelength of 8 Å to one of 100 Å.

X-ray astronomy   a new branch of astronomy using the x-ray region of the electromagnetic spectrum and made feasible by the use of orbiting spacecraft.

Z

Zodiac   the band centered on the ecliptic divided into 12 equal sections. Each section was assigned a constellation and sign by the ancients.

Appendix

ENGLISH-METRIC CONVERSION UNITS

The principal advantage of the metric system over the English system is that the metric system is based upon powers of ten. Any powers-of-ten system is as good as any other, but the metric system has the advantage that it has been adopted by more people than any other single system.

1 inch = 2.54 centimeters
1 foot = 30.48 centimeters = 0.3048 meter
1 yard = 91.44 centimeters = 0.9144 meter
1 mile = 160930 centimeters = 1609.3 meters = 1.6093 kilometers
1 ounce = 28.3495 grams = 0.0283 kilogram
1 pound = 453.6 grams = 0.4536 kilogram
1 pint (fluid) = 47.32 centiliters = 0.4732 liter
1 quart (fluid) = 94.64 centiliters = 0.9464 liter
1 kilometer = 0.6214 miles
1 meter = 1.0936 yards
1 centimeter = 0.3937 inch
1 liter = 2.1134 pints (fluid)
1 gram = 0.0353 ounce

Brief table of decimal multiples

Decimal	Power Notation	Prefix	Symbol
0.001	10^{-3}	milli-	m
0.01	10^{-2}	centi-	c
0.1	10^{-1}	deci-	d
1	10^0		
10	10	deca-	da
100	10^2	hecto-	h
1,000	10^3	kilo-	k
1,000,000	10^6	mega-	M
1,000,000,000	10^9	giga-	G

Powers-of-ten notation is extremely convenient, as can be ascertained from the last line of the table. The rules for multiplication and division in this notation are simple. For multiplication we have $10^a \times 10^b = 10^{a+b}$, thus $10^2 \times 10^5 = 10^7$. For division we have $10^c \div 10^d = 10^{c-d}$, thus $10^2 \div 10^5 = 10^{-3}$.

Greek alphabet

A	α	alpha	I		iota	P	ρ	rho
B	β	beta	K	κ	kappa	Σ	σ	sigma
Γ	γ	gamma	Λ	λ	lambda	T	τ	tau
Δ	δ	delta	M	μ	mu	Υ	υ	upsilon
E	ϵ	epsilon	N	ν	nu	Φ	ϕ	phi
Z	ζ	zeta	Ξ	ξ	xi	X	χ	chi
H	η	eta	O	o	omicron	Ψ	ψ	psi
Θ	θ	theta	Π	π	pi	Ω	ω	omega

Great reflecting telescopes (> 3 meters)

Year	Observatory and Location	Objective (meters)
1976	Zelenchukskaya Astrophysical Observatory, USSR	6.0
1948	Hale Observatory, Mt. Palomar, California	5.1
1979	Smithsonian Astrophysical Observatory, Mt. Hopkins, Arizona	4.4[a]
1973	Kitt Peak National Observatory, Kitt Peak, Arizona	4.0
1974	Cerro Tololo Inter-American Observatory, Cerro Tololo, Chile	4.0
U.C.[b]	United Kingdom Infrared Telescope, Mauna Kea, Hawaii	3.9
1975	Australian National Observatory, Siding Spring Mtn., Australia	3.8
1975	European Southern Observatories, Cerro La Silla, Chile	3.6
U.C.[b]	French, Canadian, Hawaiian Observatory, Mauna Kea, Hawaii	3.5
1959	Lick Observatory, Mount Hamilton, California	3.0
1979	NASA Infrared Telescope, Mauna Kea, Hawaii	3.0

[a] Multiple mirror telescope, six 1.8-meter mirrors.
[b] Under construction.

Great refracting telescopes (>65 cm)

Year	Optician	Observatory and Location	Objective (cm)	Focal Length (cm)
1897	Alvan Clark	Yerkes Observatory, Williams Bay, Wisconsin	102	1935
1888	Alvan Clark	Lick Observatory, Mt. Hamilton, California	91	1760
1893	Henry Brothers	Observatorie de Paris, Meudon, France	83	1615
1899	Steinheil	Astrophysikalisches Observatory, Potsdam, Germany	80	1200
1886	Henry Brothers	Bischottsheim Observatory, University of Paris, at Nice, France	76	1600
1914	Brashear	Allegheny Observatory, Pittsburgh, Pennsylvania	76	1411
1894	Howard Grubb	Royal Greenwich Observatory, Herstmonceux, England	71	850
1878	Howard Grubb	Universitäts-Sternwarte, Vienna, Austria	67	1050
1925	Howard Grubb	Union Observatory, Johannesburg, South Africa	67	1070
1883	Alvan Clark	Leander McCormick Observatory, Charlottesville, Virginia	66	1000
1873	Alvan Clark	U.S. Naval Observatory, Washington, D.C.	66	990
1953[a]	McDowell	Mount Stromlo, Canberra, Australia	66	1100
1897	Howard Grubb	Royal Greenwich Observatory, Herstmonceux, England	66	680

[a] First used in Johannesburg, South Africa, 1926.

A Brief Chronology of Astronomy

ca. 3000 B.C.	The earliest known recorded observations are made in Babylonia
ca. 1400 B.C.	Earliest known Chinese calendar
ca. 1000 B.C.	Earliest recorded Chinese, Hindu observations
ca. 800 B.C.	Earliest preserved sundial (Egyptian)
ca. 500 B.C.	Pythagorean school advances concept of celestial motions on concentric spheres
ca. 430 B.C.	Anexagoras explains eclipses and phases of the moon
ca. 400 B.C.	Philolaus speculates that the earth moves
ca. 400–300 B.C.	Several cosmological systems involving moving concentric spheres proposed by Plato, Eudoxus, and others
ca. 350 B.C.	Earliest known star catalog (Chinese)
ca. 250 B.C.	Aristarchus advances arguments favoring a heliocentric cosmology
ca. 200 B.C.	Erathosthenes measures earth's diameter
160–127 B.C.	Hipparchus develops trigonometry, analyzes generations of observational data, obtains highly accurate celestial observations
ca. A.D. 140	Ptolemy measures distance to moon; proposes geocentric cosmology involving epicycles
1054	Chinese observe supernova in Taurus
1543	Copernicus publishes *De Revolutionibus* with the heliocentric theory
1572	Tycho Brahe observes supernova; immutability of celestial sphere cast in doubt
1546–1601	Tycho accurately measures motions of the planets
1608	Lippershey invents the telescope
1609	Kepler, using Tycho's measurements, shows planets move in ellipses
1609	Galileo uses telescope to observe moons of Jupiter and crescent phase of Venus, thus lending support to Copernican hypothesis; Galileo conducts experiments in dynamics
1675	Romer measures the velocity of light
1686–1687	Newton's *Principia:* Newton combines the results of terrestrial and celestial natural philosophy to obtain the fundamental laws of motion and gravity
ca. 1690	Halley shows the great comets observed every 75 years are one and the same comet in an elliptical orbit; he discovers proper motions of stars
ca. 1690	Huygens makes estimate for distance to stars based upon assumption that the sun is a typical star
1727	Bradley observes aberration of starlight, conclusive proof of Copernican theory
ca. 1750	Wright proposes disk model for Milky Way
1755	Kant proposes nebulae are "island universes"; proposes solar system formed from rotating cloud of gas
1738–1822	W. Herschel constructs large telescopes; discovers Uranus (1781); observes gaseous nebulae
1801	First asteroid discovered
1802	Solar spectrum first viewed
1803	W. Herschel discovers binary stars
1837	First stellar parallax measured (by F. G. W. Struve)
1840	Draper produces first astronomical photograph
1843	The effect of motion on light spectra is explained by Doppler
1845	Earl of Rosse discovers spiral structure of some "nebulae"
1846	Neptune predicted independently by Leverrier and Adams and discovered by Galle
1850–1900	Development of spectrum analysis; stellar spectra used for first time to obtain temperatures and compositions of stars
1877	Schiaparelli sees "canals" on Mars
1905	Einstein's special theory of relativity
1905–1920	Einstein develops his theory of gravitation; general relativity

1914	Slipher discovers that spiral nebulae are receding from us; Lemaitre, DeSitter, and Eddington explain this phenomenon using general relativity
1915	Hooker reflecting telescope (2.5-m) constructed at Mount Wilson
1917	Shapley shows that solar system is more than 10,000 parsecs from the center of the Milky Way galaxy
1924	Hubble measures distances to spirals and confirms the viewpoint that they are galaxies in their own right
1910–1930	Russell, Eddington, and others develop the theory of stellar structure
1920–1930	Shapley, Oort, Linblad investigate rotation of Milky Way galaxy
1930–1960	Nuclear physics develops; used to explain the energy source of the stars
1930	Pluto discovered by Tombaugh
1931	Jansky discovers extra-terrestrial radio radiation
1937	Discovery of first interstellar molecule
1947–1960	Astronomical instruments sent by rocket above earth's atmosphere
1949	Great Hale reflector (5-m) went into routine operation
1951	Observation of neutral hydrogen at 21 cm wavelength
1957	Sputnik I orbits earth
1959	Space probe hits the moon
1961	Yuri Gagarin becomes first person in space
1963	Discovery of quasi-stellar objects
1965	Discovery of 3 K background radiation
1965	First close photographs of Mars by Mariner 4
1968	Discovery of pulsars
1969	Apollo 11 lands first men on the moon
1969	Discovery of first complex organic interstellar molecule (formaldehyde)
1973	First close photographs of Jupiter by Pioneer 10
1974	First close photographs of Venus and Mercury by Mariner 10
1974	Confirmation that some x-ray binary components are black holes
1975	Venera 9 lands on Venus and returns picture of surface
1975	Radio "pictures" depict certain radio galaxies having emitting regions 5,000,000 parsecs across
1976	Viking 1 & 2 place landers on Mars
1977	Discovery of rings of Uranus
1978	Discovery of Pluto's satellite
1979	Discovery of Jupiter's rings
	Discovery of volcanoes on Io
	First close photographs of Saturn by Pioneer 11

Index

A

Abell, G., 435
aberration, chromatic, 65
    of starlight, 46
    spherical, 66
absolute magnitude, 298
absorption, photographic, 349
acceleration, 134
accretion, 229
achromat, 65
achondrites, 222
active sun, 251
    galaxies, 449
Adams, J. C., 205
advance of perihelion, 138
aerolites, 222
age, cluster, 391
airglow, 13
albedo, 105, 149
Albireo, 270
Allende meteorite, 370
Almagest, 257
alphabet, Greek, 492
alpha particles, 254
Alpher, R., 454
Alphonso of Castile, 131
altitude, 10
Amalthea, 198
Ambartsumian, V., 439
amino acids, 462
Angstrom unit, 62
angular momentum, 101, 226, 285, 376
antapex, 283
antimatter, 381, 457
antiparticle, 381
aperture, 65
apex, 283
aphelion, 47
apogee, 93
Apollo spacecraft, 93
apparent distance, 9
    magnitude, 296
    orbit, 308
    place, 9
Aratus, 257
Aristarchus, 131

Arp, H., 446f
aspect, 140
association, 357, 393
asteroids, 139ff
astrographs, 68
astrolabe, 36
astrometric companions, 280
    binary, 308
astronomical latitude, 31
    unit, 47
astronomy, chronology of, 493
    gamma ray, 78
    radar, 74
    space, 469
    ultraviolet, 78
    x ray, 78
atmosphere, Jupiter's, 189
    sun's, 233
    Venus', 163
atmospheric composition, Mars, 179
    Venus, 163
atmospheric pressure, earth, 11
    Mars, 169
    Venus, 160
atom, 286
    excited, 290
    ionized, 290
    neutral, 290
    normal, 290
    primeval, 454
atomic clocks, 40
    mass unit, 361
    number, 287
    shells, 289
    weight, 287
autumnal equinox, 48
aurora, 13
    Australis, 19
    Borealis, 19
    above Jupiter, 192
azimuth, 10

B

Baade, W., 318
Balmer series, 288
background radiation, 456
Barnard's star, 281, 309
barred spiral galaxies, 420
Barringer crater, 224
Bayer, J., 260

Becker, W., 390, 409
Bell, J., 328
Bessel, F. W., 46, 279, 308
Bethe, H., 362, 454
big-bang model, 453
binary, astrometric, 308
    spectroscopic, 311, 326
    stars, 307f
    visual, 307
BL Lacertids, 430, 448
black dwarfs, 356, 360
    holes, 356, 376ff
Bode's law, 141
Bohr, N., 287
Bok, B. J., 412, 428
Bok, P., 412
Bok globules, 342
bolometric magnitude, 298, 312
Bondi, H., 453
Bonner Durchmusterung, 260
Boss, L., 387
Bradley, J., 46
Brahe, T., 132, 326
breccias, 108
bright points, 254
brightest stars, table of, 299
bulges, tidal, 99

C

Caesar, Julius, 58
calendar, 56
    Gregorian, 59
    Hebrew, 57
    Julian, 58
    lunar, 56
    lunisolar, 57
    Mayan, 57
    Mohammedan, 56
    solar, 57
Callisto, 110
Caloris Basin, 151
Cameron, A. G. W., 228
Cannon, A. J., 293
capture hypothesis, 116
carbon core, 367f
    cycle, 362
carbonaceous chondrites, 184
Cassini division, 200
catalyst, 362
celestial equator, 28
    horizon, 10

495

celestial equator *(cont.)*
　meridian, 10
　poles, 27
　sphere, 8
center of mass, 93, 136
Central Bureau
　　for Astronomical
　　Telegrams, 209
centrifugal effect, 25
cepheids, 316ff
　classical, 317
　type I, 317
　type II, 317, 395
Ceres, 184
chain craters, 111
Chandler wobble, 25
Chandrasekhar limit, 371
chaotic terrain, 169
Charon, 207
chemical elements, solar, 238
　table of, 286
chemistry, interstellar, 354
Cheseaux, J. P. L. de, 452
chromosphere, 244ff
chromospheric network, 246
chondrules, 222
chronology of astronomy, 493
circumpolar stars, 32
circumstellar rings, 340
cirrus clouds, 12
Clark, A., 309
closed universe, 457
clouds, high-velocity, 414
　interstellar, 340ff
　Magellanic, 398, 428
cluster, age, 391
　evolution of, 400
　galactic, 385
　globular, 393ff, 425
　Hyades, 273, 387ff
　open, 385f
　Pleiades, 273, 386
　star, 357, 384ff
　variables, 317
clusters of galaxies, 435
collimator, 83
color excess, 349
　index, 297
coma, 209
comet, 208ff
　cloud, 215
　Halley's, 209

comet *(cont.)*
　mass, 214
　nucleus, 213
conjunction, 140
constellations, 257ff
　table of, 258f
contacts, 126
continuum, 237
　source, 414
contraction, 357
convection zone, 236
convergence method, 388
convergent point, 387
coordinates, equatorial, 27
　galactic, 406
Copernican system, 131
Copernicus, N., 131
core, carbon, 367f
　earth's, 4f
　helium, 364
　lunar, 114
　Mercury, 153
　neutron, 375
　rigid, 366
Coriolis acceleration, 24
　effect, 22
corona, 127, 247
　F, 250
　K, 250
　L, 249
coronal holes, 250
　loops, 250
correcting plate, 69
cosmic distance scale, 433
cosmic rays, 354, 380f
　solar, 253
cosmic year, 408
cosmological expansion, 451
　principle, 451
Council of Nicaea, 59
counterglow, 220
Crab nebula, 273, 326, 369, 375,
　415, 434, 471
crape ring, 200
craters, lunar, 107
　Mercury's, 151
creation of matter, 453
crescent moon, 94
crust, 4
crystalline lattice, 373
cumulous clouds, 12
Curtis, H., 419

cusps, 95
Cyclops, project, 466

D

dark nebulae, 342
　globules, 343
day, 35
　Julian, 59
　lunar, 97
　sidereal, 35
　solar, 36
deceleration parameter, 457
declination, 29
　axis, 66
decoupling time, 455
deferent, 131
degenerate gas, 372
degeneracy, electron, 372
　neutron, 375
density, of sun, 234
　of universe, 458
density wave, 412
Descartes, R., 226
de Sitter, W., 453
diagram, *H-R,* 301, 320
diatomic molecules, 292
Deimos, 181
differential rotation, 241
　galactic, 408
differentiated, 113
diffraction limited, 82
　telescope, 473
Digges, T., 451
dimming by dust, 348
dipole field, 5
Dirac-type universe, 462
direct motion, 130
dispersion, 65
distance, look-out, 452
distance modulus, 300, 391
diurnal circles, 31
domes, 111
Doppler, C. J., 85
Doppler displacements, 281, 440,
　346
　effect, 85, 282, 311, 336
　redshift, 461
　shifts, 323, 332, 333, 380, 447
Doppler–Fizeau effect, 85
double-double, 269
double-planet hypothesis, 115

double stars, 307ff
Dreyer, J., 385
dust, dimming by, 348
  grains, 350
  interstellar, 346
dusty tails, 213
dwarf galaxies, 430
  novae, 326
  stars, 301
dynamical parallax, 313

E

earthlight, 95
earthquakes, 4
earth tides, 27, 100
earth's atmosphere, 3
  core, 5
  magnetic field, 5
Easter, 59
eccentricity, 47
eclipse, annular, 120
  family, 123
  lunar, 117
  partial, 121
  penumbral, 118
  seasons, 122
  solar, 120
  table of solar, 124, 125
  total, 120
  umbral, 118
  year, 122
ecliptic, 47
  pole, 48
eclipsing stars, 314
Einstein redshift, 373
ejecta, 108
electron, 286, 455
electron gas, 372
electromagnetic radiation, 62
elliptical galaxies, 420
elongation, 140
emission nebulae, 339
Engligh–metric units, 491
epicycle, 131
equator, earth's, 7
  galactic, 406
equatorial coordinates, 27
  mounting, 66
escape velocity, 136
  lunar, 112
Europa, 193

event horizon, 378
evolution, of clusters, 400
  of life forms, 462
  of massive stars, 368ff
  stellar, 356ff
Ewen, H. I., 348, 410
exclusion principle, Pauli, 372
exit cone, 379
excited atom, 290
expanding universe, 457
extragalactic nebulae, 432
extraterrestrial life, 464
eyepiece, 65

F

faculae, 247
fall, 223
feed, telescope, 73
Fermi, E., 289
filaments, 251
find, 223
fine structure constant, 462
fireball, 216
  primeval, 454
fission hypothesis, 115
Fizeau, H., 85
Flamsteed, J., 260
flare, solar, 253
  stars, 333, 416
flash spectrum, 244
flocculi, 247
flux units, 296
F-number, 65
focal length, 65
  ratio, 65
focus, cassegrain, 67
  coudé, 67
  Nasmyth, 67
  Newtonian, 67
  prime, 67
forbidden lines, 340
force, 134
Foucault pendulum, 22
Fraunhofer, J. von, 237
Fraunhofer lines, 237
  table of, 238
frequency, 62
Friedmann, A., 453
frigid zones, 55
full moon, 94
fundamental constants, 462

G

galactic center, 407
  coordinates, 406
  system (Milky Way), 403ff
galaxies, 418ff
  Also, see key words, e.g., active,
    dwarf, etc.
galaxy, formation, 439
  synthetic, 409
  Milky Way, 403ff
Galileo Galilei, 102, 133
Galilean satellites, 192
Galle, J. G., 205
Gamow, G., 454
Ganymede, 193
gap, Hertzsprung, 392
gas, degenerate, 372
  electron, 372
gaseous planets, 140, 461
  tails, 213
gauss, magnetic flux unit, 242
gegenschein, 19, 220
general relativity, model universe,
    456
  theory of, 377
geocentric theory, 22
geographical latitude, 31
geomagnetic pole, 5
giant elliptical galaxies, 427
  planets, 140ff
  stars, 301
gibbous phase, 94
globular clusters, 425
globules, dark, 357, 342
  Bok, 342
Gold, T., 453
Gould, B., 257
Gould's belt, 404, 413
grains, dust, 350
granules, 235
grating, normal, 82
  blazed, 82
gravitational redshift, 373
  collapse, 446
gravity research, 477
great circle, 7
Great Red Spot, 189
Great Wall of China, 2
great debate, 419
Greek alphabet, 492
  planets, 186
greenhouse effect, 12, 161

H

Halley, W., 210
Halley's comet, 209f
halo, stellar, 394
  galactic, 407, 416
  population, 399
halos, atmospheric, 17
harmonic law, Kepler's, 132, 310
Harvard classification, 293
harvest moon, 97
hat-brim effect, 422
heliocentric system, 131
heliostat, 233
helium core, 364
  flash, 366
  shell, 367
Herschel, W., 63, 204, 307, 413
Hertz, H., 63
Hertzsprung, E., 300, 301
Hertzsprung gap, 392
Hevelius, J., 103
Hewish, A., 328
high velocity clouds, 414
Hipparchus, 93, 131, 295
horizon, earth's, 9
  event, 378
Horsehead nebula, 344
H-R diagram, 320, 359, 364, 389
hot big-bang model, 456
hour angle, 30
  circles, 28, 267
Hoyle, F., 453
Hubble constant, 443, 453
  flow, 451
Hubble, E., 420, 443
humanoids, 451
Humason, M., 420
hunter's moon, 97
Huygens, C., 200
Hyades cluster, 273
hydrogen burning, 364
  shell, 367
H-I region, 347
H-II region, 347, 410

I

igneous rocks, 4
image converter, 90
  intensifier, 88
  tube, 88
index, color, 297

infrared radiation, 63, 78
  stars, 359
insolation, 54
instability strip, 369f
interactions, electromagnetic, 462
  gravitational, 462
  strong nuclear, 462
  weak nuclear, 462
interacting galaxies, 438
interferometry, long-baseline, 76
  very long baseline, 76
intergalactic medium, 439
International Astronomical Union,
  103
international date line, 42
interplanetary medium, 251
interstellar chemistry, 354
  clouds, 340
  dust, 346
  lines, 347
  magnetic field, 380
  medium, 346
  molecule, 351ff, 461
  reddening, 349
  spectra, 346
Io, 103
Io, plumes of, 104
ionized atom, 290
  helium, 369
ionosphere, 13, 254
irregular galaxies, 420
  variables, 322
island universe, 419
isotopes, 287

J

Jansky, K., 73
Jovian planets, 140ff
Julian day, 59
Jupiter, 189ff
  auroral glow, 192
  comet family, 212
  magnetic field, 190
  ring of, 191

K

Kant, I., 226, 356
Keenan, P. C., 302
Kepler, J., 132, 452
Kepler's harmonic law, 142, 310
  laws, 132

Kepler's harmonic law *(cont.)*
  second law, 379
  star, 327
Keplerian motion, 408
kimberlite pipes, 4
Kirkwood's gaps, 185
Kuiper, G. P., 226

L

Lagoon nebula, 344
Lamaitre, G., 453
Laplace, P.S. de, 356, 376
latitude, astronomical, 31
  galactic, 406
  geographical, 31
law, Bode's, 141
  harmonic, 132
  of equal areas, 132
  of gravitation, 135
  Stefan-Boltzmann, 291
  Wien displacement, 291
laws, Kepler's, 132
  Newton's, 134
leap second, 40
  year, 59
Leavitt, H., 318
length of the day, 26
Leverrier, U. J. J., 146, 205
librations, 98
life, extraterrestrial, 464
  on Mars, 180
  origin of, 461
light curve, 313
light-gathering power, 71
light, velocity of, 64
light year, 143
limb darkening, of stars, 315
  solar, 236
limit, Chandrasekhar, 371
Lin, C. C., 412
line of sight velocity, 283
Lippershey, H., 102
local group of galaxies, 431, 434
local standard of rest, 284
longitude, celestial, 49
  galactic, 406
  terrestrial, 7
longitudinal waves, 4
long-period variables, 321
look-out distance, 452
Lowell, P., 168, 205
luminosity class, 302

lunar craters, 107
  core, 114
  mantle, 113
  rays, 108
Lyman series, 288
Lyot, B., 250

M

mackerel sky, 12
Magellanic Cloud, Small, 319, 333
Magellanic Clouds, 274, 398, 428
magnetic field, earth's, 6
  interstellar, 380
  Jupiter's, 190
  Mars', 169
  Mercury's, 150
  neutron star, 330
  Saturn's, 200
  solar, 242
  Venus', 159
magnetic tail, 6
magnetosphere, Jupiter's, 191
magnifying power, 72
magnitudes, 87, 268, 296ff
  absolute, 298
  apparent, 296
  bolometric, 298, 312
  photographic, 297
  visual, 297
main sequence, 280, 300, 359
main sequence fitting, 390
  stars, 360
major axis, 47
mantle, earth's, 4
  lunar, 113
maps, star, 261ff
Mars, 165ff
  Also see key word,
  e.g., magnetic field, etc.
mascons, 115
mass, 134
  apparent, 138
  center of, 93
  loss, 315
  of comets, 214
mass–energy equivalence, 361
mass–luminosity relation, 312
massive stars, evolution of, 368
Maunder minimum, 242, 255
  butterfly diagram, 242
mean sun, 38

mechanics, Newtonian, 138
  relativistic, 138
medium, interplanetary, 251
  interstellar, 346ff
Mercury, 148ff
  Also see key word,
  e.g., magnetic field, etc.
meridians, terrestrial, 7
  standard, 41
mesons, 138
Messier, C., 385
metal deficient, 401
meteor crater, 224
meteorite, 216ff
  Allende, 370
  craters, 224
meteoroids, 215ff
  stream, 218
meteor, 215ff
  showers, 219
  streams, 218
  table of showers, 219
micrometeorites, 107, 222
microwave region, 464
midnight sun, 55
midocean ridges, 3
Milky Way, 403ff
Miller, J., 419
minor planets, 139ff
Mira, 321
Mira type variables, 321
Mizar, 307, 311
model, hot big-bang, 456
  general relativity, 456
model universe, 452
modulus, distance, 391
molecular hydrogen, 204
  spectra, 291
molecule, 286
  diatomic, 292
  interstellar, 351, 461
  organic, 351
momentum, angular, 101, 376
month, sidereal, 95
  synodic, 95
moon dogs, 17
moon, harvest, 97
  hunter's 97
moonquakes, 114
Morgan-Keenan system, 298
Morgan, W. W., 302
multiple-mirror telescope, **474**
multiple systems, 307

N

N galaxies, 430, **444**
nadir, 9
nearest stars, table of, 280
nebula, Crab, 273, 326, 330, **471**
  dark, 342
  emission, 339
  Orion, 273, 340
  planetary, 367f
  reflection, 339, 342
  Ring, 269
nebulae, extragalactic, 432
  interstellar, 340
  planetary, 334, 445
nebular hypothesis, 226
nebulium, 340
Neptune, 204f
neutral atom, 290
neutrino, 369, 381
neutron, 286, 455
  capture, 369
  core, 375
  degeneracy, 375
  stars, 329ff, 356, 369, **374ff**
Newton, I., 64, 133
Nix Olympica, 170
noctilucent clouds, 12
noise minimum, 464
nonperiodic comets, 211
nonthermal emission, 189
novae, 323, 340, 434
  dwarf, 326
  recurrent, 325
nuclear reactions, 364
nucleosynthesis, 357, 455
nucleus, 209, 286

O

objective, 65
  prism, 293
occultations, 96
Olbers, H. W. M., 452
Olbers' paradox, 451
Olympus Mons, 169
Oort constants, 408
Oort, J., 408
open universe, 457
opposition, 140
oppositions of Mars, 166
optical doubles, 307
  pairs, 307

orbit, apparent, 308
  earth's, 47
  relative, 136
orbital elements, 308
organic molecule, 351
origin of life, 461
Orion arm, 410
  nebula, 273, 340
ozone, 12

P

paradox, Olbers', 451
parallax, 45, 92
  dynamical, 313
  equatorial horizontal, 92
  lunar, 92
  solar, 233
  stellar, 279
parallels of latitude, 7
parsec, 143, 278
particle physics, 380
Paschen series, 288
peculiar galaxies, 429
  stars, 304
Peebles, P. J. E., 456
penumbra, 117, 239
Penzias, A., 456
perigee, 93
perihelion, 47
period-luminosity relation, 318
periodic comets, 212
permafrost, 171
Perseus arm, 410
phases, 3
phenomena, 257
Phobos, 181
photoelectric effect, 87
photographic absorption, 349
  magnitude, 297
  zenith tube, 35
photometric function, 105
photometry, 87
photon sphere, 379
photons, 64, 288
photosphere, 235
photosynthesis, 232
Piazzi, G., 184
Pine, M. R., 228
plages, 247
planet, 139ff
  gaseout, 140
  giant, 140

planet (cont.)
  minor, 140
  terrestrial, 140
planetarium, 130, 275
planetary nebula, 334f, 340, 367,
    445
planets, gaseous, 140, 461
  table of, 142
  terrestrial, 140, 461
plate tectonics, 4
Pleiades cluster, 273
Pluto, 205f
Pogson, N., 296
polar axis, 66
  hoods, 167
Polaris, 27
populations, stellar, 397ff
  table of, 400
position angle, 307
positron, 138, 381
powers-of-ten notation, 491
Praesepe, 268
precession, 50
  general, 51
  lunisolar, 51
  of the equinoxes, 51
prime meridian, 7
primeval atom, 454
  fireball, 454
principal planets, 139ff
principle, cosmological, 451
  Pauli exclusion, 372
  perfect cosmological, 451
Principia, 134
prism, objective, 293
prograde, 116
project Cyclops, 466
prominence, 251
  active, 251
  eruptive, 253
  loop, 251
  quiescent, 257
proper motion, 281
proton, 254, 286, 455
proton–proton reaction, 362
protoplanet hypothesis, 226
protostars, 360
protosun, 228
Proxima, 279, 310
Ptolemaic system, 131
Ptolemy, C., 131, 295
pulsar, Crab, 330
  x ray, 322

pulsars, 314, 328ff, 374
pulsating stars, 316
Purcell, E. M., 348, 410

Q

QSO, 445
quadrature, 140
quasars, 418, 445
quasistellar objects, 418, 430, 443ff
quiet sun, 251

R

radar, 106
radial velocity, 87, 282, 325
radiant, 218
radiation, electromagnetic, 62
  infrared, 78
  synchrotron, 330
  thermal, 73
  three degree background, 456
  x ray, 379
radio astronomy, 73
  galaxies, 430, 437
  source counts, 456
  sources, 415f
radiographic pictures, 437
radius, Schwarzschild, 378
rays, lunar, 108
recombination lines, 289
recurrent novae, 325
reddening, interstellar, 349
red giant, 364
redshift, absorption-line, 445
  Doppler, 461
  Einstein, 373
  emission-line, 445
  gravitational, 373
  of galaxies, 442f
reflecting telescopes, table of, 492
reflection effect, 316
  nebulae, 339, 342
reflector, 66
refracting telescopes, table of, 492
refraction, 14, 16, 64
refractor, 65, 66
regolith, 108
relation, mass–luminosity, 312
  period–luminosity, 318
  Titius–Bode, 141
relativistic mechanics, 138
relativity, general theory, 139, 377
  principle of, 137

relativity, general theory *(cont.)*
  special theory, 138
  theory of, 361
resolving power, 72
retroreflectors, 93
retrograde, 116, 130
  rotation of Venus, 155
revolution, 21
Riccioli, G. B., 103
Richter scale, 114
right ascension, 28
rigid core, 366
rilles, lunar, 111
ring, galaxies, 439
  Jupiter's, 191
  nebula, 269, 335
  Saturn's, 200
  Uranus', 204
Roberts, W. W., 412
Roman, N. G., 302
rotation, 21
  differential, 241, 408
  of galaxies, 440
  of stars, 284
  spectrum, 292
RR Lyrae variables, 317, 395
runaway stars, 391f
Russell, H. N., 300

S

Sagittarius arm, 410
Sandage, A., 443
saros, 123
satellites, table of, 145
Saturn, 199ff
  Also see key word, e.g., magnetic field, etc.
Scaliger, J., 59
Schmidt, M., 445
Schwarzschild, K., 378
Schwarzschild radius, 378
scintillation, 16
second, atomic, 39f
  leap, 40
sedimentary rocks, 4
seeing, 17
seismic waves, 4
self-gravitation, 357
semiregular variables, 323
Seyfert, C., 430
Seyfert galaxies, 430, 444
Shapley, H., 318, 396, 407, 419, 479

shell, helium, 367
  hydrogen burning, 367
  stars, 303
shells, atomic, 289
  stellar, 340
shooting star, 215
short-period variables, 318
Shu, F., 412
siderites, 222
siderolites, 222
singularity, 378
Skylab, 81
Slipher, V. M., 419, 440, 442, 449, 479
small circle, 7
solar constant, 239
  eclipse, 247
  flares, 253
  motion, 283
  parallax, 233
  wind, 6, 250
solar-terrestrial relations, 254f
space astronomy, 469ff
spacecraft, Apollo, 93
spalation, 304
spectra, interstellar, 346
  molecular, 291
  of galaxies, 440
  stellar, 285ff
spectral types, 293
spectrogram, 83
spectrograph, 82
spectroheliograph, 233
spectroscope, 83, 292
spectroscopic binary, 311, 326
spectrum, absorption-line, 83
  bright-line, 83
  continuous, 83
  dark-line, 83
  emission-line, 83
  flash, 244
  rotation, 292
  vibration-rotation, 292
sphere, celestial, 8
  photon, 379
  Strömgren, 346
spicule, 246
spin-orbital relationship of Mercury, 150
  of Venus, 159
spiral arms, 348
  galaxies, 420
  pattern, 412

spiral arms *(cont.)*
  structure, 409, 412
  tracers, 410
spirals, normal, 421
SS 433, 331, 375
star cluster, 384ff
  maps, 261ff
starlight, aberration of, 46
starquake, 376
stars, see topic, e.g., brightest stars, T Tauri stars, etc.
Stefan–Boltzmann relation, 291
stella mira, 321
stellar evolution, 356ff
  populations, 397ff
  shells, 340
  spectra, 285ff, 293
  wind, 370
stones from heaven, 221
stratosphere, 12
stratus clouds, 12
Strömgren sphere, 346
Struve, F. G. W., 46
summer solstice, 48
sun, 231ff
  atmosphere, 233
  density, 284
  dogs, 17
  evolution of, 356ff
  grazer, 213
  magnetic field, 242
  midnight, 55
  motion of, 283
sunspot cycle, 241f
sunspot, principal, 239
sunspots, 239ff
superclusters, 437
superfluid, 376
supergiant, 368
supergranules, 246
supernovae, 326, 340, 353, 369, 434
supernovae remnants, 327, 331, 369, 375, 415
supernovae, type I and II, 434
Swift, J., 181
synchrotron radiation, 190, 330
synthetic galaxy, 409
syzygy, 100

T

tail, comet, 209
technetium stars, 304

tectonic activity, 111
telescope, 61ff
  diffraction limited, 473
  Maksutov–Bouwers, 70
  multiple-mirror, 474
  radio, 73ff
  Schmidt, 69
  supersynthesis, 76
  synthesis, 76
  table of reflectors, 492
  table of refractors, 492
  Very Large Array 76, 476
telluric bands, 238
temperate zones, 55
terminator, 94, 151
terrae, 102
terrestrial planets, 140, 461
Terrell, J., 447
thermal equilibrium, 359
  radiation, 73, 189
three degree Kelvin background ra-
    diation 456
tidal bulges, 99
tidal lock, of Mercury, 150
  of Venus, 159
tides, 99ff
  earth, 100
  neap, 100
  spring, 100
time, 34ff
  apparent solar, 37
  atomic, 40
  coordinated universal, 39
  daylight saving, 42
  decoupling, 455
  dilation, 138
  ephemeris, 39
  equation of, 38
  equinoctial, 35
  Greenwich mean, 41
  local mean, 41
  mean solar, 38
  sidereal, 35
  universal, 39
  zones, 41
Titan, 203
Titius–Bode relation, 141
Tombaugh, C., 205
torrid zone, 55
total eclipse of the sun, 126
transits, of Mercury, 144
  of Venus, 146
  upper, 35

transverse waves, 4
trapezium, 340, 387
triple alpha process, 363
Triton, 205
Trojan asteroids, 186
  planets, 186
troposphere, 12
tropic of Cancer, 55
  of Capricorn, 55
T Tauri stars, 323, 343, 358, 393
twilight, 33
  astronomical, 34
  civil, 34
  nautical, 34
twinkling, 16
Tychonic model, 132
type I cepheids, 317
  population, 391, 399
type II cepheids, 317, 395
  population, 398ff

U

umbra, 117, 239
universe, closed, 457
  density of, 458
  Dirac-type, 462
  expanding, 457
  model, 452
  steady state, 453
  hot big-bang, 456
upper transit, 35
Uranus, 204f
Uranus' rings, 204

V

van Allen belts, 6
van de Hulst, H. C., 289, 347, 410
van de Kamp, P., 309
van Maanen, A., 419
variables, cepheid, 316
  cluster, 317
  irregular, 322
  long-period, 321
  RR Lyrae, 317
  semiregular, 323
  short-period, 318
variable stars, 313ff
  eclipsing, 314
  see variables
variation of latitude, 25
velocity of light, 64
velocity, radial, 87, 282

Venus, 154ff
vernal equinox, 28
vertical circles, 10
Very Large Array, 74, 476
Viking missions, 176
visible horizon, 10
visual binary, 307
  magnitude, 297
Vorontsov–Velyaminov, B. A., 438

W

wavelength, 62
weird terrain, 153
Wells, H. G., 468
white dwarf, 301, 311, 323, 356,
    371ff
Widmanstätten structure, 222
Wien displacement law, 291
Wilson, R., 456
wind, solar, 250
  stellar, 370
windows, electromagnetic, 73
winter solstice, 49
Wolf–Rayet stars, 294, 303, 335
W. Virginis stars, 317
wreath nebula, 271
wrinkle ridges, 111

X

x-ray pulsar, 332
  radiation, 379
x rays, 80

Y

year, anomalistic, 53
  cosmic, 408
  eclipse, 53, 122
  galactic, 408
  leap, 58
  of the seasons, 53
  sidereal, 53
  tropical, 53

Z

Zeeman effect, 243
zenith, 9
zenith distance, 11
zodiac, 51
zodiacal light, 19, 49
zone time, 41
Zwicky, F., 374, 435, 438

100

10 2/03